Lecture Notes in Computer Science 3990

Commenced Publication in 1973
Founding and Former Series Editors:
Gerhard Goos, Juris Hartmanis, and Jan van Leeuwen

J. Christopher Beck   Barbara M. Smith (Eds.)

Integration
of AI and OR Techniques
in Constraint Programming
for Combinatorial
Optimization Problems

Third International Conference, CPAIOR 2006
Cork, Ireland, May 31 – June 2, 2006
Proceedings

 Springer

Volume Editors

J. Christopher Beck
University of Toronto, Department of Mechanical & Industrial Engineering
5 King's College Rd, Toronto, ON M5S 3G8, Canada
E-mail: jcb@mie.utoronto.ca

Barbara M. Smith
University College Cork, Cork Constraint Computation Centre
Cork, Ireland
E-mail: b.smith@4c.ucc.ie

Library of Congress Control Number: 2006925627

CR Subject Classification (1998): G.1.6, G.1, G.2.1, F.2.2, I.2, J.1

LNCS Sublibrary: SL 1 – Theoretical Computer Science and General Issues

ISSN   0302-9743
ISBN-10   3-540-34306-7 Springer Berlin Heidelberg New York
ISBN-13   978-3-540-34306-6 Springer Berlin Heidelberg New York

Springer is a part of Springer Science+Business Media

springer.com

© Springer-Verlag Berlin Heidelberg 2006
Printed in Germany

Typesetting: Camera-ready by author, data conversion by Scientific Publishing Services, Chennai, India
Printed on acid-free paper   SPIN: 11757375   06/3142   5 4 3 2 1 0

Preface

This volume contains the proceedings of the Third International Conference on Integration of AI and OR Techniques in Constraint Programming for Combinatorial Optimization Problems (CPAIOR 2006). The conference was held in Cork, Ireland, from May 31 to June 2, 2006. Information about the conference can be found at http://tidel.mie.utoronto.ca/cpaior/. Previous meetings in this series include two international conferences held in Nice (2004) and Prague (2005) and five international workshops held in held in Ferrara (1999), Paderborn (2000), Ashford (2001), Le Croisic (2002), and Montreal (2003).

The goal of these meetings is to provide a forum for researchers to present approaches which highlight the integration of CP, AI, and OR techniques. An additional important goal is to allow researchers from diverse backgrounds to learn about techniques in other areas that are used for solving combinatorial optimization problems and therefore to encourage cross-fertilization. One measure of the success that has been enjoyed by these meetings is the number of publications outside this conference series (e.g., at the International Conference on the Principles and Practice of Constraint Programming) that directly explore integrated approaches to solving large and difficult combinatorial problems.

CPAIOR 2006 received 67 submissions. In order to streamline the reviewing process, a subcommittee of the Programme Committee, consisting of Michael Trick, Pascal Van Hentenryck, and the Programme Chairs, evaluated each submission to ensure relevance to the conference aims. The subcommittee unanimously judged that 41 of the submissions were sufficiently relevant to proceed to the full review stage. Each of these submissions received three reviews by members of the Programme Committee. The reviews were extensively discussed during an online Programme Committee meeting. As a result, the Programme Committee chose 20 (29.9%) to be included in the proceedings and presented at the conference.

The authors of the papers in this volume have been invited to submit extended versions to a special issue of the *Annals of Operations Research* entitled "Constraint Programming, Artificial Intelligence and Operations Research." All papers submitted will be subject to an additional rigorous review process and we expect the special issue to be published in early 2008.

In addition to the technical sessions, three invited talks were given by leading researchers. These diverse talks address the uses of optimization technology in visual art (Robert Bosch, Oberlin College, USA); the growing interest in the AI planning community in solving mixed discrete/continuous problems by exploiting existing CP and OR techniques (Maria Fox, University of Strathclyde, UK); and the issue of duality, a central issue in both traditional OR and CP solution approaches (John Hooker, Carnegie Mellon University, USA).

CPAIOR 2006 continued the tradition of holding a Master Class on a focused topic as part of the conference. This year's Master Class, organized by Ken Brown and Armagan Tarim, consisted of six tutorial sessions on the topic of "Modelling and Solving for Uncertainty and Change." The speakers at the Master Class were Nesim Erkip (Bilkent University, Turkey), Hélène Fargier (IRIT, Toulouse, France), Alexei Gaivoronski (Norwegian University of Science and Technology, Norway), Brahim Hnich (Izmir University of Economics, Turkey), Pascal Van Hentenryck (Brown University, USA), and Gérard Verfaillie (ONERA/CERT, France).

We would like to thank the Programme Committee for their careful work over the past few months in ensuring a high-quality programme for the conference. We would also like to thank everyone involved in the organization of the conference, notably Barry O'Sullivan, the Conference Chair; Ken Brown and Armagan Tarim, the Chairs of the Master Class; Ian Miguel, the Publicity Chair; Michela Milano, the Sponsorship Chair; Tom Carchrae, the Webmaster; and Eleanor O'Riordan for her administrative support. It would have been impossible to hold CPAIOR 2006 without their significant contributions of time and effort.

Finally, we would like to thank the institutions listed below who helped to sponsor the conference. Their generosity enabled the conference to attract invited speakers and instructors for the Master Class as well as to fund student participation. These funds, therefore, greatly contributed to the success of the conference.

March 2006

J. Christopher Beck
Barbara M. Smith
Programme Chairs
CPAIOR 2006

Organization

Conference Organization

Conference Chair	Barry O'Sullivan (Cork Constraint Computation Centre)
Programme Chairs	J. Christopher Beck (University of Toronto)
	Barbara M. Smith (Cork Constraint Computation Centre)
Master Class Chairs	Ken Brown (Cork Constraint Computation Centre)
	Armagan Tarim (Cork Constraint Computation Centre)
Publicity Chair	Ian Miguel (University of St. Andrews)
Sponsorship Chair	Michela Milano (Università di Bologna)
Webmaster	Tom Carchrae (Cork Constraint Computation Centre)

Programme Committee

Gautamkumar Appa, London School of Economics, UK
Philippe Baptiste, Ecole Polytechnique, France
Roman Bartàk, Charles University, Czech Republic
Mats Carlsson, SICS, Sweden
Ondřej Čepek, Charles University, Czech Republic
Hani El Sakkout, CISCO Systems, Inc., USA
Bernard Gendron, CRT and University of Montreal, Canada
Carmen Gervet, Imperial College London, UK / Brown University, USA
Carla Gomes, Cornell University, USA
Narendra Jussien, Ecole des Mines de Nantes, France
Stefan Karisch, Carmen Systems, Canada
François Laburthe, Bouygues, France
Andrea Lodi, University of Bologna, Italy
Gilles Pesant, CRT and Ecole Polytechnique de Montreal, Canada
Jean-François Puget, ILOG, France
Jean-Charles Régin, ILOG, France
Michel Rueher, University of Nice-Sophia Antipolis, France
Meinolf Sellmann, Brown University, USA
Helmut Simonis, CrossCore Optimization Ltd, UK
Gilles Trombettoni, University of Nice-Sophia Antipolis, France
Michael Trick, Carnegie Mellon University, USA
Pascal Van Hentenryck, Brown University, USA
Mark Wallace, Monash University, Australia
Weixiong Zhang, Washington University, USA

Additional Referees

Kai Becker
Sophie Demassey
Guy Desaulniers
Jonathan Gaudreault
Marco Gavanelli
Mattias Grönkvist
Justin W. Hart
Manuel Iori
Christophe Jermann

Martin Joborn
Yahia Lebbah
Olivier Lhomme
Vassilis Liatsos
Tomas Liden
Enrico Malaguti
Claude Michel
Ioannis Mourtos
Bertrand Neveu

Stefano Novello
Fabrizio Riguzzi
Andrea Roli
Francesca Rossi
Christine Solnon
Andrea Tramontani
Willem-Jan van Hoeve
Neil Yorke-Smith
Alessandro Zanarini

Sponsors

Association of Constraint Programming
Bouygues, France
Carmen Systems, Sweden
Cork Constraint Computation Centre, Ireland
ILOG, S.A., France
Intelligent Information Systems Institute, Cornell, USA
Science Foundation Ireland, Ireland
University College Cork, Ireland

Table of Contents

Invited Talks

Technical Papers

Opt Art

Robert Bosch

Oberlin College, Oberlin OH 44074, USA
DominoArtwork.com, Oberlin OH 44074, USA

Abstract. Optimization deals with finding the best way to complete
a task—creating a schedule for a tournament, matching professors with
courses, constructing an itinerary for a traveling salesman. It has been
applied successfully to such a great number of diverse disciplines that
one could argue that it can be put to good use in *every* imaginable field.
In this talk, we will showcase its amazing utility by describing some
applications in the area of art: portraits constructed out of complete sets
of dominoes (via integer programming) mosaics comprised of abstract
geometric tiles (via integer programming and various heuristics), and
continuous line drawings (via the "solution" of large-scale instances of
the traveling salesman problem).

Fig. 1. Examples of Opt Art

J.C. Beck and B.M. Smith (Eds.): CPAIOR 2006, LNCS 3990, p. 1, 2006.

Planning for Mixed Discrete Continuous Domains

Maria Fox

Department of Computer Science and Information Systems
University of Strathclyde, U.K.

Abstract. Mixed discrete-continuous systems are hybrid systems that exhibit both discrete changes of state, describable in terms of their logical and metric properties, and continuous numeric change describable in terms of differential equations. Continuous change occurs within a state as a consequence of one or more continuous processes being active in that state, whilst discrete change results in state transitions. Such hybrid systems are well-understood in the formal verification and real-time control communities.

Many real planning problems involve interaction with continuously changing values that directly affect both the validity and efficiency of plans. The problem of planning with continuous effects is harder than planning under the assumption of discrete change. The planner must be capable of reasoning about the evolution of continuous processes and their interactions with discrete state changes. For this reason, the standard approach to handling complex continuous effects in planning is to abstract them out of the domain model by lifting the representation to a level where all change can be seen as discrete.

In this talk we discuss progress we have made towards planning in mixed discrete-continuous domains. We begin by arguing that there are problems of critical interest to potential users of planning technology that cannot be adequately modelled under the assumption of discreteness. We then discuss an approach to planning in these domains that relies on the integration of a discrete planner with a continuous non-linear constraint solver. We present some results taken from a range of planning domains featuring continuous change. We discuss the future of this branch of planning and relate our work to the AI and OR literature.

J.C. Beck and B.M. Smith (Eds.): CPAIOR 2006, LNCS 3990, p. 2, 2006.

Duality in Optimization and Constraint Satisfaction

J.N. Hooker

Carnegie Mellon University, Pittsburgh, USA
john@hooker.tepper.cmu.edu

Abstract. We show that various duals that occur in optimization and constraint satisfaction can be classified as inference duals, relaxation duals, or both. We discuss linear programming, surrogate, Lagrangean, superadditive, and constraint duals, as well as duals defined by resolution and filtering algorithms. Inference duals give rise to nogood-based search methods and sensitivity analysis, while relaxation duals provide bounds. This analysis shows that duals may be more closely related than they appear, as are surrogate and Lagrangean duals. It also reveals common structure between solution methods, such as Benders decomposition and Davis-Putnam-Loveland methods with clause learning. It provides a framework for devising new duals and solution methods, such as generalizations of mini-bucket elimination.

1 Two Kinds of Duals

Duality is perennial theme in optimization and constraint satisfaction. Well-known optimization duals include the linear programming (LP), Lagrangean, surrogate, and superadditive duals. The constraint satisfaction literature discusses constraint duals as well as search methods that are closely related to duality.

These many duals can be viewed as falling into two classes: inference duals and relaxation duals [12]. The two classes represent quite different concepts of duality. This is perhaps not obvious at first because the traditional optimization duals just mentioned can be interpreted as both inference and relaxation duals.

Classifying duals as inference or relaxation duals reveals relationships that might not otherwise be noticed. For instance, the surrogate and Lagrangean duals do not seem closely related, but by viewing them as inference duals rather than relaxation duals, one sees that they are identical except for a slight alteration in the type of inference on which they are based.

A general analysis of duality can also unify some existing solution methods and suggest new ones. Inference duals underlie a number of nogood-based search methods and techniques for sensitivity analysis. For instance, Benders decomposition and Davis-Putnam-Loveland methods with clause learning, which appear unrelated, are nogood-based search methods that result from two particular inference duals. Since any inference method defines an inference dual, one can

J.C. Beck and B.M. Smith (Eds.): CPAIOR 2006, LNCS 3990, pp. 3–15, 2006.
© Springer-Verlag Berlin Heidelberg 2006

in principle devise a great variety inference duals and investigate the nogood-based search methods that result. For example, filtering algorithms can be seen as inference methods that define duals and give rise to new search methods, such as decomposition methods for planning and scheduling.

Relaxation duals underlie a variety of solution methods that are based on bounding the objective function. A relaxation dual solves a class of problem relaxations that are parameterized by "dual variables," in order to obtain a tight bound on the objective function value. The LP, surrogate, Lagrangean, and superadditive duals familiar to the optimization literature are relaxation duals. A constraint dual is not precisely a relaxation dual but immediately gives rise to one that generalizes mini-bucket elimination methods.

Inference and relaxation duals are precise expressions of two general problem-solving strategies. Problems are often solved by a combination of search and inference; that is, by searching over values of variables, which can yield a certificate of feasibility for the original ("primal") problem, and by simultaneously drawing inferences from constraints, which can yield a certificate of optimality by solving the dual problem. A problem belongs to NP when the primal solution has polynomial size and to co-NP when the dual solution has polynomial size.

Problems can also be solved by a combination of search and relaxation; that is, by enumerating relaxations and solving each. The relaxation dual is one way of doing this, since it searches over values of dual variables and solves the relaxation corresponding to each value.

2   Inference Duality

An *optimization problem* can be written

$$\min_{x \in D} \{f(x) \mid \mathcal{C}\} \tag{1}$$

where $f(x)$ is a real-valued function, $\mathcal{C}$ is a constraint set containing variables $x = (x_1, \ldots, x_n)$, and D is the *domain* of x. A solution $\bar{x} \in D$ is *feasible* when it satisfies $\mathcal{C}$ and is *optimal* when $f(\bar{x}) \leq f(x)$ for all feasible x. If there is no feasible solution, the optimal value of (1) is ∞. If there is no lower bound on $f(x)$ for feasible x, the problem is *unbounded* and the optimal value is $-\infty$.[1] A value $\bar{z}$ is a feasible value of (1) if $f(x) = \bar{z}$ for some feasible x, or if $\bar{z} = \infty$, or if $\bar{z} = -\infty$ and the problem is unbounded.

A *constraint satisfaction problem* can be viewed the problem of determining whether the optimal value of

$$\min_{x \in D} \{0 \mid \mathcal{C}\}$$

is 0 or ∞.

---

[1] We exclude problems that have no optimal value, such as $\min_{x \in \mathbb{R}} \{x \mid x > 0\}$.

2.1 The Inference Dual

The *inference dual* of (1) is the problem of finding the greatest lower bound on $f(x)$ that can be inferred from $\mathcal{C}$ within a given proof system. The inference dual can be written

$$\max_{P \in \mathcal{P}} \left\{ v \;\middle|\; \mathcal{C} \overset{P}{\vdash} (f(x) \geq v) \right\} \tag{2}$$

where $\mathcal{C} \overset{P}{\vdash} (f(x) \geq v)$ indicates that proof P deduces $f(x) \geq v$ from $\mathcal{C}$. The domain of variable P is a family $\mathcal{P}$ of proofs. A pair $(\bar{v}, \bar{P})$ is a feasible solution of (2) if $\bar{P} \in \mathcal{P}$ and $\mathcal{C} \overset{\bar{P}}{\vdash} (f(x) \geq \bar{v})$, and it is optimal if $\bar{v} \geq v$ for all feasible (v, P). If $f(x) \geq v$ cannot be derived from $\mathcal{C}$ for any finite v, the problem is infeasible and has optimal value $-\infty$. If for any v there is a feasible (v, P), the problem is unbounded and has optimal value ∞. A value $\bar{v}$ is a feasible value of (2) if $(\bar{v}, P)$ is feasible for some $P \in \mathcal{P}$, or if $\bar{v} = -\infty$, or if $\bar{v} = \infty$ and (2) is unbounded.

The original problem (1) is often called the *primal* problem. Any feasible value of the dual problem is clearly a lower bound on any feasible value of the primal problem, a property known as *weak duality*. The difference between the optimal value of the primal and the optimal value of the dual is the *duality gap*.

The constraint set $\mathcal{C}$ *implies* $f(x) \geq v$ when $f(x) \geq v$ for all $x \in D$ satisfying $\mathcal{C}$. The proof family $\mathcal{P}$ is *complete* if for any v such that $\mathcal{C}$ implies $f(x) \geq v$, there is a proof $P \in \mathcal{P}$ that deduces $f(x) \geq v$ from $\mathcal{C}$. If $\mathcal{P}$ is complete, then there is no duality gap. This property is known as *strong duality*.

Solution of the inference dual for a complete proof family $\mathcal{P}$ solves the optimization problem (1), in the sense that a solution $(\bar{v}, \bar{P})$ of the dual provides a proof that $\bar{v}$ is the optimal value of (1). If $\bar{P}$ always has polynomial size, then the dual belongs to NP and the primal problem belongs to co-NP. Solution of the inference dual for an incomplete proof family may not solve (1) but may be useful nonetheless, for instance by providing nogoods and sensitivity analysis.

2.2 Nogood-Based Search

Nogoods are often used to exclude portions of the search space that have already been explicitly or implicitly examined. The inference dual can provide a basis for a nogood-based search.

Suppose we are solving problem (1) by searching over values of x in some manner. The search might proceed by splitting domains, fixing variables, or by adding constraints of some other sort. Let $\mathcal{B}$ be the set of constraints that have been added so far in the search. The constraint set has thus been enlarged to $\mathcal{C} \cup \mathcal{B}$. The inference dual of this restricted problem is

$$\max_{P \in \mathcal{P}} \left\{ v \;\middle|\; \mathcal{C} \cup \mathcal{B} \overset{P}{\vdash} (f(x) \geq v) \right\} \tag{3}$$

If $(\bar{v}, \bar{P})$ solves the dual, we identify a subset $\mathcal{N}$ of constraints that include all the constraints actually used as premises in the proof $\bar{P}$. That is, $\bar{P}$ remains a valid proof when $\mathcal{C} \cup \mathcal{B}$ is replaced by $\mathcal{N}$. Then by weak duality we can infer the *nogood*

$$\mathcal{N} \rightarrow (f(x) \geq \bar{v})$$

This nogood is a valid constraint and can be added to $\mathcal{C}$, which may accelerate the search. For instance, if $\mathcal{N}$ contains only a few variables, then restricting or fixing only a few variables may violate the nogood, allowing us to avoid a dead end earlier in the search process.

An important special case of this idea identifies a subset $\bar{\mathcal{B}} \subset \mathcal{B}$ of the search constraints that preserves the validity of $\bar{P}$. That is, $\bar{P}$ remains a proof when $\mathcal{C} \cup \bar{\mathcal{B}}$ replaces $\mathcal{C} \cup \mathcal{B}$. Then we can use the nogood

$$\bar{\mathcal{B}} \rightarrow (f(x) \geq \bar{v})$$

as a side constraint that guides the search, rather than adding it to $\mathcal{C}$. Suppose for example that the search proceeds by splitting domains; that is, by adding bounds of the form $L_j \leq x_j \leq U_j$ to $\mathcal{B}$. Suppose further than at some point in the search we obtain a solution $(\bar{v}, \bar{P})$ of the inference dual and find that the only bounds used as premises in $\bar{P}$ are $L_j \leq x_j$ and $x_k \leq U_k$. Then we can write the nogood

$$(L_j \leq x_j, \; x_k \leq U_k) \rightarrow (f(x) \geq \bar{v})$$

To obtain a solution value better than $\bar{v}$, we must avoid all future branches in which $x_j < L_j$ and $x_k > U_k$.

We can equally well apply this technique when we branch by fixing a variable x_j to each of the values in its domain. Suppose that at some point in the search the variables in x_F have been fixed to values $\bar{x}_F$, and the variables in x_U remain unfixed, where $x = (x_F, x_U)$. Thus $\mathcal{B} = \{x_F = \bar{x}_F\}$. We obtain a solution $(\bar{v}, \bar{P})$ of the inference dual and identify a subset x_J of variables in x_F such that $\bar{P}$ is still valid when $x_F = \bar{x}_F$ is replaced by $x_J = \bar{x}_J$. The resulting nogood

$$(x_J = \bar{x}_J) \rightarrow (f(x) \geq \bar{v})$$

tells us that if we want a solution value better than $\bar{v}$, the remainder of the search should exclude solutions x in which $x_J = \bar{x}_J$.

2.3   Sensitivity Analysis

Sensitivity analysis determines the sensitivity of the optimal value of (1) to perturbations in the problem data. Suppose that we have solved (1) and found its optimal value to be z^*. A simple form of sensitivity analysis relies on an optimal solution $(\bar{v}, \bar{P})$ of the inference dual [11]. Let $\bar{\mathcal{C}}$ be a subset of $\mathcal{C}$ for which $\bar{P}$ remains a valid proof of $f(x) \geq \bar{v}$. Then changing or removing the premises in $\mathcal{C} \setminus \bar{\mathcal{C}}$ has no effect on the bound and therfore cannot reduce the optimal value of (1) below $\bar{v}$. If there is no duality gap, then $z^* = \bar{v}$, and changing or removing these constraints has no effect on the optimal value of (1).

A sharper analysis can often be obtained by observing how much the individual constraints in $\mathcal{C}$ can be altered without invalidating the proof $\bar{P}$. One can also observe whether a proof having the same form as $\bar{P}$ would deduce $f(x) \geq v'$ for some $v' < v$ when the constraints in $\mathcal{C}$ are altered in certain ways. Both of these strategies have long been used in linear programming, for example. They can be applied to integer and mixed integer programming as well [6].

From here out we focus on nogood-based search rather than sensitivity analysis.

2.4   Linear Programming Dual

A linear programming (LP) problem has the form

$$\min_{x \geq 0} \{cx \mid Ax \geq b\} \tag{4}$$

The inference dual is

$$\max_{P \in \mathcal{P}} \left\{ v \; \Big| \; Ax \geq b \overset{P}{\vdash} (cx \geq v) \right\} \tag{5}$$

The proofs in family $\mathcal{P}$ are based on nonnegative linear combination and domination. Let a *surrogate* of a system $Ax \geq b$ be any linear combination $uAx \geq ub$, where $u \geq 0$. An inequality $ax \geq \alpha$ *dominates* $bx \geq \beta$ when $a \leq b$ and $\alpha \geq \beta$. There is a proof $P \in \mathcal{P}$ of $f(x) \geq v$ when some surrogate of $Ax \geq b$ dominates $f(x) \geq v$. The proof P is encoded by the vector u of *dual multipliers*. Due to the classical Farkas lemma, the proof family $\mathcal{P}$ is complete, which means that strong duality holds.

The inference dual (5) of (4) is essentially the same as the classical LP dual of (4). A solution (v, P) is feasible in the dual problem (5) when some surrogate $uAx \geq ub$ dominates $cx \geq v$, which is to say $uAx \leq c$ and $ub \geq v$. So when the dual is bounded (i.e., the primal is feasible), it can be seen as maximizing v subject to $uAx \leq c$, $ub \geq v$, and $u \geq 0$, or equivalently

$$\max_{u \geq 0} \{ub \mid uA \leq c\} \tag{6}$$

which is the classical LP dual. Strong duality holds for the classical dual unless both the primal and dual are infeasible.

When the LP dual is used in nogood-based search, the well-known method of *Benders decomposition* results [2]. It is applied to problems that become linear when certain variables x_F are fixed:

$$\min_{x_U \geq 0} \{f(x_F) + cx_U \mid g(x_F) + Ax_U \geq b\} \tag{7}$$

Suppose that when x_F is fixed to $\bar{x}_F$, (7) has optimal value $\bar{z}$ and optimal dual solution $\bar{u}$. By strong duality $\bar{v} - f(\bar{x}_F) = \bar{u}(b - g(\bar{x}_F))$, which means that

$$f(x_F) + \bar{u}(b - g(x_F)) \tag{8}$$

is the largest possible lower bound on the optimal value of (7) when $x_F = \bar{x}_F$. But since $\bar{u}$ remains dual feasible when $\bar{x}_F$ in (7) is replaced by any x_F, weak duality implies that (8) remains a valid lower bound for any x_F. This yields the nogood

$$z \geq f(x_F) + \bar{u}(b - g(x_F)) \tag{9}$$

where z represents the objective function of (7). This nogood is known as a *Benders cut*. If the dual of (7) is unbounded, there is a direction or ray $\bar{u}$ along which its solution value can increase indefinitely. In this case the Benders cut (9) simplifies to $\bar{u}(b - g(x_F)) \leq 0$.

In the Benders algorithm, the set x_F of fixed variables is static. The algorithm searches over values of x_F by solving a *master problem* in each iteration of the search. The master problem minimizes z subject to the Benders cuts obtained in previous iterations. The optimal solution of the master problem becomes the next $\bar{x}_F$. The search terminates when the optimal value of the master problem is equal to the previous $\bar{z}$. The master problem can be solved by any desired method, such as branch and bound if it is mixed integer programming problem.

2.5   Surrogate Dual

The surrogate dual results when one writes the inference dual of an inequality-constrained problem, again using nonnegative linear combination and domination as an inference method. When the inequalities and objective function are linear, the surrogate dual becomes the linear programming dual. When a slightly stronger form of domination is used, we obtain the Lagrangean dual, as is shown in the next section.

The surrogate dual [10] is defined for a problem of the form

$$\min_{x \in D} \{f(x) \mid g(x) \leq 0\} \tag{10}$$

where $g(x)$ is a vector of functions. A surrogate of $g(x) \leq 0$ is any linear combination $ug(x) \leq 0$ with $u \geq 0$. Let $P \in \mathcal{P}$ deduce $f(x) \geq v$ from $g(x) \leq 0$ when some surrogate $ug(x) \leq 0$ dominates $f(x) \geq v$. We will use the weakest possible form of domination: $ug(x) \leq 0$ dominates $f(x) \geq v$ whenever the former implies the latter. This family $\mathcal{P}$ of proofs is generally incomplete.

Under this definition of $\mathcal{P}$, the inference dual of (10) finds the largest v such that $ug(x) \leq 0$ implies $f(x) \geq v$ for some $u \geq 0$. The inference dual therefore becomes the *surrogate dual*

$$\max_{u \geq 0} \left\{ \min_{x \in D} \{f(x) \mid ug(x) \leq 0\} \right\} \tag{11}$$

A difficulty with the surrogate dual is that it is generally hard to solve. Yet if the problem (10) has special structure that allows easy solution of the dual, the resulting nogoods could be used in a search algorithm.

2.6   Lagrangean Dual

Like the surrogate dual, the Lagrangean dual is defined for inequality-constrained problems of the form (10). Again the proofs in $\mathcal{P}$ consist of nonnegative linear combination and domination, but this time a stronger form of domination is used.

In the surrogate dual, $ug(x) \leq 0$ dominates $f(x) \geq v$, which can be written $v - f(x) \leq 0$, when $ug(x) \leq 0$ implies $v - f(x) \leq 0$. In the Lagrangean dual, $ug(x) \leq 0$ dominates $v - f(x) \leq 0$ when $ug(x) \geq v - f(x)$ for all $x \in D$.

Under this definition of $\mathcal{P}$, the inference dual of (10) finds the largest v such that $ug(x) \geq v - f(x)$ for some $u \geq 0$. Since $ug(x) \geq v - f(x)$ can be written $f(x) + ug(x) \geq v$, the inference dual becomes the *Lagrangean dual*

$$\max_{u \geq 0} \left\{ \min_{x \in D} \{ f(x) + ug(x) \} \right\} \tag{12}$$

The Lagrangean dual has the nice property that the optimal value $\theta(u)$ of the minimization problem in (12) is a concave function of u. This means that $\theta(u)$ can be maximized by a hill-climbing search. Subgradient optimization techniques are often used for this purpose [1, 19].

When the inequalities $g(x) \leq 0$ and the objective function $f(x)$ are linear, both the surrogate and Lagrangean duals become the linear programming dual, since the two types of domination collapse into implication.

Nogoods can be obtained from the Lagrangean dual much as from the LP dual. If at some point in the search x_F is fixed to $\bar{x}_F$ and $\bar{u}$ solves the Lagrangean dual problem

$$\max_{u \geq 0} \left\{ \min_{x \in D} \{ f(\bar{x}_F, x_U) + ug(\bar{x}_F, x_U) \} \right\} \tag{13}$$

then we have the nogood

$$z \geq f(\bar{x}_F, x_U) + \bar{u}g(\bar{x}_F, x_U) \tag{14}$$

Nogoods of this sort could be generated in a wide variety of search methods, but they have apparently been used only in the special case of *generalized Benders decomposition* [9]. This method can be applied when there is a set x_F of variables for which (13) has no duality gap when x_F is fixed. The set x_F is therefore static, and x_F is fixed by solving a master problem that contains the Benders cuts (14). In practice the classical method obtains the multipliers $\bar{u}$ by solving

$$\min_{x \in D} \{ f(\bar{x}_F, x_U) \mid g(\bar{x}_F, x_U) \leq 0 \}$$

as a nonlinear programming problem and letting $\bar{u}$ be the Lagrange multipliers that correspond to the optimal solution. However, the multipliers could be obtained by solving the Lagrangean dual directly. as a nonlinear programming problem and letting $\bar{u}$ be the Lagrange multipliers that correspond to the optimal solution. However, the multipliers could be obtained by solving the Lagrangean dual directly.

2.7 Superadditive/Subadditive Dual

The *subadditive dual* [17] has been studied in connection with integer programming problems, which can be written

$$\min_{x \in D} \{ cx \mid Ax \geq b \} \tag{15}$$

where D is the set of n-tuples of nonnegative integers. (The *superadditive dual* is used when one maximizes cx subject to $Ax \leq b$.) The subadditive dual can be viewed as an inference dual, using a form of inference that generalizes inference by nonnegative linear combination and domination. Let a real-valued function $h(\cdot)$ be *subadditive* when $h(d + d') \leq h(d) + h(d')$ for all d, d'. We will say that a proof in $\mathcal{P}$ derives $cx \geq v$ from $Ax \geq b$ when $h(Ax) \geq h(b)$ dominates $cx \geq v$ for some nondecreasing, subadditive function h, and that $h(Ax) \geq h(b)$ dominates $cx \geq v$ when $h(b) \geq v$ and $h(Ax) \leq cx$ for all $x \in D$. This inference method can be shown to be complete for linear integer inequalities, based on cutting plane theory developed in [3, 4, 16, 20].

When $\mathcal{P}$ is defined in this way, the inference dual (2) becomes

$$\max_{h \in H} \{ g(b) \mid g(Ax) \leq cx, \text{ all } x \in D \}$$

where H is the set of subadditive, nondecreasing functions. This is the subadditive dual of (15). Since $\mathcal{P}$ is complete, it is a strong dual.

The subadditive dual has been used primarily for sensitivity analysis in integer programming (e.g. [5]). It has apparently not been used in the context of nogood-based search. Since the form of domination used to define $\mathcal{P}$ is that used in the surrogate dual, one could obtain a Lagrangean analog of the subadditive dual by substituting the form of domination used in the Lagrangean dual.

2.8 Duals for Propositional Satisfiability

Propositional satisfiability (SAT) problems are often solved by a *Davis-Putnam-Loveland* (DPL) method with *clause learning* (e.g., [18]). These methods can be seen as nogood-based search methods derived from an inference dual.

The SAT problem can be written

$$\min_{x \in \{0,1\}^n} \{ 0 \mid \mathcal{C} \} \tag{16}$$

where $\mathcal{C}$ is a set of logical clauses. To formulate an inference dual, let $\mathcal{P}$ consist of *unit resolution* proofs (i.e., repeated elimination of variables that occur in unit clauses until no unit clauses remain). The dual problem (2) has optimal value ∞ when unit resolution proves unsatisfiability by deriving the empty clause. Since unit resolution is not a complete inference method, there may be a duality gap: the dual may have optimal value zero when the primal is unsatisfiable.

Now suppose we solve (16) by branching on the propositional variables x_j. At each node of the search tree, certain variables x_F are fixed to $\bar{x}_F$. Let $\mathcal{U}$ contain the unit clause x_j when $\bar{x}_j = 1$ and $\neg x_j$ when $\bar{x}_j = 0$. We now solve the inference dual of (16) with the clause set $\mathcal{C} \cup \mathcal{U}$. If the optimal value is ∞, we generate a nogood and backtrack; otherwise we continue to branch. To generate the nogood, we identify a subset $\bar{\mathcal{U}}$ of $\mathcal{U}$ for which some portion of the unit resolution proof obtains the empty clause from $\mathcal{C} \cup \bar{\mathcal{U}}$. Then

$$\bigvee_{x_j \in \bar{\mathcal{U}}} \neg x_j \vee \bigvee_{\neg x_j \in \bar{\mathcal{U}}} x_j$$

is a nogood or "learned" clause that can be added to $\mathcal{C}$ before backtracking. This results in a basic DPL algorithm with clause learning.

Similar algorithms can be developed for other types of resolution, including full resolution, which could be terminated if it fails to derive the empty clause in a fixed amount of time.

2.9 Domain Filtering Duals

A domain filtering algorithm can be viewed as an inference method and can therefore define an inference dual. For concreteness suppose that the domain of each variable is an interval $[L_j, U_j]$, and consider a filtering algorithm that tightens lower and upper bounds. The inference dual is most straightforward when we assume the objective function $f(x)$ is monotone nondecreasing. In this case we can let $\mathcal{P}$ contain all possible proofs consisting of an application of the filtering algorithm to obtain reduced bounds $[L_j, U_j]$, followed by an inference that $f(x) \geq f(L_1, \ldots, L_2)$. This defines an inference dual (2).

If the filtering method achieves bounds consistency, $f(L_1, \ldots, L_n)$ is the optimal value of (1), there is no duality gap, and the problem is solved.

If bounds consistency is not achieved, the dual can be useful in nogood-based search. Suppose we search by domain splitting, and let $\mathcal{B}$ contain the bounds currently imposed by the branching process. We can examine the filtering process to identify a subset $\bar{\mathcal{B}} \subset \mathcal{B}$ of bounds that are actually used to obtain the lower bounds L_j that affect the value of $f(L_1, \ldots, L_n)$. The resulting nogood

$$\bar{\mathcal{B}} \rightarrow (f(x) \geq f(L_1, \ldots, L_n))$$

can be used as described earlier.

A related idea has proved very useful in planning and scheduling [12, 15, 13]. Let each variable x_j in x_F indicate which facility will process job j. The jobs assigned to a facility i must be scheduled subject to time windows; the variables in x_U indicate the start times of the jobs. The processing times of job j may be different on the different facilities. For definiteness, suppose the objective $f(x)$ is to minimize the latest completion time over all the jobs (i.e., minimize makespan). We solve each scheduling problem with a constraint programming method that combines branching with edge finding. This can be viewed as a complete inference method that defines an inference dual with no duality gap. Let $\bar{v}_i$ be the minimum makespan obtained on facility i for a given assignment $\bar{x}_F$. By examining the edge finding and branching process, we identify a subset x_J of job assignments for which the minimum makespan on each facility i is still $\bar{v}_i$. Then we have the nogood

$$(x_F = \bar{x}_F) \rightarrow \left(f(x) \geq \max_i \{\bar{v}_i\} \right)$$

Nogoods of this sort are accumulated in a master problem that is solved to obtain the next $\bar{x}_F$, thus yielding a generalized form of Benders decomposition [14]. The assignments in x_J are identified by noting which jobs play a role in the edge finding at each node of the search tree; details are provided in [13].

3   Relaxation Duality

A *parameterized relaxation* of the optimization problem (1) can be written

$$\min_{x \in D} \{f(x, u) \mid \mathcal{C}(u)\} \tag{17}$$

where $u \in U$ is a vector of *dual variables*. The constraint set $\mathcal{C}(u)$ is a relaxation of $\mathcal{C}$, in the sense that every $x \in D$ that satisfies $\mathcal{C}$ satisfies $\mathcal{C}(u)$. The objective function $f(x, u)$ is a lower bound on $f(x)$ for all x feasible in (1); that is, $f(x, u) \leq f(x)$ for all $x \in D$ satisfying $\mathcal{C}$.

Clearly the optimal value of the relaxation (17) is a lower bound on the optimal value of (1). The *relaxation dual* of (1) is the problem of finding the parameter u that yields the tightest bound [12]:

$$\max_{u \in U} \left\{ \min_{x \in D} \{f(x, u) \mid \mathcal{C}(x, u)\} \right\} \tag{18}$$

Let z^* be the optimal value of (1), and $\theta(u)$ be the optimal value of the minimization problem in (18). Since $\theta(u)$ is a lower bound on z^* for every $u \in U$, we have weak duality: the optimal value $\bar{v}$ of the relaxation dual (18) is a lower bound on z^*.

The lower bound $\bar{v}$ can abbreviate the search, as for example in a branch-and-relax (branch-and-bound) scheme. The parameterized relaxation is chosen so that $\theta(u)$ is easy to compute. The dual problem of maximizing $\theta(u)$ over $u \in U$ may be solved by some kind of search procedure, such as subgradient optimization in the case of Lagrangean relaxation. The maximization problem need not be solved to optimality, since any $\theta(u)$ is a valid lower bound.

3.1   Equivalence to an Inference Dual

Although inference and relaxation duality are very different concepts, a relaxation dual is always formally equivalent to an inference dual, provided there exists a solution algorithm for the parameterized relaxation. There does not seem to be a natural converse for this proposition.

To formulate the relaxation dual (18) as an inference dual, suppose that an algorithm $P(u)$ is available for computing $\theta(u)$ for any given $u \in U$. We can regard $P(u)$ as a proof that $f(x) \geq \theta(u)$ and let $\mathcal{P} = \{P(u) \mid u \in U\}$. The resulting inference dual (2) is $\max_{u \in U} \{\theta(u)\}$, which is identical to the relaxation dual (18).

3.2   Linear Programming and Surrogate Duals

A simple parameterized relaxation for the inequality-constrained problem (10) uses a surrogate relaxation of the constraints but leaves the objective function unchanged. The relaxation therefore minimizes $f(x)$ subject to $ug(x) \leq 0$, where $u \geq 0$. The resulting relaxation dual is the surrogate dual (11) of (1). Since the surrogate dual of an LP problem is the LP dual, the relaxation dual of an LP problem is likewise the LP dual.

3.3 Lagrangean Dual

Another parameterized relaxation for (10) removes the constraints entirely but "dualizes" them in the objective function. The parameterized relaxation minimizes $f(u, x) = f(x) + ug(x)$ subject to $x \in D$. The function $f(x, u)$ is a lower bound on $f(x)$ for all feasible x since $ug(x) \leq 0$ when $u \geq 0$ and $g(x) \leq 0$. The resulting relaxation dual is precisely the Lagrangean dual (13).

The close connection between surrogate and Lagrangean duals, conceived as inference duals, is much less obvious when they are reinterpreted as relaxation duals.

3.4 Superadditive/Subadditive Dual

The subadditive dual discussed earlier can be viewed as a relaxation dual that generalizes the surrogate dual. We can give the integer programming problem (15) a relaxation parameterized by subadditive, nondecreasing functions h, in which we minimize cx subject to $h(Ax) \geq h(b)$ and $x \in D$. (In the surrogate dual, the function h is multiplication by a vector u of nonnegative multipliers.) This yields the relaxation dual

$$\max_{h \in H} \left\{ \min_{x \in D} \left\{ cx \mid h(Ax) \geq h(b) \right\} \right\}$$

which is equivalent to the subadditive dual.

3.5 Constraint Dual

The constraint dual is related to a relaxation dual. More precisely, the constraint dual can be given a parameterized relaxation that yields a relaxation dual. A special case of the relaxation has been applied in mini-bucket elimination and perhaps elsewhere.

It is convenient to let x_J denote the tuple of variables x_j for $j \in J$. Given a constraint set $\mathcal{C}$, the *constraint dual* of $\mathcal{C}$ is formed by "standardizing apart" variables that occur in different constraints and then equating these variables. So if $x_{j_1}, \ldots, x_{j_{n_i}}$ are the variables in constraint $C_i \in \mathcal{C}$, let $y^i = (y_1^i, \ldots, y_{n_i}^i)$ be a renaming of these variables. Also let J_{ik} be the index set of variables that occur in both C_i and C_k. The constraint dual associates the dual variable y^i with each constraint C_i, where the domain D_i of y^i is the set of tuples that satisfy C_i. The dual constraint set consists of the binary constraints $y_{J_{ik}}^i = y_{J_{ik}}^k$ for each pair i, k.

The constraint dual can be relaxed by replacing each $y_{J_{ik}}^i = y_{J_{ik}}^k$ with $y_{J'_{ik}}^i = y_{J'_{ik}}^k$ where $J'_{ik} \subset J_{ik}$. It is helpful to think about the constraint graph G corresponding to the dual, which contains a vertex for each variable y_j^i and an edge between two variables when they occur in the same tuple y^i or in the same equality constraint. Removing equality constraints deletes the corresponding edges from G, resulting in a sparser graph $G(E)$, where E is the set of edges corresponding to the equality constraints that remain. The relaxation is therefore parameterized by the subset E of edges that defines $G(E)$. This relaxation also serves as a parameterized relaxation $\mathcal{C}(E)$ of the original constraint set $\mathcal{C}$. Thus if the constraint satisfaction problem is written

$$\min_{y^i \in D_i, \text{ all } i} \{0 \mid \mathcal{C}\} \tag{19}$$

then we can write the relaxation dual

$$\max_{E \in \mathcal{E}} \{0 \mid \mathcal{C}(E)\}$$

where $\mathcal{E}$ is some (generally incomplete) family of subsets E. To solve the dual, we check the feasibility of $\mathcal{C}(E)$ for each $E \in \mathcal{E}$. The family $\mathcal{E}$ normally would be chosen so $G(E)$ has small induced width for $E \in \mathcal{E}$, since in this case $\mathcal{C}(E)$ is easier to solve by nonserial dynamic programming.

One way to construct $\mathcal{E}$ is to define sets of "mini-buckets" [7, 8]. We consider various partitions of the constraints in $\mathcal{C}$, where the kth partition defines disjoint subsets or mini-buckets $\mathcal{C}_{k1}, \ldots, \mathcal{C}_{km_k}$. For each k and each $t \in \{1, \ldots, m_k\}$ we let E_{kt} contain the edges corresponding to equality constraints between variables occurring in $\mathcal{C}_{kt}$, so that $\mathcal{C}(E_{kt}) = \mathcal{C}_{kt}$. Now $\mathcal{E}$ is the family of all sets E_{kt}. Thus, rather than solve the relaxations $\mathcal{C}_{k1}, \ldots, \mathcal{C}_{km_k}$ corresponding to a single set of mini-buckets as in [7], we solve relaxations $\mathcal{C}_{kt}$ for all E_{kt}. Other relaxation duals based on reducing the induced width are discussed in [12]. All of these approaches can be applied to problems (19) with a general objective function $f(x)$, as is done in mini-bucket elimination schemes.

References

1. F. Barahona and R. Anbil. The volume algorithm: Producing primal solutions with a subgradient algorithm. *Mathematical Programming*, 87:385–399, 2000.
2. J. F. Benders. Partitioning procedures for solving mixed-variables programming problems. *Numerische Mathematik*, 4:238–252, 1962.
3. C. E. Blair and R. G. Jeroslow. The value function of a mixed integer program. *Mathematical Programming*, 23:237–273, 1982.
4. V. Chvátal. Edmonds polytopes and a hierarchy of combinatorial problems. *Discrete Mathematics*, 4:305–337, 1973.
5. W. Cook, A. M. H. Gerards, A. Schrijver, and E. Tardos. Sensitivity results in integer programming. *Mathematical Programming*, 34:251–264, 1986.
6. M. Dawande and J. N. Hooker. Inference-based sensitivity analysis for mixed integer/linear programming. *Operations Research*, 48:623–634, 2000.
7. R. Dechter. Mini-buckets: A general scheme of generating approximations in automated reasoning. In *Proceedings of the 15th International Joint Conference on Artificial Intelligence (IJCAI 97)*, pages 1297–1302, 1997.
8. R. Dechter and I. Rish. Mini-buckets: A general scheme for bounded inference. *Journal of the ACM*, 50:107–153, 2003.
9. A. M. Geoffrion. Generalized benders decomposition. *Journal of Optimization Theory and Applications*, 10:237–260, 1972.

10. F. Glover. Surrogate constraint duality in mathematical programming. *Operations Research*, 23:434–451, 1975.
11. J. N. Hooker. Inference duality as a basis for sensitivity analysis. *Constraints*, 4:104–112, 1999.
12. J. N. Hooker. *Logic-Based Methods for Optimization: Combining Optimization and Constraint Satisfaction*. Wiley, New York, 2000.
13. J. N. Hooker. A hybrid method for planning and scheduling. *Constraints*, 10:385–401, 2005.
14. J. N. Hooker and G. Ottosson. Logic-based Benders decomposition. *Mathematical Programming*, 96:33–60, 2003.
15. V. Jain and I. E. Grossmann. Algorithms for hybrid MILP/CP models for a class of optimization problems. *INFORMS Journal on Computing*, 13:258–276, 2001.
16. R. G. Jeroslow. Cutting plane theory: Algebraic methods. *Discrete Mathematics*, 23:121–150, 1978.
17. E. L. Johnson. Cyclic groups, cutting planes and shortest paths. In T. C. Hu and S. Robinson, editors, *Mathematical Programming*, pages 185–211. Academic Press, 1973.
18. M. Moskewicz, C. F. Madigan, Ying Zhao, Lintao Zhang, and S. Malik. Chaff: Engineering an efficient SAT solver. In *Proceedings of the 38th Design Automation Conference (DAC01)*, pages 530–535, 2001.
19. A. Nedic and D. P. Bertsekas. Incremental subgradient methods for nondifferentiable optimization. *SIAM Journal on Optimization*, 12:109–138, 2001.
20. L. A. Wolsey. The b-hull of an integer program. *Discrete Applied Mathematics*, 3:193–201, 1981.

A Totally Unimodular Description of the Consistent Value Polytope for Binary Constraint Programming

Ionuţ D. Aron, Daniel H. Leventhal, and Meinolf Sellmann

Brown University, Department of Computer Science
115 Waterman Street, Providence, RI 02912, U.S.A.
{ia, dleventh, sello}@cs.brown.edu

Abstract. We present a theoretical study on the idea of using mathematical programming relaxations for filtering binary constraint satisfaction problems. We introduce the consistent value polytope and give a linear programming description that is provably tighter than a recently studied formulation. We then provide an experimental study that shows that, despite the theoretical progress, in practice filtering based on mathematical programming relaxations continues to perform worse than standard arc-consistency algorithms for binary constraint satisfaction problems.

Keywords: Cost-based filtering, hybrid methods, mathematical programming.

1 Introduction

As a result of the growing interaction between the mathematical programming and constraint programming communities, it has now become standard to use mathematical programming tools to derive information useful both for domain filtering and for guiding the search. On real-world constraint satisfaction problems (CSPs), and especially optimization problems, hybrid methods have been shown to outperform pure solution approaches. As a result of a decade long research, a rich tool-box for hybridization is now available: from the idea of optimization constraints [7, 14, 17] and associated notions of relaxed or approximated consistency [5, 19], reduced-cost filtering [16], to sophisticated problem-dependent techniques based on Bender's decomposition [9], Lagrangian decomposition [6, 18, 20, 21], or column generation [4, 11]. Also, specialized hybrid approaches have been developed for special problems like computing orthogonal Latin squares [2] or to solve the social golfer problem [22].

Despite these successes, in the past hybridization on binary constraint satisfaction problems (BCSPs) has been nothing less than disappointing. Many approaches that looked very promising on paper have failed to give real benefits. While this is common knowledge in the research community and has lead to the common belief that mathematical programming techniques only pay off when a problem contains constraints that contain large numbers of variables where constraint programming (CP) propagation is weak, we are not aware of any paper that would state such a negative result. Consequently, we frequently see that, despite prior experience that developing hybrid

J.C. Beck and B.M. Smith (Eds.): CPAIOR 2006, LNCS 3990, pp. 16–28, 2006.

methods for BCSPs is not a promising research avenue, the undoubtedly tempting idea lures researchers into developing new hybrid filtering approaches for BCSPs.

A recent approach regarding hybrid filtering for BCSPs is presented in [12]. The authors of that paper suggest to use a relaxation of an equivalent integer programming (IP) formulation of a given BCSP for domain filtering. Two ideas were novel in that contribution: first, the idea to use a Lagrangian relaxation instead of the commonly used linear relaxations for filtering. And second, to use a formulation that specifically targets individual assignments.

We were intrigued by those two ideas and decided to investigate them further. We address two questions: First, can the Lagrangian relaxation suggested in [12] yield to more effective filtering than standard linear programming (LP) relaxations? And second, does it pay off to focus on individual assignments for filtering in a tree search where what matters is the trade-off between filtering effectiveness and filtering time?

In order to answer those two questions, we start out in Section 2 by discussing different models for BCSPs and how they can be translated into integer programs. Based on those models, in Section 3, we develop an LP relaxation that is *provably tighter* than the Lagrangian relaxation developed in [12]. While offering the prospect of more effective filtering, that LP relaxation can also be computed much faster than Lagrangian relaxations when using standard LP software like Cplex.

In Section 4, we then present numerical results on various CSP and BCSP benchmark classes. The experiments show that, once again, mathematical programming techniques are inferior to standard arc-consistency on feasibility problems.

2 CSP and IP Models

2.1 Positive and Negative Representations of BCSPs

A binary constraint satisfaction problem (BCSP) consists of a finite set of *variables* $\mathcal{V} = \{V_1, \ldots, V_n\}$, a finite *domain* $D_i = \{v_1^i, v_2^i, \ldots, v_{l_i}^i\}$ for each variable V_i, and a finite collection of *constraints* $\mathcal{C} = \{C_1, \ldots, C_m\}$. Each constraint C is a constraint over two variables $\mathrm{Vars}(C) \subseteq \mathcal{V}$. Every constraint C can be viewed as a subset of the Cartesian product of the domains of the variables in $\mathrm{Vars}(C)$ (i.e. the set of tuples that *satisfy* the constraint). Alternatively, C could also be viewed as the *complement* of this product (i.e. the set of tuples that *do not satisfy* the constraint, which are commonly referred to as *no-goods*). As we will see later, although equivalent, these two views of constraints lead to very different linear models.

Let $y_{iu} \in \{true, false\}$ represent the truth value of assignment $V_i = u$ (i.e. $y_{iu} = true$ iff $V_i = u$). The two representations of C_{ij}, as described above, become:

1. Positive Representation: Tuples that satisfy C_{ij}.

$$(P_{CSP})\; C_{ij} ::= \mathcal{R}_{ij} = \{(u, v) \in D_i \times D_j : (u, v) \text{ satisfies } C_{ij}\}$$

For any value $u \in D_i$, the set of tuples $\{(u, v) : (u, v) \in \mathcal{R}_{ij}\}$ can be seen as the logical implication:

$$y_{iu} \rightarrow \bigvee_{v:(u,v)\in\mathcal{R}_{ij}} y_{jv} \tag{1}$$

This states that once we have assigned value u to variable V_i, we must also assign (at least) one of the values v to variable V_j. For this reason, we will call this representation the *positive representation of BCSPs*.

2. Negative Representation: Tuples that violate C_{ij}.

$$(D_{CSP}) \; C_{ij} ::= \overline{\mathcal{R}_{ij}} = \{(u,v) \in D_i \times D_j : (u,v) \text{ violates } C_{ij}\}$$

In this case, for any value $u \in D_i$, the set of tuples $\{(u,v) : (u,v) \in \overline{\mathcal{R}_{ij}}\}$ can be seen as the logical implication:

$$y_{iu} \rightarrow \bigwedge_{v:(u,v)\in\overline{\mathcal{R}_{ij}}} \neg y_{jv} \tag{2}$$

This states that once we have assigned value u to variable V_i, we cannot assign to variable V_j any of the values v. We therefore refer to this representation as the *negative representation of BCSPs*.

Note that, when written as logical implications, there is nothing in the positive representation that prevents us from assigning multiple values to a variable (i.e. $y_{iu} = y_{iv} = true$ for $u \neq v$), just as there is nothing in the negative representation that says that we must assign values to variables (i.e. $y_{iu} = true$ for some $u \in D_i$). However, once we enforce the implicit constraints that each variable V_i must take one and only one value $u \in D_i$, it is not hard to see that:

Lemma 1. *In a BCSP, positive and negative constraint representations are equivalent.*

Proof. Let $s(P_{CSP}) = (y_{iu} \mid 1 \le i \le n, u \in D_i)$ denote a solution of the positive BCSP. If $y_{iu} = true$ in $s(P_{CSP})$, then by (1) for any j there exists a value v such that $(u,v) \in \mathcal{R}_{ij}$ and $y_{jv} = true$. Since V_j can only take one value, it means that for any other value $v_k \in D_j$, $y_{jv_k} = false$. In particular, for all v_k such that $(u,v_k) \notin \mathcal{R}_{ij}$, we have $y_{jv_k} = false$, which means that (2) also holds. If on the other hand $y_{iu} = false$ in $s(P_{CSP})$ then obviously (2) holds as well. Thus, $s(P_{CSP})$ is also a solution for the negative BCSP. Conversely, let $s(D_{CSP}) = (y_{iu} \mid 1 \le i \le n, u \in D_i)$ denote a solution of the negative BCSP. If $y_{iu} = true$ in $s(D_{CSP})$, then for any j, by (2), there exists no value v such that $(u,v) \in \overline{\mathcal{R}_{ij}}$ and $y_{jv} = true$. Since V_j must take at least one value, it means that there exists a value $v_k \in D_j$, with $(u,v_k) \notin \overline{\mathcal{R}_{ij}}$ such that $y_{jv_k} = true$. In other words, (1) also holds. If $y_{iu} = false$ in $s(D_{CSP})$, then (1) holds as well. Thus, $s(D_{CSP})$ is also a solution for the positive BCSP. $\square$

2.2 Linear Models of BCSPs

Our discussion of the two representations for BCSPs in Section 2.1, and in particular the formulation of constraints as logical implications provides the basis to model BCSPs as 0-1 integer linear programs: A logical formula written in *conjunctive normal form* (CNF) can be easily modeled as a set of inequalities involving 0-1 variables. Using the fact that $a \rightarrow b \equiv \neg a \vee b$, and that $a \vee (b \wedge c) \equiv (a \vee b) \wedge (a \vee c)$, we can write (1) and (2) in CNF in the following way:

$$y_{iu} \rightarrow \bigvee_{v:(u,v)\in\mathcal{R}_{ij}} y_{jv} \equiv \neg y_{iu} \vee \left(\bigvee_{v:(u,v)\in\mathcal{R}_{ij}} y_{jv} \right) \tag{3}$$

and

$$y_{iu} \rightarrow \bigwedge_{v:(u,v)\in\overline{\mathcal{R}_{ij}}} \neg y_{jv} \equiv \neg y_{iu} \vee \left(\bigwedge_{v:(u,v)\in\overline{\mathcal{R}_{ij}}} \neg y_{jv} \right) \equiv \bigwedge_{v:(u,v)\in\overline{\mathcal{R}_{ij}}} (\neg y_{iu} \vee \neg y_{jv}) \tag{4}$$

Let $x_{iu} \in \{0,1\}$, $x_{iu} = 1$ iff $y_{iu} = true$. This allows us to rewrite (3) and (4) as linear inequalities in terms of x:

$$(1 - x_{iu}) + \sum_{v:(u,v)\in\mathcal{R}_{ij}} x_{jv} \geq 1 \tag{5}$$

and

$$(1 - x_{iu}) + (1 - x_{jv}) \geq 1, \forall v : (u,v) \in \overline{\mathcal{R}_{ij}} \tag{6}$$

Based on these formula, we are now ready to give the two IP formulations resulting from the positive and negative representations of a BCSP.

Positive IP model (P_{IP})

$$
\begin{aligned}
\max \quad & 0 \\
\text{s.t.} \quad & x_{iu} \leq \sum_{v:(u,v)\in\mathcal{R}_{ij}} x_{jv} & & \forall i, \forall j, \forall u : (u,v) \in \mathcal{R}_{ij} & (7) \\
& \sum_{u\in D_i} x_{iu} = 1 & & \forall i \in \{1,\ldots,n\} & (8) \\
& x_{iu} \in \{0,1\} & & \forall i \in \{1,\ldots,n\}, \forall u \in D_i & (9)
\end{aligned}
$$

Negative IP model (N_{IP})

$$
\begin{aligned}
\max \quad & 0 \\
\text{s.t.} \quad & x_{iu} + x_{jv} \leq 1 & & \forall i, \forall j, \forall (u,v) \in \overline{\mathcal{R}_{ij}} & (10) \\
& \sum_{u\in D_i} x_{iu} = 1 & & \forall i \in \{1,\ldots,n\} & (11) \\
& x_{iu} \in \{0,1\} & & \forall i \in \{1,\ldots,n\}, \forall u \in D_i & (12)
\end{aligned}
$$

The first set of constraints, (7) and (10) encode the constraints of the positive and negative BCSP and are equivalent to the inequalities (5) and (6), respectively. Constraints (8) and (11) state that each variable V_i must take one and only one value from its corresponding domain D_i. They are the same as the implicit constraints we discussed in Section 2.1 and recall that they are the ones that ensure the equivalence of the two models. The last set of constraints (9) and (12) forbid solutions in which x_{iu} take fractional values. These are of course the constraints that make solving both these IPs difficult and are commonly the ones that are relaxed first to solve such problems in operations research. The purpose of the following study is to show that the linear relaxation derived from the positive formulation is strictly stronger than the weakened Lagrangian relaxation of the negative formulation which is used in [12].

3   Integer Programming Relaxations and Filtering

Based on the positive and negative IP models developed in the previous section, we now investigate how they could be used for filtering. In [12] the research is based on the (N_{IP}) model. Two relaxation steps are taken: First, constraints (10) are aggregated and thereby weakened since new fractional solutions are introduced. This was done because the authors felt that the number of constraints in (10) were too many. Then, for a given potential assignment $V_p = q$, a Lagrangian relaxation is considered where all constraints in (10) that do not affect x_{pq} are softened by penalizing a violation rather than enforcing the constraints.

Without the first aggregation step, let us study the polytope of feasible solutions to the Lagrangian subproblem that evolves when we relax all constraints in (10) that do not affect x_{pq}. We call the LP relaxation of the following IP the *consistent value polytope*. Since it can be viewed as derived from the negative formulation, we denote it with $(CV - N)$:

$$(CV - N) : (1) \sum_{k:v_k \in D_i} x_{ik} = 1 \qquad \forall\, 1 \le i \le n$$
$$(2)\ x_{pq} + x_{jl} \le 1 \qquad \forall\, 1 \le j \le n, l \in D_j, (q,l) \in \overline{R_{pj}}$$
$$(3)\ x \in \{0,1\}^n$$

In [12], a large number of aggregated versions of these IPs with changing objectives need to be solved in order to compute the Lagrangian relaxation value. In (CV-N) we did not aggregate any constraints, therefore we achieve a tighter relaxation. What is even more important, there exists a reformulation of (CV-N) that is totally unimodular which allows us to solve these IPs by means of linear programming. Consider

$$(CV - P) : (1) \sum_{k:v_k \in D_i} x_{ik} = 1 \qquad \forall\, 1 \le i \le n$$
$$(2) \sum_{l:(q,l) \in R_{pj}} x_{jl} \ge x_{pq} \qquad \forall\, 1 \le j \le n,\ R_{pj} \in C$$
$$(3)\ x \in \{0,1\}^n$$

Of course, the reformulation of (CV-N) above was motivated by what we called the positive formulation of BCSPs earlier. Formally, we can show:

Lemma 2. *The integer programs (CV-N) and (CV-P) are equivalent.*

Proof. When removing all constraints in the corresponding BCSP that do not involve V_p, then (CV-N) and (CV-P) are exactly the IPs that evolve from the negative and positive formulations of the resulting BCSP. Therefore, the proof of Lemma 1 shows that both IPs are indeed equivalent.

Even though the reformulation from a negative representation of constraints to their positive formulation appears academic at first, it has a very important consequence when the IP model is considered:

Theorem 1. *The integer program (CV-P) is totally unimodular.*

Proof. After eliminating all duplicate and all unit-vector columns from the constraint matrix, neither of which affect total unimodularity, we get the following structure:

$$
\begin{array}{c}
(1) \\[2em] \hline \\ (2)
\end{array}
\left|
\begin{array}{ccccccc}
1 & 0 & \dots & \dots & & & 0 \\
| & | & & & & & | \\
0 & \dots & \dots & & & 0 & 1 \\
\hline
-1 & 1 & 0 & \dots & 0 & & \\
| & | & & & & & | \\
-1 & 0 & \dots & & 0 & 1
\end{array}
\right|
$$

We note that part (1) is now an identity matrix, so it does not affect total unimodularity and can also be eliminated. Then, in part (2), we can eliminate all unit vectors again and we are left with just one column where all entries are -1, i.e., every square submatrix of this column matrix is -1.[1] □

As a consequence of Lemma 2 and Theorem 1, the linear relaxation of (CV-P) describes exactly the convex hull of feasible solutions to (CV-N). Consequently, the Lagrangian subproblem can be solved in polynomial time. This is hardly surprising from a CP perspective: the Lagrangian relaxation rids ourselves of all constraints that do not incorporate variable V_p. Consequently, polynomial arc-consistency methods perform perfectly in terms of filtering effectiveness on the relaxed BCSP.

From an IP perspective, the fact that we found a totally unimodular description of the polytope of the Lagrangian subproblems enables us now to solve a tighter Lagrangian relaxation than the one proposed in [12] simply by means of linear programming: It is a well-known fact that if the Lagrangian subproblem is totally unimodular (it is then sometimes also referred to as exhibiting the integrality property), then the Lagrangian relaxation and the linear continuous relaxation have the same value [1]. To make this point very clear: Theorem 1 states that the *Lagrangian subproblem* is TU. We can therefore solve the Lagrangian relaxation by means of linear programming. Then, the overall linear relaxation is of course not TU (which would indeed come as a big surprise as then the NP-hard BCSP was solvable in P).

In summary, we have shown that the linear relaxation on (P_{IP}), while much easier to solve, is equivalent to the Lagrangian relaxation of (N_{IP}). Consequently, it is *strictly better* than a Lagrangian relaxation on an aggregated version of (N_{IP}). Therefore, the filtering algorithms that we derive from the relaxation based on the positive model are more effective and faster than the one that is considered in [12]. Note that this improvement does not restrict the choice of objective function. We can, as it was suggested in [12], investigate specific assignments by maximizing different specific variables x_{pq} in turn, or we could choose a more global objective function and perform reduced cost filtering.

What we view as even more important here is that in the positive model we have found a way to formulate *binary constraints as collection of integer constraints with tighter linear programming relaxations*. Consequently, we have found an improved formulation that we can use when binary constraints constitute a part of the constraint

---

[1] We owe the idea to this simplified proof to an anonymous referee.

structure of an optimization problem, where it is well-known that it is essential to exploit tight global bounds on the objective.

4   Experimental Evaluation

In our experimental study, we focus purely on feasibility problems and the idea presented in [12] to base an efficient filtering algorithm for BCSPs on mathematical programming methods. In order to base a filtering algorithm on the relaxations that we studied in the previous section, first we follow the second main contribution that was made in [12]. It consists in the introduction of an objective function that is assignment specific. In [12], the authors compute upper bounds on (N_{IP}) augmented by an objective that tries to maximize the value of one single variable x_{pq}. Clearly, if that upper bound drops below 1, then this implies that V_p cannot consistently take value q, and the value is removed from D_p.

In our first series of experiments, we try to reproduce the results reported in [12]. We follow their approach and solve a series of linear relaxations of (P_{IP}) with changing objectives to maximize x_{pq} for the different variables. As a result of Section 3, we know that this filtering technique is at least as effective as the one presented in [12].

The following first set of experiments was run on a 2.4 GHz Intel Hyperthreading processor with 2 GB RAM. In order to provide a close comparison with [12], we use randomly generated BCSPs as our benchmark set. The problems were generated using the random uniform BCSP generator available at [3]. For each experiment, we generated 200 random instances. The test programs were implemented using ILOG Concert 2.0 to interface with Solver 6.0 and CPLEX 9.0 [10]. We generated problems of comparable size and structure: 16 variables, 8 values per domain, and 32 allowed pairs per constraint. We varied the number of constraints from 10 to 120 in increments of 10. In order to verify the validity of the observed trends, we also used a second class of problems, smaller in size, with the following characteristics: 10 variables, 10 values per domain, 32 allowed pairs. For these problems, we varied the number of constraints from 5 to 50 in increments of 5.

To assure that our experimentation is correct, first we solved all problems in our test set to completion using ILOG Solver 6.0 and looked at the distribution of feasible instances. The results are shown in Figure 1, and it is clear that, as the number of constraints approaches 120 for the larger problems and 50 for the smaller ones, the number of feasible instances drops sharply. This is a typical phase transition phenomenon, and an easy-hard-less hard partition is visible by the time required by the solution algorithm on these instances. Figure 2 shows the time needed by a standard CP solver for the large benchmark. It is clearly visible that the hardest instances are those around the phase transition.

The percentage of values filtered using the relaxation of (P_{IP}) at the root node is plotted in Figure 3, for both sets of problem instances (small and large). This confirms the results reported in [12] where it was found that hybrid filtering is *far more effective* than standard arc-consistency algorithms *at the root node*. On our problems, at the root node arc-consistency is unable to filter any substantial number of values, which is why the corresponding line runs close to the 0% horizontal.

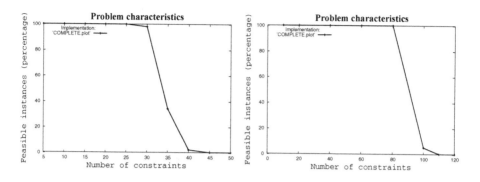

Fig. 1. Percent of solvable instances over the number of constraints for the small (left) and large (right) instances

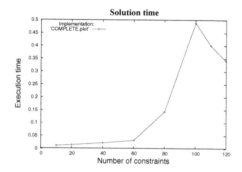

Fig. 2. Time (sec) required by a pure CP solver to solve the BCSPs

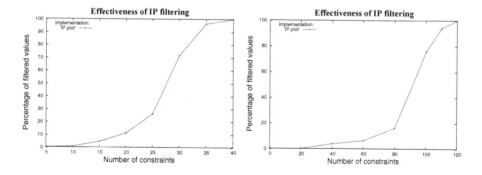

Fig. 3. Percentage of filtered values using the relaxation of (P_{IP}) for the small (left) and large (right) instances

However, what is not made explicit in [12] is that the high percentage of filtered values when the number of constraints gets closer to 120 is actually due to the fact that most problems in that range are *infeasible* and that relaxation-based filtering is able to

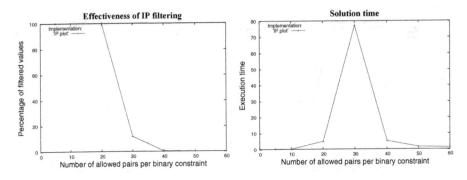

Fig. 4. Effect of the number of allowed pairs on the performance of $\mathbf{P}_{IP_1}$ on the percentage of solvable instances (left) and the propagation time at the root node for the relaxation based filtering method (right)

detect that at the root node! On infeasible problems, the filtering algorithm naturally reports 100% removal of values. It is solely due to this effect that the time per filtered value decreases so massively as it was reported in [12].

It is also important to mention that this performance is obtained only if we iterate the filtering process (i.e. solve relaxations of (P_{IP}) as long as we have at least one filtered value in a round). If we perform a single round of IP filtering (i.e. solve the relaxation of (P_{IP}) once for each variable x_{iu}), the number of filtered values grows only to about 30% as we approach 120 constraints. The large difference can be explained by the fact that, for most problems, the LP relaxation is unable to detect infeasibility in only one round. It typically does so after 4-5 rounds, and then the percentage of filtered values reported jumps to 100%. Obviously, an iterated application of the filtering algorithm increases the effectiveness — but of course it comes at the cost of more cpu time, which, as we will see shortly, is too much to make this kind of filtering worthwhile in the context of random BCSPs.

We also studied the effect of constraining the problem in a different way: namely by varying the number of allowed pairs per constraint instead of varying the number of constraints. For this experiment, we generated problems with 16 variables, 8 values per domain, 60 constraints and varied the number of valid pairs from 5 to 60 in increments of 5. The results are shown in Figure 4. Again, we observe the a clear phase transition, which happens at around 30 pairs per constraint, and that is supported by the problem characteristics observed in Figure 5.

So far we have been able to confirm the results reported in [12]. Now, we were of course curious to see whether the idea of iterated LP-based filtering with assignment specific objectives actually pays off within a tree search. After all, while the improvements in filtering effectiveness at the root node are quite good, what we are ultimately interested in is of course the time that it takes to complete the search and actually solve instances. Therefore, we study how fast the LP filtering is compared to that of the constraint solver. While for virtually every instance that we studied, the first propagation step of the constraint solver failed to remove *any* values from the domains of the variables, the performance of arc-consistency techniques *within a tree search* is far better: When comparing the time the constraint solver took to solve the *entire problem* with

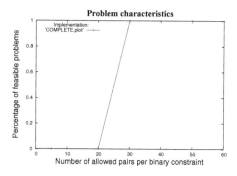

Fig. 5. Characteristics of the instances where we varied the number of pairs per constraint

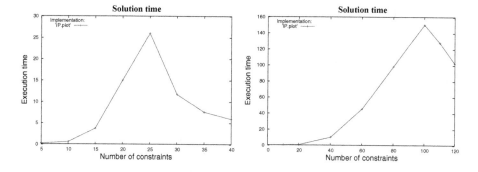

Fig. 6. Time (sec) required by the relaxation of (P_{IP}) to complete filtering

the time it took the LP approach just to filter *at the root node*, we see that the difference is hugely in favor of the constraint solver, by orders of magnitude. While at the
phase transition (where more effective filtering should be of most importance) the time
to filter according to (P_{IP}) only at the root-node peaks at around 150 seconds, standard arc-consistency algorithms complete the entire search in half a second on average
(compare Figures 6 and 2). Consequently, despite the far more effective filtering that
they offer, the algorithms published in [12] are just not worthwhile to solve random
BCSPs.

To summarize our findings to this point: filtering binary constraints based on mathematical programming relaxations is more effective than standard arc-consistency
methods, thanks to the global view on the problem that the relaxation provides. However, even despite our strengthening the relaxation and speeding up its computation
time by showing that an alternative LP relaxation dominates the Lagrangian relaxation
introduced in [12], the idea to use an iterated procedure to filter every domain value
individually is just far too costly to pay off within a tree search — no matter whether
we compare at the under-constrained, over-constrained, or critically constrained region.

In order to improve the efficiency of LP-based filtering, we need to make it less expensive, even at the cost of losing some of its vast effectiveness. Therefore, we tried out

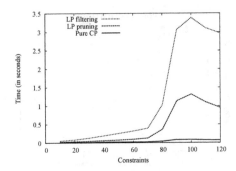

Fig. 7. Comparison of pure CP, LP-pruning, and LP-filtering on random BCSP instances

two different kinds of weakened approaches: The first computes an initial LP-solution to the problem, then it chooses those assignments $X_p = q$ for which the continuous value of x_{pq} is lower than some threshold value $\epsilon > 0$, and finally it sets up a new objective for each of those variable with one filtering iteration only. We refer to this approach as *LP-filtering*. The second approach sacrifices even more effectiveness by using the LP-relaxation just for pruning purposes. It just solves the initial LP once and backtracks if and only if that LP turns out to be infeasible. We denote this second approach with *LP-pruning*.

We performed a second set of experiments to compare the performance of LP-based BCSP propagation and pure CP. The following test results show the averages[2] over 30 runs per data point on a 2 GHz AMD Athlon processor with 512 MB RAM. Figure 7 visualizes the results of our experiments on the large benchmark of random BCSP instances with 16 variables, 8 domain values per variable, and 32 allowed pairs per binary constraint. Again, we see a clear easy-hard-less hard pattern. The comparison shows that a pure CP solver is orders of magnitude faster than LP-filtering and LP-pruning, whereby the latter, despite its weaker effectiveness, is still about twice as fast.

For our last experiment, we were curious whether the good efficiency of pure arc-consistency methods was maybe caused by the unstructured character of our benchmark set. Therefore, we repeated the experiment in Figure 7 on a benchmark set that contains 13-queens instances with additional random binary constraints on the queens. We use the standard CP model where we add one queen-variable for each column and the values that they take correspond to the row index that the queen takes. Alldifferent constraints on rows, columns, and diagonals enforce the 13-queen problem. In Figure 8 we plot the percentage of feasible instances and the solution time by our three solvers over the number of (additional) binary constraints added to the problem.

We see that LP-filtering is able to catch up with LP-pruning, but the comparison with the pure CP solver is devastating. We conclude that the idea of basing a BCSP filtering algorithm on mathematical programming just does not pay off within a tree search.

---

[2] Although we can only visualize averages in our plots, we would like to mention that we also checked the medians and variances to eliminate the possibility that some extreme outliers disproportionally bias the comparison.

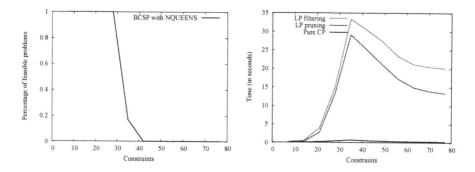

Fig. 8. Comparison of pure CP, LP-pruning, and LP-filtering on random BCSP instances

Of course, it is well-known fact that the use of relaxations is essential for many optimization problems. For this case, when binary constraints are part of the problem, we have introduced a linear programming formulation that approximates the convex hull of feasible integer solutions better than previously studied relaxations. However, for pure feasibility problems, we find that pure CP is the method of choice.

5 Conclusion

We presented a filtering algorithm based on linear programming (LP) models for BC-SPs. The LP relaxations that we used are provably stronger than those developed in [12]. At the same time, filtering can now be based on standard linear programming technology which reduces the programming effort and speeds up the filtering process considerably. Our numerical results show that LP-based filtering for BCSPs leads to more effective filtering. In so far, we can confirm the findings in [12]. However, ultimately we are interested in solving BCSPs by search methods. And in the realm of search, what matters is not so much the effectiveness of filtering methods, but the trade-off between effectiveness and time, i.e. efficiency. Our experiments on random instances show clearly that the additional time for filtering based on mathematical programming does not pay off for BCSPs when compared with standard CP arc-consistency techniques. We therefore reconfirm the common (yet to the best of our knowledge unpublished) belief that hybrid methods perform very poorly on BCSPs: for these problems, leaner and faster inference continues to be the right way to go.

References

1. R.K. Ahuja, T.L. Magnati, J.B. Orlin. *Network Flows*. Prentice Hall, 1993.
2. G. Appa, D. Magos, I. Mourtos, An LP-based proof for the non-existence of a pair of Orthogonal Latin Squares for n=6. *OR Letters*, 32(4): 336–344, 2004.
3. C. Bessiere. Random Uniform CSP Generators. *http://www.lirmm.fr/~bessiere/generator.html*.
4. T. Fahle, U. Junker, S.E. Karisch, N. Kohl, M. Sellmann, B. Vaaben. Constraint programming based column generation for crew assignment. *Journal of Heuristics*, 8(1):59-81, 2002.

5. T. Fahle, M. Sellmann. Cost-Based Filtering for the Constrained Knapsack Problem. *Annals of Operations Research*, 115:73–93, 2002.
6. F. Focacci, A. Lodi, M. Milano. Cutting Planes in Constraint Programming: An Hybrid Approach. *Proceedings of CP-AI-OR'00*, Paderborn Center for Parallel Computing, Technical Report tr-001-2000:45–51, 2000.
7. F. Focacci, A. Lodi, M. Milano. Cost-Based Domain Filtering. *Principles and Practice of Constraint Programming (CP)* Springer LNCS 1713:189–203, 1999.
8. J.N. Hooker. A hybrid method for planning and scheduling. *Proceedings of Principles and Practice of Constraint Programming (CP 2004)*, Springer LNCS 3258:305–316, 2004.
9. J.N. Hooker and G. Ottosson. Logic-based Benders decomposition. *Mathematical Programming*, 96:33–60, 2003.
10. ILOG SA. ILOG Concert 2.0. *http://www.ilog.com*.
11. U. Junker, S.E. Karisch, N. Kohl, B. Vaaben, T. Fahle, M. Sellmann. A Framework for Constraint programming based column generation. *Principles and Practice of Constraint Programming (CP)*, Springer LNCS 1713:261–274, 1999.
12. M.O.I. Khemmoudj, H. Bennaceur, A. Nagih. Combining Arc-Consistency and Dual Lagrangean Relaxation for Filtering CSPs. Proceedings of CPAIOR'05, LNCS 3524:258–272, 2005.
13. H-J. Kim and J. N. Hooker. Solving fixed-charge network flow problems with a hybrid optimization and constraint programming approach. *Annals of Operations Research* 115:95–124, 2002.
14. M. Milano. *Integration of Mathematical Programming and Constraint Programming for Combinatorial Optimization Problems*, Tutorial at CP2000, 2000.
15. G.L. Nemhauser and L.A. Wolsey. *Integer and Combinatorial Optimization*. Wiley, 1988.
16. G. Ottosson, E.S. Thorsteinsson. Linear Relaxation and Reduced-Cost Based Propagation of Continuous Variable Subscripts. *CP-AI-OR'00*, Paderborn Center for Parallel Computing, Technical Report tr-001-2000:129–138, 2000.
17. J.-C. Régin. Cost-Based Arc Consistency for Global Cardinality Constraints. *Constraints*, 7(3-4):387–405, 2002.
18. Meinolf Sellmann. Theoretical Foundations of CP-based Lagrangian Relaxation. *Proceedings of the 10th intern. Conference on the Principles and Practice of Constraint Programming (CP)*, Springer LNCS 3258:634-647, 2004.
19. M. Sellmann. Approximated Consistency for Knapsack Constraints. *CP*, Springer LNCS 2833: 679–693, 2003.
20. M. Sellmann and T. Fahle. Constraint Programming Based Lagrangian Relaxation for the Automatic Recording Problem. *Annals of Operations Research*, 118:17-33, 2003.
21. M. Sellmann and T.Fahle. Coupling Variable Fixing Algorithms for the Automatic Recording Problem. *Annual European Symposium on Algorithms (ESA)*, Springer LNCS 2161: 134–145, 2001.
22. M. Sellmann and W. Harvey. Heuristic Constraint Propagation. *Proceedings of the 8th intern. Conference on the Principles and Practice of Constraint Programming (CP)*, Springer LNCS 2470: 738–743, 2002.

Undirected Forest Constraints

Nicolas Beldiceanu[1], Irit Katriel[2,*], and Xavier Lorca[1]

[1] LINA FRE CNRS 2729, École des Mines de Nantes, FR-44307 Nantes Cedex 3, France
{Nicolas.Beldiceanu, Xavier.Lorca}@emn.fr
[2] BRICS[**], University of Aarhus, Åbogade 34, Århus, Denmark
irit@daimi.au.dk

Abstract. We present two constraints that partition the vertices of an undirected n-vertex, m-edge graph $\mathcal{G} = (\mathcal{V}, \mathcal{E})$ into a set of vertex-disjoint trees. The first is the *resource-forest* constraint, where we assume that a subset $R \subseteq \mathcal{V}$ of the vertices are *resource* vertices. The constraint specifies that each tree in the forest must contain at least one resource vertex. This is the natural undirected counterpart of the *tree* constraint [1], which partitions a directed graph into a forest of directed trees where only certain vertices can be tree roots. We describe a hybrid-consistency algorithm that runs in $\mathcal{O}(m + n)$ time for the *resource-forest* constraint, a sharp improvement over the $\mathcal{O}(mn)$ bound that is known for the directed case. The second constraint is *proper-forest*. In this variant, we do not have the requirement that each tree contains a resource, but the forest must contain only *proper* trees, i.e., trees that have at least two vertices each. We develop a hybrid-consistency algorithm for this case whose running time is $\mathcal{O}(mn)$ in the worst case, and $\mathcal{O}(m\sqrt{n})$ in many (typical) cases.

1 Introduction

Constraints that describe graph properties were considered from an early stage of constraint programming research. Some examples are the *Hamiltonian circuit* and *spanning tree* constraints of ALICE [2] that were later followed by the *cycle* [3] and *path* constraints [4], which were, respectively, introduced in later versions of CHIP [5] and Ilog Solver [6]. A more recent example is the *tree*(NTREE, VER) constraint [1], which receives an integer variable NTREE and a graph described by the vertex-list VER. Some of the vertices are specified as "possible roots" and the constraint determines that the graph consists of NTREE directed trees, each of which is rooted at a "possible root".

A natural network design problem is the following. We are given an undirected graph $\mathcal{G} = (\mathcal{V}, \mathcal{E})$ where $R \subseteq \mathcal{V}$ is a set of vertices that correspond to a certain resource, e.g., a printer. The remaining vertices represent the tasks (clients/users). The problem is to cover the vertices of the graph with trees (networks) such that every tree contains at least one vertex from R (every network has a printer). We could replace each undirected edge by two anti-parallel directed arcs and then use the *tree* constraint which can be filtered in $\mathcal{O}(mn)$ time. However, undirected graphs are often much simpler than directed graphs. Indeed, we will show that the *resource-forest* constraint, the undirected counterpart of the *tree* constraint, can be filtered in $\mathcal{O}(m + n)$ time.

---
[*] Supported by the Danish Research Agency (grant # 272-05-0081).
[**] Basic Research in Computer Science, funded by the Danish National Research Foundation.

J.C. Beck and B.M. Smith (Eds.): CPAIOR 2006, LNCS 3990, pp. 29–43, 2006.

We then turn to another variant of the problem, the *proper-forest*(NTREE, VER) constraint which specifies that the graph is a forest of NTREE *proper trees* as defined by A. CAYLEY in 1889 [7]: A proper tree is a connected, cycle-free graph with at least two vertices. Note that with the *proper-forest* constraint the issue of resources does not exist (or, equivalently, all vertices are resource vertices). It can apply to the design of fault-tolerant networks, where each network needs to contain at least two computers so that each computer can back up the other. The *proper-forest* variant appears to be more complex than the *resource-forest* one; we show a filtering algorithm for *proper-forest* whose running time is $\mathcal{O}(mn)$ in the worst case, and is dominated by the complexity of determining which edges of the graph belong to at least one maximum cardinality matching. As we will see, the worst case occurs when the domain of NTREE is ground and contains a certain value. In all other cases, the algorithm's bottleneck is finding a maximum matching in the graph, which can be done in $\mathcal{O}(m\sqrt{n})$ time.

Since both constraints involve integer and set variables, our filtering algorithms achieve *hybrid-consistency*, which is a type of consistency suitable for this context, introduced by Bessière *et al.* [8]. It will be formally defined in Section 2, but intuitively hybrid-consistency means that every integer variable is arc-consistent and every set variable is bound-consistent.

The rest of the paper is organized as follows. Section 2 provides the necessary background on constraint programming and graph theory. Section 3 introduces the *resource-forest* and *proper-forest* constraints. Sections 4 and 5, respectively, present filtering algorithms for the *resource-forest* and *proper-forest* constraints. Finally, Section 6 contains a summary of the known results on filtering tree-partitioning constraints.

2   Preliminaries

In this section we recall some of the constraint programming and graph theory terminology that we use in the rest of the paper.

Definition 1. *An integer variable V ranges over a finite set of integers denoted by $\mathcal{D}(V)$. The extremal values in $\mathcal{D}(V)$ are denoted by $\min(V)$ and $\max(V)$.*

Definition 2. *The domain of a set variable V is a set of sets of integers. It is specified by its lower bound $\underline{V}$ and its upper bound $\overline{V}$ and contains all sets that contain $\underline{V}$ and are contained in $\overline{V}$. When the set variable V is ground we have that $\underline{V} = \overline{V}$. The values in $\underline{V}$ are the mandatory values of V and the values in $\overline{V} \setminus \underline{V}$ are its potential values.*

Definition 3 (Hybrid-consistency [8]). *A constraint C defined on the integer variables $V_1^d, \ldots, V_l^d$ and the set variables $V_{l+1}^s, \ldots, V_n^s$ is hybrid-consistent iff:*

1. *For every pair (V^d, v) such that V^d is an integer variable of C and $v \in \mathcal{D}(V^d)$, there exists at least one solution to C in which V^d is assigned the value v.*
2. *For every pair (V^s, v) such that V^s is a set variable of C, if $v \in \underline{V^s}$ then v belongs to the set assigned to V^s in all solutions to C and if $v \in \overline{V^s} \setminus \underline{V^s}$ then v belongs to the set assigned to V^s in at least one solution and is excluded from this set in at least one solution.*

Definition 4. *Graph theoretic terms [9] . Let $\mathcal{G} = (\mathcal{V}, \mathcal{E})$ be an undirected graph. A path in $\mathcal{G}$ is a sequence of vertices, such that every two consecutive vertices are joined* by an edge. A path is simple *if every vertex appears on it at most once.* A bridge *in $\mathcal{G}$ is an edge $e \in \mathcal{E}$ whose removal increases the number of maximal connected components of $\mathcal{G}$.* A matching *in $\mathcal{G}$ is a set $M \in \mathcal{E}$ of edges such that every vertex in $\mathcal{V}$ is incident on at most one edge from M.*

3 The *resource-forest* and *proper-forest* Constraints

In this section, we define and motivate the *resource-forest* and the *proper-forest* constraints, introduce their corresponding graphs, define them formally and provide examples that illustrates the semantics of the constraint as well as the problem of filtering them to hybrid-consistency.

In many graph-partitioning problems, the vertex set of the graph is the union of a set of *resource* vertices and a set of *task* vertices. Independently of the pattern used to cover the graph, this distinction between the two types of vertices comes from the need that each partition has to contain at least one resource vertex. This distinction between resource and task vertices was already introduced in the *cycle* constraint [3]. An example of application for the *cycle* constraint is the vehicle routing problem which consists in allocating a set of trucks (resources) to deliver goods to a set of shops (tasks). The *resource-forest* constraint, on the other hand, can be used to model the problem of allocating hardware resources in a network. Here, a resource represents a piece of hardware (e.g., a printer) and a task represents a computer. The solution (forest) is a network in which each computer is connected with at least one printer.

In 1889, A. CAYLEY [7] introduced the definition of a *tree* as a connected graph without cycles which contains at least two vertices. We will call Cayley's tree a *proper tree*. A *proper forest*, then, is a set of proper trees. The *proper-forest* constraint partitions the vertices of an undirected graph into a set of vertex-disjoint proper trees.

Formally, each of the *resource-forest*(NTREE, VER) and *proper-forest*(NTREE, VER) constraints is defined on an integer variable NTREE and an array VER which is essentially an adjacency-list representation of a graph. Each item $v \in$ VER has the following attributes, which complete the description of the graph:

- I is an integer between 1 and n, which can be interpreted as the *label* of v.
- N is a set variable whose elements are integers (vertex labels) between 1 and n. The lower and upper bounds of N can respectively be interpreted as the *set of mandatory neighbors* and the *set of mandatory or potential neighbors* of v.
- R (only for the *resource-forest* constraint) is a boolean flag which is true if the vertex is a resource vertex and false if it is a task vertex.

Notation: For each $1 \leq i \leq n$, VER[i] is the i-th item of the VER collection, while VER[i].I, VER[i].N, and VER[i].R, respectively, denote the I, N and R attributes of VER[i].

When speaking of global constraints, it is often convenient to reason about a graph that models the constraint rather than directly about the constraint (see, e.g., the *cycle* [3], *path* [10,4], and *alldifferent* [11] constraints). In the case of the *resource-forest* and *proper-forest* constraints, the graph model is obvious: It is the undirected graph $\mathcal{G} = (\mathcal{V}, \mathcal{E})$ in which the vertices represent the elements of VER and the edges represent the neighborhood relations between them. Each edge of the graph has a *type* (solid or dotted) which indicates whether it represents a mandatory (solid) or a potential (dotted) neighborhood relation.

Since it can be easily achieved by a linear-time preprocessing step, we will assume in the rest of this paper that the associated graph does not contain loops and that the N sets of the vertices are symmetric, i.e., $i \in \text{VER}[j].\underline{\text{N}} \Leftrightarrow j \in \text{VER}[i].\underline{\text{N}}$ (in this case we will say that i and j are *mandatory neighbors*) and $i \in \text{VER}[j].\overline{\text{N}} \Leftrightarrow j \in \text{VER}[i].\overline{\text{N}}$ (in this case, if i and j are not mandatory neighbors then they are *possible neighbors*). Note that the preprocessing step may find that the constraint has no solution. This can happen if $i \in \text{VER}[i].\underline{\text{N}}$ (there is a mandatory loop) or $\exists i, j : i \in \text{VER}[j].\underline{\text{N}} \wedge j \notin \text{VER}[i].\overline{\text{N}}$ (i is a mandatory neighbor of j but j is not a possible neighbor of i). Formally, the graph associated with a *resource-forest* or a *proper-forest* constraint is defined as follows.

Definition 5. *For a* resource-forest(NTREE, VER) *or a* proper-forest(NTREE, VER) *constraint, the associated graph is the undirected graph* $\mathcal{G} = (\mathcal{V}, \mathcal{E})$ *where* $\mathcal{V} = \{v_i : i \in [1, n]\}$ *and* $(v_i, v_j) \in \mathcal{E}$ *iff* $i \in \text{VER}[j].\overline{\text{N}} \wedge j \in \text{VER}[i].\overline{\text{N}}$. *We distinguish between solid and dotted edges: The edge* $(v_i, v_j) \in \mathcal{E}$ *is solid if* i *and* j *are mandatory neighbors and dotted if* i *and* j *are potential neighbors. Finally, we denote the number of edges in the graph,* $|\mathcal{E}|$, *by* m.

In the case of the resource-forest *constraint, we distinguish between resource vertices and task vertices; the set* R *of resource vertices is* $\{v_i : \text{VER}[i].\text{R} = \text{true}\}$. *All vertices in* $\mathcal{V} \setminus$ R *are task vertices.*

The *resource-forest* constraint specifies that its associated graph is a forest where each tree contains at least one resource and the *proper-forest* constraint specifies that its associated graph is a proper forest. Formally:

Definition 6. *A ground* resource-forest(NTREE, VER) *constraint is satisfied iff the following conditions hold:*

(1) $\forall i \in [1, n] : \text{VER}[i].\text{I} = i$,
(2) $\forall i, j \in [1, n] : i \in \text{VER}[j].\text{N} \Leftrightarrow j \in \text{VER}[i].\text{N}$ *(i.e., the neighborhood relation is symmetric),*
(3) *The associated graph* $\mathcal{G}$ *consists of* NTREE *maximal connected components such that each component contains at least one vertex from* R *and does not contain any cycles.*

Definition 7. *A ground* proper-forest(NTREE, VER) *constraint is satisfied iff the following conditions hold:*

(1) $\forall i \in [1, n] : \text{VER}[i].\text{I} = i$,
(2) $\forall i, j \in [1, n] : i \in \text{VER}[j].\text{N} \Leftrightarrow j \in \text{VER}[i].\text{N}$ *(i.e., the neighborhood relation is symmetric),*
(3) *The associated graph* $\mathcal{G}$ *is a forest of* NTREE *(vertex-disjoint) proper trees.*

The following example will be used throughout the paper.

Example 1. Part (A) of Figure 1 shows the input graph $\mathcal{G}$, where the mandatory edges are solid and the rest are dotted. Parts (B) and (C) of the figure show two possible solutions to the *resource-forest* constraint on this graph, one with two trees and the other with three trees. Parts (B) and (D) show two solutions to the *proper-forest* constraint on this graph, with two and seven proper trees, respectively.

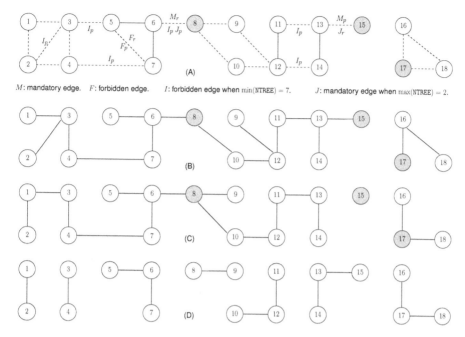

Fig. 1. (A) An undirected graph with 3 (grayed) resource vertices (B) a solution with 2 trees for the *resource-forest* and *proper-forest* constraints (C) a solution with 3 trees for the *resource-forest* constraint (D) a solution with 7 proper trees for the *proper-forest* constraint. Notice that each kind of edges (M, F, I and J) are indiced by p in the case of the *proper-forest* and by r in the case of the *resource-forest*.

A hybrid-consistency algorithm for the *resource-forest* constraint on $\mathcal{G}$ should discover that regardless of the contents of the domain of NTREE, the edge marked by M_r, i.e., the edge $(6, 8)$, is mandatory and that the edge $(5, 7)$ (marked by F_r) is forbidden. Furthermore, it should set the domain of NTREE to be the intersection of its previous value with $\{2, 3\}$. If, in the input, $\mathcal{D}(\text{NTREE}) = \{2\}$, the algorithm should also discover that the edge marked by J_r, i.e., the edge $(13, 15)$, is mandatory. Section 4 will justify this pruning.

A hybrid-consistency algorithm for the *proper-forest* constraint on $\mathcal{G}$ should discover that the edge $(13, 15)$ (marked by M_p) is mandatory and that the edge $(5, 7)$ (marked by F_p) is forbidden. Next, it should set the domain of NTREE to be the intersection of its previous value with $\{2, 3, 4, 5, 6, 7\}$. If, in the input, $\mathcal{D}(\text{NTREE}) = 2$, the algorithm should discover that the edge marked by J_p, i.e., the edge $(6, 8)$, is mandatory. Finally, if $\mathcal{D}(\text{NTREE}) = \{7\}$ in the input, the algorithm should discover that the edges marked by I_p, i.e., $(2, 3)$, $(3, 5)$, $(4, 7)$, $(6, 8)$, $(11, 13)$, and $(12, 14)$, are forbidden. Section 5 will justify this pruning.

Before we can describe the filtering algorithms for the *resource-forest* and *proper-forest* constraints, we need to define the mandatory graph $\mathcal{G}_{TRUE}$ and the possible graph $\mathcal{G}_{MAYBE}$ associated with a graph $\mathcal{G}$. An example appears in Figure 2.

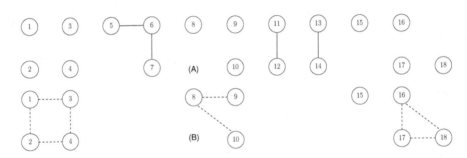

Fig. 2. The graphs (A) $\mathcal{G}_{TRUE}$ and (B) $\mathcal{G}_{MAYBE}$ associated with the graph of Figure 1, Part (A)

Definition 8. *(Mandatory graph) Given a* resource-forest *or* proper-forest *constraint and its associated graph* $\mathcal{G}$, *the graph* $\mathcal{G}_{TRUE}$ *contains all edges that must be in the forest. Formally,* $\mathcal{G}_{TRUE} = (V, \mathcal{E}_{TRUE})$, *where* $\mathcal{E}_{TRUE}$ *is the set of solid edges in* $\mathcal{G}$.

Definition 9. *(Possible graph) Given a* resource-forest *or* proper-forest *constraint and its associated graph* $\mathcal{G}$, *the graph* $\mathcal{G}_{MAYBE}$ *contains the subgraph induced by the vertices that are not incident on mandatory edges. Formally,* $\mathcal{G}_{MAYBE} = (V_{MAYBE}, \mathcal{E}_{MAYBE})$ *where* V_{MAYBE} *contains all vertices that are isolated in* $\mathcal{G}_{TRUE}$ *and* $\mathcal{E}_{MAYBE} = \mathcal{E} \cap (V_{MAYBE} \times V_{MAYBE})$.

4 Filtering the *resource-forest* Constraint

4.1 Checking Feasibility of the *resource-forest* Constraint

Theorem 1 specifies necessary and sufficient conditions for the existence of a solution to a *resource-forest* constraint. The first two conditions ensure that it is possible to partition the graph into a forest with a resource in every tree and the third ensures that the number of trees in the forest is within the domain of NTREE.

Theorem 1. *There is a solution to the* resource-forest(NTREE, VER) *constraint iff the following conditions hold:*

(1) $\mathcal{G}_{TRUE}$ *does not contain any cycles.*
(2) *Every maximal connected component of* $\mathcal{G}$ *contains at least one resource vertex.*
(3) $\mathcal{D}(\text{NTREE}) \cap [\text{MINTREE}, \text{MAXTREE}] \neq \emptyset$, *where* MINTREE *is the number of maximal connected components in* $\mathcal{G}$ *and* MAXTREE *is the number of maximal connected components of* $\mathcal{G}_{TRUE}$ *that contain at least one resource vertex.*

Proof. **Sufficiency:** To prove that the three conditions are sufficient, we assume that they hold and show that for every value $k \in [\text{MINTREE}, \text{MAXTREE}]$, we can construct a spanning forest of $\mathcal{G}$ with k trees, each of which contains a resource vertex.

Case 1: $k = \text{MAXTREE}$. Let $\mathcal{T} = \{C_1, \cdots, C_p\}$ be the maximal connected components of $\mathcal{G}_{TRUE}$. By definition, exactly MAXTREE of them contain at least one resource vertex. By Condition (2), every component which does not contain a resource vertex is connected by a path of $\mathcal{G}$ to a component which does. To obtain a solution of k trees, we merge every component that does not contain any resource vertices with one that does, and output a spanning tree of each component.

Case 2: $k <$ MAXTREE. We first construct a forest of MAXTREE trees as in Case 1 and then merge trees until there are k trees: While there are too many trees, select two trees which are connected by an edge e and merge them by including e in the forest. Since MINTREE is the number of maximal connected components in $\mathcal{G}$, as long as $k >$ MINTREE we are guaranteed to find two trees that can be merged.

Necessity: Clearly, if $\mathcal{G}_{TRUE}$ contains a cycle, the solution cannot be a forest. If $\mathcal{D}(\text{NTREE}) \cap [\text{MINTREE}, \text{MAXTREE}] = \emptyset$ then we have $\max(\text{NTREE}) <$ MINTREE or $\min(\text{NTREE}) >$ MAXTREE. In either case, the constraint is infeasible: We cannot create less than MINTREE trees because a tree must be connected. To see that we cannot create more than MAXTREE trees, note a connected component of $\mathcal{G}_{TRUE}$ cannot be broken, so every component can contribute at most one tree. Furthermore, the vertices of a component that does not contain a resource vertex must belong to the same tree as the vertices of a component that does contain a resource vertex. In other words, a component cannot contribute a tree to the forest if it does not contain a resource vertex.   □

4.2   Hybrid-Consistency of the *resource-forest* Constraint

Figure 3 shows the algorithm for filtering a *resource-forest* constraint to hybrid-consistency. First, it verifies that the constraint has at least one solution, using the characterization of Theorem 1. Lines 3 to 7 prune with respect to the fact that a solution must be a forest (while ignoring the cardinality of the forest and the condition on the resources). In Line 7 the algorithm removes any dotted edge (u, v) where u and v are connected by solid edges; since the solid edges must be in the solution, this dotted edge would create a cycle (e.g., the edge $(5, 7)$ in Part (A) of Figure 1). In Lines 8 to 10 it identifies dotted edges which must be in the solution because removing them would separate one or more vertices from the resources, and makes them solid (e.g., the edge $(6, 8)$ in Part (A) of Figure 1). Line 11 narrows the domain of NTREE.

Lines 12 to 17 are executed only when the domain of NTREE is ground. In this case, the number of trees in a solution is fixed and if it is equal to MINTREE (as defined in the statement of Theorem 1), all bridges of $\mathcal{G}$ are mandatory and are turned solid, because otherwise the number of connected components of the graph, and therefore also the number of trees in any solution, is strictly larger than MINTREE (e.g., the edge $(13, 15)$ in Part (A) of Figure 1). On the other hand, if the value of NTREE is equal to MAXTREE, then every maximal connected component of $\mathcal{G}_{TRUE}$ which contains a resource should contribute a tree to the solution, so a dotted edge between two such components must not be in the forest and is removed.

4.3   Correctness

In the full version of this paper, we prove that the algorithm achieves hybrid-consistency by showing that:

1. We did not remove an edge from the graph or a value from NTREE that belong to a solution.

2. Every remaining edge in $\mathcal{G}$ and value in $\mathcal{D}(\text{NTREE})$ participates in at least one solution, and every remaining dotted edge is excluded from at least one solution.
3. Every edge that we turned from dotted to solid participates in all solutions.

```
 1. if the constraint has no solution (see Theorem 1) then
 2.    report failure and exit;
 3. Compute the maximal connected components (CCs) of G_TRUE.
 4. foreach v ∈ V do
 5.    C(v) ← the CC of G_TRUE that contains v;
 6. foreach dotted edge (u,v) ∈ E do
 7.    if C(u) = C(v) then remove (u,v) from the graph;
 8. foreach dotted edge e do
 9.    if removing e creates a CC of G without resource vertices then
10.       Turn e into a solid edge;
11. D(NTREE) ← D(NTREE) ∩ [MINTREE, MAXTREE];
12. if D(NTREE) = {MINTREE} then
13.    foreach dotted edge e that is a bridge of G do
14.       Turn e into a solid edge;
15. if D(NTREE) = {MAXTREE} then
16.    foreach dotted edge e = (u,v) where C(u) ≠ C(v) and
              both C(u) and C(v) contain a resource vertex do
17.          Remove e from the graph;
```

Fig. 3. A hybrid-consistency algorithm for the *resource-forest* constraint

4.4 Complexity

The steps of the algorithm excluding Lines 8 to 10 require cycle detection, computing maximal connected components and identifying bridges, all of which can be done in linear time [12, p.18]. We now show that Lines 8 to 10 also take linear time. Clearly, an edge whose removal creates a maximal CC without a resource is a bridge of $\mathcal{G}$. But not all bridges have this property. We create a reduced graph $\mathcal{H}$ by contracting every biconnected component of $\mathcal{G}$ into a single vertex. The graph $\mathcal{H}$ is a tree whose edges are exactly the bridges of $\mathcal{G}$. We will say that a vertex of $\mathcal{H}$ is a resource vertex if one of the vertices of $\mathcal{G}$ that were contracted into it is a resource vertex, i.e., if the biconnected component it represents has a resource. We need to identify which of the edges of $\mathcal{H}$ are edges whose removal would create a connected component of $\mathcal{H}$ without resources. In other words, we have reduced our problem to the same problem on trees. This holds for an edge if one of its endpoints is the root of a subtree without resources. We select an arbitrary resource vertex of $\mathcal{H}$ and perform a DFS traversal of $\mathcal{H}$ starting at this vertex. Whenever we backtrack from a vertex v, we communicate to its parent p whether a resource was encountered in the subgraph rooted at v. If not, then the edge (p, v) is turned into a solid edge (if it is not solid already).

Thus, we have shown:

Theorem 2. *The algorithm of Figure 3 filters the* resource-forest *constraint to hybrid-consistency in* $\mathcal{O}(m + n)$ *time.*

5   Filtering the *proper-forest* Constraint

5.1   Checking Feasibility of the *proper-forest* Constraint

Theorem 3 specifies the conditions for the existence of a solution to a *proper-forest* constraint. The first two conditions ensure that it is possible to partition the graph into a proper forest and the third ensures that the number of proper trees in the proper forest is within the domain of NTREE.

Theorem 3. *There is a solution to the* proper-forest(NTREE, VER) *constraint iff the following conditions hold:*

(1) $\mathcal{G}$ *does not have isolated vertices,*
(2) $\mathcal{G}_{TRUE}$ *does not contain any cycles,*
(3) $\mathcal{D}(\text{NTREE}) \cap [\text{MINTREE}, \text{MAXTREE}] \neq \emptyset$, *where* MINTREE *is the number of maximal connected components in* $\mathcal{G}$ *and* MAXTREE *is the number of maximal connected components of cardinality at least two in* $\mathcal{G}_{TRUE}$ *plus the cardinality of a maximum cardinality matching in* $\mathcal{G}_{MAYBE}$.

Proof. **Sufficiency:** To prove that the three conditions are sufficient, we assume that they hold and show that for every value $k \in [\text{MINTREE}, \text{MAXTREE}]$, we can construct a spanning forest of $\mathcal{G}$ with k proper trees. We begin with $k = \text{MAXTREE}$ and proceed to an arbitrary $k \in [\text{MINTREE}, \text{MAXTREE}]$.

- Let $\mathcal{T} = \{T_1, \cdots, T_p\}$ be a maximum spanning forest of $\mathcal{G}_{TRUE}$, i.e., each T_i is an isolated vertex from $\mathcal{G}_{TRUE}$ or a spanning tree of a maximal connected component of $\mathcal{G}_{TRUE}$. Observe that a tree T_i of cardinality one is also a vertex of $\mathcal{G}_{MAYBE}$.
- To construct a spanning forest of cardinality MAXTREE, compute a maximum cardinality matching $\mathcal{M}$ in $\mathcal{G}_{MAYBE}$ and modify $\mathcal{T}$ as follows:
  - By definition of $\mathcal{G}_{MAYBE}$, each matching edge connects two singletons T_i and T_j of $\mathcal{T}$. Merge them into a tree of cardinality two.
  - For every vertex $u \in V_{MAYBE}$, corresponding to a tree T_i of cardinality one which is unmatched by $\mathcal{M}$, select a neighbor v of u and include the edge (u, v) in the spanning forest. In other words, merge the tree T_i with the tree T_v to which v belongs. Condition 1 guarantees that this is possible. It is easy to see that the forest consists of exactly MAXTREE trees.
- If $k < \text{MAXTREE}$, merge proper trees until there are k proper trees: While there are too many proper trees, select two proper trees that are connected by an edge e from $\mathcal{G}_{MAYBE}$ and merge them by including e in the proper forest. Since MINTREE is the number of maximal connected components in $\mathcal{G}$, as long as $k > \text{MINTREE}$ we are guaranteed to find two proper trees that can be merged.

Necessity: It remains to show that all three conditions are necessary. Clearly, if $\mathcal{G}$ contains an isolated vertex v, then v does not belong to a subgraph of $\mathcal{G}$ which is a proper tree and if $\mathcal{G}_{TRUE}$ contains a cycle, the solution must contain a cycle so it cannot be a proper forest. Finally, if $\mathcal{D}(\text{NTREE}) \cap [\text{MINTREE}, \text{MAXTREE}] = \emptyset$ then we have $\max(\text{NTREE}) < \text{MINTREE}$ or $\min(\text{NTREE}) > \text{MAXTREE}$. In either case, the constraint does not have a solution: We cannot create less than MINTREE proper trees because a proper tree must be connected. To see that we cannot create more than MAXTREE proper trees, note that the number of proper trees that $\mathcal{G}_{TRUE}$ can contribute is at most the number of connected components it has (we cannot break a connected component of $\mathcal{G}_{TRUE}$) and that a vertex of $\mathcal{G}_{MAYBE}$ can either form a new proper tree with another vertex from $\mathcal{G}_{MAYBE}$, or be merged into a previously existing proper tree (and not contribute to the tree-count). Clearly, a maximum cardinality matching contributes the largest possible number of proper trees from $\mathcal{G}_{MAYBE}$.   □

5.2 Hybrid-Consistency of the *proper-forest* Constraint

Figure 4 shows the algorithm for filtering a *proper-forest* constraint to hybrid-consistency. First, it verifies that the constraint has at least one solution, using the characterization of Theorem 3, and exits if the constraint is inconsistent. Lines 3 to 7 and 18 to 19 prune with respect to the fact that a solution must be a proper forest (while ignoring the cardinality of the forest). In Line 7 the algorithm removes any dotted edge (u, v) (e.g., the edge $(5, 7)$ in Part (A) of Figure 1) where u and v are connected by a path of solid edges; since the solid edges must be in the solution, this dotted edge would create a cycle. Line 19 identifies dotted edges which must be in the solution because removing them would isolate a vertex, and makes them solid (e.g., the edge $(13, 15)$ in Part (A) of Figure 1). Line 8 narrows the domain of NTREE.

Lines 9 to 17 are executed only when the domain of NTREE is ground. In this case, the number of trees in a solution is fixed and if it is equal to either MINTREE or MAXTREE (these values are defined in the statement of Theorem 3), more filtering is possible: If $\mathcal{D}(\text{NTREE}) = \{\text{MINTREE}\}$ then all bridges of $\mathcal{G}$ are mandatory and are turned solid, because otherwise the number of connected components of the graph, and therefore also the number of trees in any solution, is strictly larger than MINTREE. In the example in Part (A) of Figure 1, the edge $(6, 8)$ (which is marked by J_p) is mandatory when $\mathcal{D}(\text{NTREE}) = \{2\}$. If $\mathcal{D}(\text{NTREE}) = \{\text{MAXTREE}\}$ then three sets of edges are forbidden and are removed from the graph; we will show that including any one of these edges in a solution would reduce the number of trees we can construct to strictly less than MAXTREE. For example, if $\mathcal{D}(\text{NTREE}) = \{7\}$, the edges marked by I_p in Part (A) of Figure 1 are removed: $(11, 13)$ and $(12, 14)$ in Line 15, $(2, 3)$ in Line 16, and $(3, 5)$, $(4, 7)$ and $(8, 6)$ in Line 17.

5.3 Correctness

To prove correctness of the algorithm, we will show that:

1. We did not remove an edge from the graph or a value from NTREE that belong to a solution (Lemma 1).
2. Every remaining edge in $\mathcal{G}$ and value in $\mathcal{D}(\text{NTREE})$ participates in at least one solution, and every remaining dotted edge is excluded from at least one solution (Lemma 2).
3. Every edge that we turned from dotted to solid participates in all solutions (Lemma 3).

Lemma 1. *The algorithm in Figure 4 never removes an edge from the graph or a value from $\mathcal{D}(\text{NTREE})$ that belongs to a solution.*

Proof. Let (u, v) be an edge that was removed by the algorithm and assume that there is a solution S that contains (u, v). If (u, v) was removed in Line 7, then there is a path of solid edges from u to v, and these edges are also in the solution. But then the forest S contains a cycle, a contradiction. So we must assume that (u, v) was removed in Lines 13 to 17. In this case, we know that the number of trees in S is equal to MAXTREE (the number of CCs of $\mathcal{G}_{TRUE}$ of cardinality at least two plus the cardinality of a maximum matching in $\mathcal{G}_{MAYBE}$.) We will show that if there still is a solution S' when the constraint is on the graph $\mathcal{G}'$, which is obtained from $\mathcal{G}$ by turning (u, v) into a solid edge, then this violates Theorem 3 because S' has more than MAXTREE$'$ trees (where MAXTREE$'$ is the MAXTREE value for $\mathcal{G}'$). If (u, v) was removed in Line 15, then MAXTREE$'$ = MAXTREE − 1 because u and v are not in $\mathcal{G}_{MAYBE}$ and two non-trivial connected components of $\mathcal{G}_{TRUE}$ have been merged. The solution $S' = S$ has MAXTREE (i.e., > MAXTREE$'$) trees. If (u, v) was removed in Line 16, then u and v form a size-2 maximal connected component in $\mathcal{G}'_{TRUE}$. This increases MAXTREE by one. On the other hand, the cardinality of a maximum matching in $\mathcal{G}'_{MAYBE} = \mathcal{G}_{MAYBE} \setminus \{u, v\}$ is two less than that of $\mathcal{G}_{MAYBE}$, because otherwise (u, v) belongs to a maximum matching in $\mathcal{G}_{MAYBE}$. So MAXTREE$'$ = MAXTREE − 1 and $S' = S$ is a solution with MAXTREE trees. Finally, if (u, v) was removed in Line 17, then turning (u, v) into a solid edge inserts v into the CC of u in $\mathcal{G}'_{TRUE}$. Since v is not in $\mathcal{G}'_{MAYBE}$, the cardinality of a maximum matching in $\mathcal{G}'_{MAYBE}$ is one less than that of $\mathcal{G}_{MAYBE}$ (otherwise $\mathcal{G}_{MAYBE}$ has a maximum matching in which v is not saturated, a contradiction). Again, we get that MAXTREE$'$ = MAXTREE − 1 and $S' = S$ is a solution with MAXTREE trees. If the algorithm removes a useful value from $\mathcal{D}$(NTREE), this clearly contradicts Theorem 3.   □

```
 1. if the constraint has no solution (see Theorem 3) then
 2.    report failure and exit;
 3. Compute the maximal connected components (CCs) of G_TRUE.
 4. foreach v ∈ V do
 5.    C(v) ← the CC of G_TRUE that contains v;
 6. foreach dotted edge (u,v) ∈ E do
 7.    if C(u) = C(v) then remove (u,v) from the graph;
 8. D(NTREE) ← D(NTREE) ∩ [MINTREE, MAXTREE];
 9. if D(NTREE) = {MINTREE} then
10.    foreach dotted edge (u,v) that is a bridge of G do
11.       Turn (u,v) into a solid edge;
12. if D(NTREE) = {MAXTREE} then
13.    foreach dotted edge (u,v) do
14.       remove (u,v) from G if one of the following holds:
15          a. |C(u)| > 1, |C(v)| > 1, and C(u) ≠ C(v).
16.         b. (u,v) ∈ E_MAYBE but does not belong to any
                maximum matching in G_MAYBE.
17.         c. |C(u)| > 1 and v is saturated in every maximum
                cardinality matching of G_MAYBE.
18. foreach dotted edge (u,v) ∈ E do
19.    if u is a leaf in G then turn (u,v) into a solid edge;
```

Fig. 4. A hybrid-consistency algorithm for the *proper-forest* constraint

Lemma 2. *After applying the algorithm of Figure 4, every remaining edge in $\mathcal{G}$ and value in $\mathcal{D}(\text{NTREE})$ participates in at least one solution, and every remaining dotted edge is excluded from at least one solution.*

Proof. We have already shown in the proof of Theorem 3 that every value in $[\text{MINTREE}, \text{MAXTREE}]$ participates in a solution. We now show that every remaining edge (u, v) belongs to the forest in at least one solution. First, we construct a solution S with MAXTREE trees as before. If (u, v) belongs to the forest, we are done. Otherwise, let $S' = S \cup (u, v)$. If S' is not a solution to the constraint, it can be either because it is not a forest or because the number of trees in S' is not in $\mathcal{D}(\text{NTREE})$. If it is not a forest, it is because inserting (u, v) creates a cycle, but in this case (u, v) should have been removed in Line 7. So the number of trees, which is $\text{MAXTREE} - 1$, is not in $\mathcal{D}(\text{NTREE})$. If there is a value in $\mathcal{D}(\text{NTREE})$ which is smaller than the number of trees in S', we can merge trees as in the proof of Theorem 3 until we have a solution. Otherwise, it must be that $\mathcal{D}(\text{NTREE}) = \{\text{MAXTREE}\}$. But in this case, (u, v) should have been removed in Line 15.

It remains to show that every dotted edge is excluded from at least one solution. Let (u, v) be a dotted edge. We have shown above that there exists a solution S that uses (u, v). Let $S' = S \setminus \{(u, v)\}$. If S' is a solution for the *proper-forest* constraint, we are done. Assume, then, that S' is not a solution. This can be either because it contains an isolated vertex or because it consists of too many trees.

Assume that the removal of (u, v) created a singleton tree with the vertex u. Since (u, v) was not turned into a solid edge in Line 19, we know that u has another neighbor $n_u \neq v$, to which it is linked by a dotted edge. If v did not become a singleton tree or the number of trees was strictly larger than MINTREE, we can merge u into a neighboring tree and obtain a new solution that does not use the edge (u, v).

However, if the number of trees was exactly MINTREE and both u and v were isolated, merging each of them into a pre-existing tree would leave us with $\text{MINTREE} - 1$ trees. Fortunately, this case is not possible. Indeed, assume that it has occurred. We know that n_u belongs to a tree of the solution which contains at least two vertices and which does not contain u or v. So the connected component of u (and v) in $\mathcal{G}$ contributes at least two trees to the solution, and the solution must have more than MINTREE trees.

Finally, assume that after the removal of (u, v) we do not have any singleton trees, but we do have one tree too many. If possible, merge two trees by a dotted edge other than (u, v). If this is not possible, then it is because the number of trees is equal to the number of connected components in $\mathcal{G} \setminus \{e\}$, i.e., to MINTREE. If (u, v) was a bridge, it would have turned into a solid edge in Line 11. So there is a cycle that contains (u, v). Since we are not able to merge two trees after the deletion of (u, v), it must be that for every dotted edge on the cycle, both endpoints belong to the same tree. This implies that u and v are in the same tree after the deletion of the edge (u, v), which means that there is a cycle in the forest S, a contradiction. $\square$

Lemma 3. *Every edge that the algorithm of Figure 4 turned from dotted to solid participates in all solutions.*

Proof. Assume otherwise, i.e., there exists an edge (u, v) that was turned from dotted to solid but which is excluded from a solution S. If (u, v) turned into a solid edge in Line 19 of the algorithm then $\mathcal{G}$ has an isolated vertex, so S cannot be a proper forest.

So the transformation must have occurred in Line 11. In this case, we know that the domain of NTREE is ground and contains only MINTREE, i.e., the number of trees in S is equal to the number of connected components of $\mathcal{G}$. But then any bridge of $\mathcal{G}$, and hence the edge (u, v), must belong to S; a contradiction.   □

5.4   Complexity

The complexity of checking whether the constraint has a solution is dominated by the complexity of computing MAXTREE (all the rest can be done in linear time). To find MAXTREE we need to find the cardinality of a maximum matching in $\mathcal{G}_{MAYBE}$ and the best known upper bound for this is $\mathcal{O}(m\sqrt{n})$ time [13]. Lines 3 to 9 take linear time: We need to compute the connected components of $\mathcal{G}_{TRUE}$ and traverse the dotted edges, spending a constant amount of time for each edge. Finding all bridges of $\mathcal{G}$ in Line 10 and detecting leaves in Line 19 also takes linear time.

So far, the bottleneck is the feasibility test which takes $\mathcal{O}(m\sqrt{n})$ time. If the domain of NTREE is ground and contains only the value MAXTREE, we need to execute also Lines 13 to 17. Line 15 is trivial. For Line 16, we need to determine which edges of the graph belong to at least one maximum cardinality matching. For bipartite graphs, this can be done in linear time once a single maximum matching is known [11]. However, we need to perform this task on arbitrary graphs. In the full version of this paper, we describe an algorithm that does this in $\mathcal{O}(mn)$ time.

Finally, for Line 19 we need an algorithm that receives a graph and a maximum matching and detects which vertices of the graph are saturated in every maximum matching. We show a linear-time solution in the full version of the paper.

Theorem 4. *The algorithm of Figure 4 filters the* proper-forest *constraint to hybrid-consistency in* $\mathcal{O}(mn)$ *time if* $\mathcal{D}(\text{NTREE}) = \{\text{MAXTREE}\}$ *and in* $\mathcal{O}(m\sqrt{n})$ *time otherwise.*

6   A Summary of Known Results on *tree* Covering Constraints

This section highlights the commonalities and differences between the constraints *tree* [1], *resource-forest* and *proper-forest*. All three constraints are defined on a graph $\mathcal{G} = (\mathcal{V}, \mathcal{E})$, directed or undirected, with $|\mathcal{V}| = n$ and $|\mathcal{E}| = m$.

Figure 5 summarizes the best known running times for checking feasibility and for achieving hybrid-consistency for each constraint. Figure 6 summarizes the main graph properties used to determine relevant bounds on the number of trees allowed to cover a given graph as well as the conditions for the existence of well-formed trees according to the definition of each constraint. The last table indicates that four basic graph properties completely define these constraints: connected components (in undirected graphs), strongly connected components (in digraphs), maximum matchings, existence of cycles. For each of the constraints, necessary conditions and filtering rules were deduced with known algorithms (e.g. dfs, maximum matching, connected component detection, etc.) as well as new algorithms (e.g., identifying vertices which are saturated in any maximum matching in an undirected graph). Observe that the lower and upper bounds

Graph Pattern	Trees	Undirected trees	
	tree	proper-forest	resource-forest
Checking feasibility	$\mathcal{O}(n+m)$	$\mathcal{O}(m\sqrt{n})$	$\mathcal{O}(n+m)$
Hybrid-consistency	$\mathcal{O}(mn)$	$\mathcal{O}(mn)$ [worst], $\mathcal{O}(m\sqrt{n})$ [typical]	$\mathcal{O}(n+m)$

Fig. 5. Best known upper bounds for three *tree* covering constraints

Graph Pattern	Trees	Undirected trees	
	tree	proper-forest	resource-forest
MINTREE	$|SCC_{sink}(G)|$	$|CC(\mathcal{G})|$	$|CC(\mathcal{G})|$
MAXTREE	$|R_{potential}(\mathcal{G})|$	$|CC(\mathcal{G}_{TRUE})|+$ $\mu(\mathcal{G}_{MAYBE})$	$|CC(\mathcal{G}_{TRUE})$ with at least one *resource*|
Well-formed trees	at least one potential root in each SCC of $\mathcal{G}$	no cycle in $\mathcal{G}_{TRUE}$ no isolated vertex in $\mathcal{G}$	no cycle in $\mathcal{G}_{TRUE}$ one *resource* vertex in each $CC(\mathcal{G}_{TRUE})$
Compatible number of trees	$\mathcal{D}(\mathtt{NTREE}) \cap [\mathtt{MINTREE}, \mathtt{MAXTREE}] \neq \emptyset$		

Fig. 6. Graph properties characterizing solutions to the three *tree* covering constraints

MINTREE and MAXTREE for the *proper-forest* constraint exactly correspond to the lower and upper bounds on the number of connected components that appears in [14].

Notation: For a graph H, the number of maximal CCs in H is $|CC(H)|$, the maximum cardinality of a matching in H is $\mu(H)$, the number of sink SCCs of H is $|SCC_{sink}(H)|$ and the number of potential roots in H is $|R_{potential}(H)|$.

References

1. N. Beldiceanu, P. Flener, and X. Lorca. The *tree* Constraint. In *CP-AI-OR'05*, volume 3524 of *LNCS*, pages 64–78. Springer-Verlag, May 2005.
2. J.-L. Laurière. A language and a program for stating and solving combinatorial problems. *Artificial Intelligence*, 10:29–127, 1978.
3. N. Beldiceanu and E. Contejean. Introducing global constraint in CHIP. *Mathl. Comput. Modelling*, 20(12):97–123, 1994.
4. M. Sellmann. Cost-based filtering for shortest path constraints. In *CP 2003*, volume 2833 of *LNCS*, pages 694–708. Springer-Verlag, 2003.
5. M. Dincbas, P. Van Hentenryck, H. Simonis, A. Aggoun, T. Graf, and F. Berthier. The Constraint Logic Programming Language CHIP. In *Int. Conf. on Fifth Generation Computer Systems (FGCS'88)*, pages 693–702, Tokyo, Japan, 1988.
6. J.-F. Puget. A C++ Implementation of CLP. In *Second Singapore International Conference on Intelligent Systems (SPICIS)*, pages 256–261, Singapore, November 1994.
7. A. Cayley. A theorem on trees. *Quart. J. Math.*, 23:376–378, 1889.
8. C. Bessière, E. Hebrard, B. Hnich, Z. Kızıltan, and T. Walsh. The *range* and *roots* Constraints: Specifying Counting and Occurrence Problems. In *IJCAI-05*, pages 60–65, 2005.
9. C. Berge. *Graphes*. Dunod, New York, 2nd edition, 1985. In French.

10. M. Sellmann. *Reduction techniques in Constraint Programming and Combinatorial Optimization.* PhD thesis, University of Paderborn, 2002.
11. J.-C. Régin. A filtering algorithm for constraints of difference in CSP. In *AAAI-94*, pages 362–367, 1994.
12. M. Gondran and M. Minoux. *Graphes et algorithmes.* Eyrolles, Paris, 2nd edition, 1985. In French.
13. S. Micali and V. V. Vazirani. An $\mathcal{O}(\sqrt{|V|} \cdot |E|)$ algorithm for finding maximum matching in general graphs. In *FOCS 1980*, pages 17–27, New York, 1980. IEEE.
14. N. Beldiceanu, T. Petit, and G. Rochart. Bounds of Graph Characteristics. In P. van Beek, editor, *CP 2005*, volume 3709 of *LNCS*, pages 742–746. Springer-Verlag, 2005.

Allocation, Scheduling and Voltage Scaling on Energy Aware MPSoCs

Luca Benini[1], Davide Bertozzi[2], Alessio Guerri[1], and Michela Milano[1]

[1] DEIS, University of Bologna
V.le Risorgimento 2, 40136, Bologna, Italy
{lbenini, aguerri, mmilano}@deis.unibo.it
[2] Dipartimento di Ingegneria, University of Ferrara
V. Saragat 1, 41100, Ferrara, Italy
dbertozzi@ing.unife.it

Abstract. In this paper we introduce a complex allocation and scheduling problem for variable voltage Multi-Processor System-on-Chip (MP-SoC) platforms. We propose a methodology to formulate and solve to optimality the allocation, scheduling and discrete voltage selection problem, minimizing the system energy dissipation and the overhead for frequency switching. Our approach is based on the Logic Benders decomposition technique where the allocation is solved through an Integer Programming solver, and the scheduling through a Constraint Programming solver. The two solvers are interleaved and their interaction regulated by cutting plane generation. The objective function depends on both master and sub-problem variables. We demonstrate the efficiency of our approach on a set of realistic instances.

1 Introduction

As silicon technology keeps scaling, it is becoming technically feasible to integrate entire and complex systems on the same silicon die. This solution provides scalable computation power, and it is expected that hundreds of processor cores will be integrated on these Multi-Processor Systems-on-Chip (MPSoCs) in future technologies. MPSoCs are widely used in embedded systems (such as cellular phones, automotive control engines, etc.) where, once deployed in field, they always run the same set of applications. Since for many multimedia and signal processing applications the workload is highly predictable at design time, with minimum run-time fluctuations, an optimal allocation and scheduling for such applications can be statically derived off-line.

A critical task for recent MPSoCs is the minimization of the energy consumed since the speed of each processor can be tuned by changing its frequency. We start from a well-characterized task graph, a directed acyclic graph representing a functional abstraction of the application that will run on the MPSoCs. Each task is characterized by the number of clock cycles used for its execution. Clearly the duration of each task and the energy spent for running it depends on the clock frequency used during the task execution. In addition, tasks connected

J.C. Beck and B.M. Smith (Eds.): CPAIOR 2006, LNCS 3990, pp. 44–58, 2006.

by arcs in the task graph communicate and if they are allocated to different processors, additional communicating tasks are created for reading and writing data on a shared memory.

Defining the optimal allocation, scheduling and voltage scaling for minimizing energy in MPSoCs is the aim of this paper. Energy is consumed during task execution, task communication and for switching between two voltages (setup costs).

The problem we face is very complex. It has never been solved to optimality by the system design community and it cannot be solved by any complete commercial solver that models the problem as a whole. The method we use is the Logic Based Benders Decomposition [8], an extension of the well known OR Benders Decomposition [1] approach for dealing with solvers of any kind. In this setting, we allocate tasks to processors and decide their execution frequency in the master problem, while the subproblem schedules tasks with a fixed duration and static resource assignment. The interaction between the master and the subproblem is regulated via cutting planes generation.

The approach has been followed several times for similar problems, but never applied to scheduling for minimizing costs and setup costs. In particular, there are a number of papers using Benders Decomposition in a CP setting. [12] proposes the branch and check framework using Benders Decomposition (BD). [4] embeds BD in the CP environment ECLiPSe and shows that it can be useful in practice. [5] applied Benders decomposition to minimum cost planning and scheduling problems; in this work the objective function involves only master problem variables, while the subproblem is simply a feasibility problem. [6] and [7] used Benders decomposition for Planning and Scheduling problems with several objective functions: either minimizing the cost (involving only master problem variables), or minimizing the makespan or the tardiness or the number of late tasks (involving the last three cases only subproblem variables); here the objective function involves both master problem and subproblem variables since the execution energy is minimized by the allocation problem solver while the setup cost due to frequency switches can be minimized only at scheduling time.

2   Problem Description

The new MPSoC paradigm for hardware platform design is pushing the parallelization of applications, so that instead of running them at a high frequency on a single monolithic core, they can be partitioned into a set of parallel tasks, which are mapped and executed on top of a set of parallel processor cores operating at lower frequencies. Power minimization is a key design objective for MPSoCs to be used in portable, battery-operated devices. This goal can be pursued by means of low power design techniques at each level of the design process, from physical-level techniques (e.g., low swing signaling) up to application optimization for low power. In this paper, we focus on system-level design, where the main knobs for tuning power dissipation of an MPSoC are: allocation and scheduling of a multi-task application onto the available parallel processor cores, voltage and frequency setting of the individual processor cores. For those systems

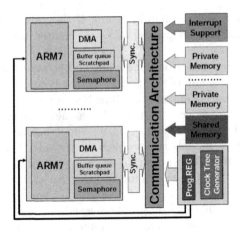

Fig. 1. Distributed MPSoC architecture

where the workload is largely predictable and not subject to run-time fluctuations (e.g., signal processing or some multimedia applications), the above design parameters can be statically set at design time. Traditional ways to tackle the mapping and configuration problem either incur overly large computation times already for medium-size task sets, or are inaccurate (e.g., use of heuristics and problem modelling with highly simplifying assumptions on system operation). Therefore, design technology for MPSoCs strongly needs accurate, scalable and composable modelling and solving frameworks.

In this paper we consider a reference template [10] for a distributed MPSoC architecture. The platform consists of computation tiles, a shared bus for inter-tile communication and a shared memory. The computation tiles are supposed to be homogeneous and consist of ARM7 processor cores (including instruction and data caches) and of tightly coupled software-controlled scratchpad memories. These latter devices can be viewed as local, low access cost memories (see Fig. 1). Messages can be exchanged by tasks through communication queues [9], which can be allocated at design time either in scratch-pad memory or in remote shared memory, depending on whether tasks are mapped onto the same processor or not.

In this architecture, each processor core can run at different clock frequencies. The frequency of each processor core is derived from a baseline system frequency by means of integer dividers. Moreover, a synchronization module must be inserted between the bus and the processor cores to allow frequency decoupling (usually a dual-clock FIFO). The bus operates at the maximum frequency (e.g., 200 MHz). For each processor core, a set of voltage and frequency couples is specified, since the feasible operating points for these cores are not continuous but rather discrete. For modern variable voltage/variable frequency cores, this set is specified in the data-sheet.

Finally, in real-life MPSoC platforms, switching voltage and frequency of a processor core is not immediate nor costless, therefore the switching overhead in terms of switching delay (referred to as setup times) and energy overhead (referred to as setup costs) must be carefully considered when selecting the optimal configuration of a system. In practice, interesting trade-offs have to be studied. On one hand, tasks can be spread across a large number of processor cores, so that these cores can operate at lower frequencies, but more communication arises and the energy cost of many running cores has to be compensated by a more energy-efficient execution of tasks. On the other hand, tasks have to be grouped onto the processor cores and scheduled taking care of minimizing the number of frequency switchings. It must be observed that application real-time requirements play a dominant role in determining solutions for the MPSoC mapping and configuration problem. A good methodology should be conservative with respect to task deadlines, so to minimize the probability of timing violations in the real system.

3   Dynamic Voltage Scaling Problem – DVSP: The Model

We consider a directed acyclic task graph G whose nodes represent a set of T tasks, are annotated with their deadline dl_t and with the worst case number of clock cycles WCN_t. Arcs represent dependencies/communications among tasks. Each arc is annotated with the amount of data two dependent tasks should exchange, and therefore the number of clock cycles for exchanging (reading and writing) these data WCN_R and WCN_W. Tasks are running on a set of processors P. Each processor can run with M energy/speed modes and has a maximum load constraint dl_p. Each task spends energy both in computing and in communicating. In addition, when the processor switches between two modes it spends time and energy. We have energy overhead E_{ij} for switching from frequency i to frequency j, and time overhead T_{ij} for switching from frequency i to j.

The Dynamic Voltage Scaling Problem is the problem of allocating tasks to processors, define the running speed of each task and schedule each of them minimizing the total energy consumed.

The method we use for handling the DVSP uses the logic-based Benders decomposition technique [8]. Similarly to [2], the problem is decomposed into two parts: the first, called Master Problem, is the allocation of processors and frequencies to tasks and the second, called Subproblem, is the scheduling of tasks given the static allocation and frequency assignments provided by the master. Note that the frequency assignment could be done in the subproblem. However, the scheduling part becomes extremely slow and performances highly decrease. In addition, the relaxation of the subproblem (introduced in section 4.1) become extremely loose. Differently from [2], the objective function depends on master and subproblem variables. In fact, the master problem minimizes the communication and execution energy, while only during the scheduling phase we could minimize the switching energy overhead.

The master problem is tackled by an Integer Programming solver (through a traditional Branch and Bound) while the subproblem through a Constraint Programming solver. The two solvers interact via no-good and cutting planes generation. The solution of the master is passed to the subproblem. We have two possible cases: (1) there is no feasible schedule: we have to compute a no-good avoiding the same allocation to be found again; (2) there is a feasible and optimal schedule minimizing the second component of the objective function: here we cannot simply stop the iteration since we are not sure we have the optimal solution overall. We have to generate a cut saying that this is the optimal solution unless a better one can be computed with a different allocation.

The procedure converges when the master problem produces a solution with the same objective function of the previous one.

4   Example

As an example, let consider 5 tasks and 5 communications, with the precedence constraints as described in Figure 2. Table 1 shows the duration (in clock cycles) of execution and communication tasks (the durations of the reading and the writing phase R_i and W_i of each communication Com_i are the half of these values). We have 2 processors, running at 2 different frequencies, 200MHz and 100MHz (so, e.g. $Task_1$ will last 500ns if runs at 200MHz and 1μs if runs at 100MHz). The processors waste 10mW when running at 200MHz and 3mW when running at 100MHz. Switching from the higher frequency to the lower needs 2ns and wastes 2pJ, while the contrary needs 3ns and wastes 3pJ. The realtime requirement settles the processor deadline at 2μs.

Table 1. Activities durations for the example

Nome	Task1	Task2	Task3	Task4	Task5	Com1	Com2	Com3	Com4	Com5
Clock	100	54	134	24	10	20	10	8	8	8

The first allocation found tries to assign the lower frequency to the third task, being the longest one and thus the most power consuming one; this solution is however not schedulable due to the deadline constraint. The second allocation found is schedulable and is also the optimal one w.r.t. the power consumption minimization (the total power consumption is 13502mW). The first two tasks are allocated on the first processor at the higher frequency and the other three tasks on the second processor: here only $Task_5$ runs at the higher frequency. The Gantt chart in Figure 2 shows the schedule of this solution.

4.1   The Master Problem Model

We model the allocation problem with binary variables X_{ptm} which take value 1 if task t is mapped on the processor p and runs in mode m, 0 otherwise. Since we also take into account communication, we assume that two communicating tasks running on the same processor do not consume any energy and do not spend any time (indeed the communication time and energy spent are included

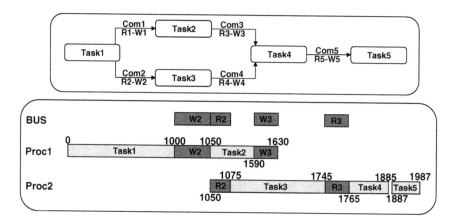

Fig. 2. Task graph and schedule for the example in Table 1

in the execution time and energy), while if they are allocated on two different processors, they both consume energy and spend time. The first task spends time and energy for writing data on a shared memory. This operation makes the duration of the task becoming longer: it increases of a quantity WCN_W/f_m where WCN_W is the number of clock cycles for writing data (it depends on the amount of data we should write), and f_m is the frequency of the clock when task t is performed. The second task should read data from the shared memory. Again its duration increases of a quantity WCN_R/f_m where WCN_R is the number of clock cycles for reading data (it depends on the amount of data we should read), and f_m is the frequency of the clock when task t is performed.

Both the read and write activities are performed at the same speed of the task and use the bus (which instead works at the maximum speed). For modelling this aspect, we introduce in the model two variables $R_{pt_1t_2m}$ and $W_{pt_1t_2m}$ taking value 1 if the task t_1 running on processor p reads (resp. writes) data at mode m from (resp. for) a task t_2 not running on p.

Any task can be mapped on only one processor and can run at only one speed. This translates in the following constraints:

$$\sum_{p=1}^{P}\sum_{m=1}^{M} X_{ptm} = 1 \quad \forall t$$

Also the communication between two tasks happens at most once:

$$\sum_{p=1}^{P}\sum_{m=1}^{M} R_{pt_1t_2m} \leq 1 \quad \forall t_1, t_2$$

$$\sum_{p=1}^{P}\sum_{m=1}^{M} W_{pt_1t_2m} \leq 1 \quad \forall t_1, t_2$$

The objective function is to minimize the energy consumption of the task execution, and of the task communication (read and write)

$$E_{comp} = \sum_{p=1}^{P} \sum_{m=1}^{M} \sum_{t=1}^{T} X_{ptm} WCN_t t_{clock_m} P_{tm}$$

$$E_{Read} = \sum_{p=1}^{P} \sum_{m=1}^{M} \sum_{t,t_1=1}^{T} R_{ptt_1m} WCN_{Rtt_1} t_{clock_m} P_{tm}$$

$$E_{Write} = \sum_{p=1}^{P} \sum_{m=1}^{M} \sum_{t,t_1=1}^{T} W_{ptt_1m} WCN_{Wtt_1} t_{clock_m} P_{tm}$$

where P_{tm} is the power consumed in a clock cycle (lasting t_{clock_m}) by the task t at mode m.

$$OF = E_{comp} + E_{Read} + E_{Write}$$

The objective function defined up to now depends only on master problem variables. However, switching from one speed to another introduces transition costs, but their value can be computed only at scheduling time. In fact, they are not constrained in the master problem original model. They are constrained by Benders Cuts instead, after the first iteration. We will present Benders Cuts in section 4.3. Therefore, in the master problem the objective function is:

$$OF_{Master} = OF + Setup$$

$$Setup = \sum_{p=1}^{P} Setup_p$$

It is worth noting that this contribution should be added to the master problem objective function, but, being the $Setup_p$ variables not constrained at the first iteration in the master problem, they are all forced to be 0. From the second iteration, instead, cuts are produced constraining variables $Setup_p$ and this contribution could be no longer 0.

This formulation will result in tasks that are potentially running initially with lower frequencies on the same processor (thus avoiding communication). A measure of control is provided by constraints on deadlines in order to prevent the blind selection of the lowest frequencies and the allocation of all tasks on the same processor. The timing is not yet known in this phase, but we can introduce some constraints that represent a relaxation of the subproblem and will reduce the solution space. For each processor, only a certain load is allowed. Therefore, on each processor the sum of the time spent for computation, plus the time spent for communication (read and write) should be less than or equal to the processor deadline dl_p:

$$T^p_{comp} = \sum_{t=1}^{T} \sum_{m=1}^{M} X_{ptm} \frac{WCN_t}{f_m}$$

$$T^p_{read} = \sum_{t=1}^{T} \sum_{m=1}^{M} \sum_{t_1=1}^{T} R_{ptt_1m} \frac{WCN_{Rtt_1}}{f_m}$$

$$T^p_{write} = \sum_{t=1}^{T} \sum_{m=1}^{M} \sum_{t_1=1}^{T} W_{ptt_1m} \frac{WCN_{Wtt_1}}{f_m}$$

$$T^p_{comp} + T^p_{read} + T^p_{write} \leq dl_p \quad \forall p \tag{1}$$

These relaxations can be tightened by considering chains of tasks in the task graphs instead of groups of tasks running on the same processor. For example consider tasks t_1, t_2, t_3, t_4 linked by precedence constraints so that $t_1 \rightarrow t_2$, $t_2 \rightarrow t_3$ and $t_3 \rightarrow t_4$. Now suppose that t_1 and t_4 are allocated on processor 1 and t_2 and t_3 on other processors. Instead of summing only the durations of t_1 and t_4 that should be less than or equal to the processor deadline, one could add also the duration of t_2 and t_3 since they should be executed before t_4. The chains in a graph can be many, we added only some of them.

Finally, task deadlines can be captured:

$$\sum_{p=1}^{P} \sum_{m=1}^{M} \left[X_{ptm} \frac{WCN_t}{f_m} + \sum_{t1=1}^{T} \left(R_{ptt_1m} \frac{WCN_{Rtt_1}}{f_m} + W_{ptt_1m} \frac{WCN_{Wtt_1}}{f_m} \right) \right] \leq dl_t \ \forall t$$

There are several improvements we have introduced in the master problem model. In particular we have removed many symmetries leading the solver to explore the same configurations several times.

4.2   The Sub-problem Model

Once allocation and voltage selection have been solved optimally, for the scheduling part each task t has an associated variable representing its starting time $Start_i$. The duration is fixed since the frequency is decided, i.e., $duration_i = WCN_i/f_i$. In addition, if two communicating tasks t_i and t_j are allocated on two different processors, we should introduce two additional activities (one for writing data on the shared memory and one for reading data from the shared memory). We model the starting time of these activities $StartWrite_{ij}$ and $StartRead_{ji}$. These activities are carried on at the same frequency of the corresponding task. If t_i writes and t_j reads data, the writing activity is performed at the same frequency of t_i and its duration $dWrite_{ij}$ depends on the frequency and on the amount of data t_i writes, i.e., WCN_{Wij}/f_i. Analogously, the reading activity is performed at the same frequency of t_j and its duration $dRead_{ji}$ depends on the frequency and on the amount of data t_j reads, i.e., $WCN_{R_{ji}}/f_j$. Clearly the read and write activities are linked together and to the corresponding task:

$$StartWrite_{ij} + dWrite_{ij} \leq StartRead_{ji} \quad \forall i, j \text{ s.t. i communicates with j}$$

$$Start_i + duration_i \leq StartWrite_{ij} \quad \forall i, j \text{ s.t. i communicates with j}$$

$$StartRead_{ji} + dRead_{ji} \leq Start_j \quad \forall i, j \text{ s.t. i communicates with j}$$

In the subproblem, we model precedence constraints in the following way: if task t_i should precede task t_j and they run on the same processor at the same frequency the precedence constraint is simply:

$$Start_i + duration_i \leq Start_j$$

If two tasks run on different processors and should communicate we should add the time for communicating.

$$Start_i + duration_i + dWrite_{ij} + dRead_{ji} \leq Start_j$$

Deadline constraints are captured stating that each task must end its execution before its deadline and, on each processor, all the tasks (and in particular the last one) running on it must end before the processor deadline.

$$Start_i + duration_i \leq dl_{t_i} \quad \forall \text{ tasks } t_i$$

$$Start_i + duration_i \leq dl_p \quad \forall i \in p, \forall p$$

Resources are modelled as follows. We have a unary resource constraint for each processor, modelled through a cumulative constraint having as parameters a list of all the variables representing the starting time of the activities (tasks, readings, writings) sharing the same resource p, their durations, their resource consumption (which is a list of 1) and the capacity of the processor which is 1.

$$cumulative(StartList_p, DurationList_p, [1], 1) \quad \forall p$$

We model the bus through an additive model we have already validated in [11]. We have an activity on the bus each time a task writes or reads data to or from the shared memory. The bus is modelled as an additive resource and several activities can share the bus, each one consuming a fraction of it until the total bandwidth is reached. The cumulative constraint used to model the bus is:

$$cumulative(StartReadWriteList, DurationList, Fraction, TotBWidth)$$

where $StartReadWriteList$ and $DurationList$ are lists of the starting times and durations of all read and write activities needing the bus, $Fraction$ is the amount of bandwidth granted to any activity when accessing the bus[1] and $TotBWidth$ is total bandwidth available of the bus.

---

[1] This value was experimentally tuned to 1/4 of the total bus bandwidth.

To model the setup time and cost for frequency switching we take advantage of the classes defined by ILOG Scheduler to manage transitions between activities. It is possible to associate a label to each activity and to define a transition matrix that specifies, for each couple of labels l_1 and l_2, a setup time and a setup cost that must be paid to schedule, on the same resource, an activity having the label l_1 just before an activity having the label l_2. When, during the search for a solution, two activities with labels l_1 and l_2 are scheduled one just after the other on the same resource, the solver will satisfy the additional constraint:

$$Start_{l_1} + duration_{l_1} + TransTime_{l_1 l_2} \leq Start_{l_2}$$

where $TransTime_{l_1 l_2}$ is the setup time specified in the transition matrix. Likewise, the solver introduces $TransCost_{ij}$ in the objective function. If S_p is the set of all the tasks scheduled on processor p, the objective function we want to minimize is:

$$OF = \sum_{p=1}^{P} \sum_{(i,j) \in S_p | next(i)=j} TransCost_{ij}$$

4.3   Generation of Logic-Based Benders Cuts

Once the subproblem has been solved, we generate Benders Cuts. The cuts are of two types:

- if there is no feasible schedule given an allocation, the cuts are the same we computed for the single voltage problem and depend on variables X_{ptm}.
- if the schedule exists, we cannot simply stop the iteration since the objective function depends also on subproblem variables. Therefore, we have to produce cuts saying that the one just computed is the optimal solution unless a better one exists with a different allocation. These cuts produce a lower bound on the setup of single processors.

The first type of cuts are no-good: we call J_p the set of couples (Task, Frequency) allocated to processor p. We impose

$$\sum_{(t,m) \in J_p} X_{ptm} \leq |J_p| - 1 \quad \forall p$$

Let us concentrate on the second type of cuts. The cuts we produce in this case are bounds on the variable $Setup$ previously defined in the Master Problem.

Suppose the schedule we find for a given allocation has an optimal setup cost $Setup^*$. It is formed by independent setups, one for each processor $Setup^* = \sum_{p=1}^{P} Setup_p^*$.

We have a bound on the setup LB_{Setup_p} on each processor and therefore a bound on the overall setup $LB_{Setup} = \sum_{p=1}^{P} LB_{Setup_p}$.

$$Setup_p \geq 0$$

$$Setup_p \geq LB_{Setup_p}$$

$$LB_{Setup_p} = Setup_p^* - Setup_p^* \sum_{(t,m)\in J_p} (1 - X_{ptm})$$

These cuts remove only one allocation. Indeed, we have also produced cuts that remove some symmetric solutions.

We have devised tighter cuts removing more solutions. Intuitively, each time we consider a solution of the problem overall, we generate an optimal setup cost $Setup^*$ for the given allocation. In the current solution, we know the number of frequency switches producing $Setup^*$. We can consider each processor independently since the frequency switches on one processor are independent from the other. We can impose cuts that say that $Setup^*$ is bound for all solutions with the same set of frequency switches of the last one found or a superset of it. To do that we have to introduce in the model variables $Next_{t_1 t_2 f_1 f_2 p}$, which complicate the model too much. In fact, our experimental results show that these cuts, even if tighter, do not lead to any advantage in terms of computational time.

4.4 Relaxation of the Subproblem

The iterative procedure presented so far can be improved by adding a bound on the setup cost and setup time in the master problem based only on information derived from the allocation.

Suppose we have five tasks running on the same processors using three different frequencies. So for instance, tasks t_1, t_3 and t_5 run at frequency f_1, t_2 runs at frequency f_2 and t_4 runs at frequency f_3. Since we have to compute a bound, we suppose that all tasks running at the same speed go one after the other. We can have six possible orders of these frequencies leading to different couples of frequency switches. A bound on the sum of the energy spent during the frequency switches is the minimal sum between two switches, i.e., the sum of all possible switches minus the maximum switch. This bound is extremely easy to compute and does not enlarge the allocation problem model.

Let us introduce in the model variables Z_{pf} taking value 1 if the frequency f is allocated at least once on the processor p, 0 otherwise. Let us call E_f the minimum energy for switching to frequency f, i.e. $E_f = min_{i,i\neq f}\{E_{if}\}$.

$$Setup_p \geq \sum_{f=1}^{M} (Z_{pf}E_f - max_f\{E_f|Z_{pf} = 1\})$$

This bound helps in reducing the number of iterations between the master and the subproblem.

Similarly, we can compute the bound on the setup time given an allocation. Let us consider $T_f = min_{i,i\neq f}\{T_{if}\}$. Therefore, we can compute the following bound.

$$SetupTime_p \geq \sum_{f=1}^{M} (Z_{pf}T_f - max_f\{T_f|Z_{pf} = 1\})$$

This bound can be used to tighten the constraint (1) in section 4.1 in the following way.

$$T^p_{comp} + T^p_{read} + T^p_{write} + SetupTime_p \leq dl_p \quad \forall p$$

so that solutions provided by the master problem are more likely to be feasible for the subproblem.

A tighter bound on the setup time and cost could be achieved by introducing in the allocation problem model variables $Next$, but as explained in section 4.3 they complicate too much the model and are not worth using.

5   Experimental Results

We have generated 500 realistic instances, with the number of tasks varying from 4 to 10 and the number of processors from 2 to 10. We assume that each processor can run at three different frequencies. We consider, as in [2], applications with a pipeline workload. Therefore we refer to the number of tasks to be allocated and we schedule a larger number of tasks corresponding to many iterations of the pipeline. We also have generated 27 realistic instances with the number of tasks varying from 8 to 14 and the number of processors from 2 to 6, with generic task graphs. The generic task graph complicates the problem since it increases the parallelism degree. We assume that each processor can run at six different frequencies. All the considered instances are solvable and we found the proved optimal solution for each of them. Experiments were performed on a 2.4GHz Pentium 4 with 512 Mb RAM. We used ILOG CPLEX 8.1, ILOG Solver 5.3 and ILOG Scheduler 5.3 as solving tools.

5.1   Comparison with Pure Approaches

In [2], we compared a solving tool based on Benders Decomposition for a similar problem with pure CP or IP based solving tools. Results shown that the pure approaches were not comparable with the hybrid one, being the search times for finding a solution to a relaxed (thus easier) problem order of magnitude higher. The problem we are facing in this paper is much more complex then the one presented in [2], since we consider also frequency switching. We developed a CP and an IP-based approach to solve allocation, scheduling and voltage selection, but not even a single (feasible) solution was found within 15 minutes, while the hybrid approach, within 4 minutes, finds the optimal solution and proves optimality for all the pipelined instances considered.

5.2   Experimental Results

In this section we show the results obtained solving the problem instances using the model described in section 3. We consider first the instances with task graphs representing a pipeline workflow. Note that here, since we are considering applications with pipeline workload, if n is the number of tasks to be allocated,

Table 2. Search time and number of iterations for instances with pipelined task graphs

Tasks				
Alloc	Sched	Procs	Time(s)	Iters
4	16	2	1,73	1,98
4	16	3	1,43	2,91
4	16	4	2,24	3,47
5	25	2	2,91	2,36
5	25	3	4,19	4,12
5	25	4	5,65	4,80
5	25	5	6,69	3,41
6	36	2	3,84	2,90
6	36	3	10,76	2,17
6	36	4	15,25	4,66
6	36	5	23,17	4,50
6	36	6	26,14	3,66
7	49	2	4,67	1,75
7	49	3	5,90	1,90
7	49	7	34,53	6,34
8	64	2	4,09	3,28
8	64	3	10,99	1,83
8	64	4	12,34	4,45
8	64	5	22,65	10,53
8	64	7	51,07	6,98
9	81	2	1,79	1,12
9	81	5	60,07	7,15
9	81	6	70,40	9,20
10	100	2	5,52	1,83
10	100	3	3,07	1,96
10	100	6	120,02	6,23
10	100	10	209,35	10,65

Table 3. Search time and number of iterations for instances with generic task graphs

Tasks				
Alloc	Sched	Procs	Time(s)	Iters
8	8	2	1,57	1
8	8	3	1,48	2
8	8	3	0,81	1
8	8	3	4,26	6
8	8	4	0,86	1
9	9	2	2,51	1
9	9	2	1,11	1
9	9	2	2,73	3
9	9	3	35,95	43
9	9	3	2,51	1
9	9	3	6,62	2
9	9	4	1,40	3
9	9	4	2,14	5
9	9	4	2,60	4
9	9	4	29,59	26
9	9	4	4,84	6
9	9	6	158,43	39
10	10	2	5,90	1
10	10	3	2,12	1
10	10	3	12,81	3
10	10	4	0,37	1
10	10	4	13,92	14
10	10	4	4,18	5
10	10	4	11,50	27
12	12	5	551,92	213
14	14	2	14,11	1
14	14	6	3624,81	2

Table 4. Number of iterations distribution ratio

Iter	1	2	3	4	5	6	7	8	9	10	11+
%	50,20	18,51	7,11	4,52	4,81	2,88	2,46	2,05	1,64	1,64	4,11

the number of scheduled tasks is n^2. Results are summarized in Table 2. The first three columns contain the number of allocated and scheduled tasks and the number of processors considered in the instances (we remind that each processor can run at three different frequencies). The last two columns represent respectively the search time and the number of iterations. Each value is the mean over all the instances with the same number of tasks and processors. We can see that for all the instances the optimal solution can be found within four minutes. The number of iterations is typically low. Table 4 shows the percentage of occurrence of a given number of iterations. We can see that the optimal solution can be found at the first step in one half of the cases and the number of iterations is at most 5 in almost the 90% of cases. This result is due to the tight relaxations added to the master problem model. We tried to remove these relaxations and we found that the search time and the number of iterations rise, in the average case, up to 1 order of magnitude and, in the worst cases, the solution cannot be found within two hours.

We extended our analysis to instances where the task graph is a generic one, so an activity can possibly read data from more than one preceding activity and possibly write data that will be read by more than one following activity, so the number of reading and writing activities can be considerably higher, being higher the number of edges in the task graph. We remind that each processor can run at six different frequencies, so the number of alternative resources a task can use is six times the number of processors. Differently from the pipelined instances, here we schedule a single repetition of each task. Table 3 summarizes the results. Each instance presented has been solved optimally. Columns have the same meaning as those already described in Table 2. We can see that typically the behaviors are similar to those found when solving the pipelined instances, but sometimes the number of iterations, and thus the search time is notably higher. This is due to the particular structure of the task graph; in fact it can happens that a high degree of parallelism between the tasks, that is a high number of tasks that can execute only after a single task, leads to allocations that are not schedulable. The master problem solver thus looses time proposing to the scheduler a high number of unfeasible allocation. Introducing in the master problem model some relaxations coming from an analysis of the task graph structure, and in particular from the precedence constraints, can lead to better results.

6   Conclusion and Future Research

An exact algorithm for allocation, scheduling and voltage selection has been proposed exploiting the method of Logic-based Benders Decomposition. Experimental results show that the approach using CP and IP for the problem as a whole cannot solve any of the instances considered, while our approach solves them all to optimality. A number of improvements can be conceived the most important concerning the use of a column generation approach for the master problem would most probably lead to a significant speed up. As a second improvement cutting planes that can be derived from [3] and integrated in the master problem model. In addition, we are investigating tighter cutting planes based on information derived from the precedence graph.

Acknowledgement. This work has been partially supported by MIUR under the COFIN2005 project *Mapping di applicazioni multi-task basate su Programmazione a vincoli e intera.*

References

1. J. F. Benders. Partitioning procedures for solving mixed-variables programming problems. *Numerische Mathematik*, 4:238–252, 1962.
2. D. Bertozzi, L. Benini, A. Guerri, and M. Milano. Allocation and scheduling for mpsocs via decomposition and no-good generation. In *Procs. of the 11th Intern. Conference on Principles and Practice of Constraint Programming - CP 2005*, pages 107–121, Sites, Spain, Sept. 2005. Springer.
3. M. Fischetti E. Balas and W. Pulleyblank. The precedence constrained asymmetric travelling salesman problem. *Mathematical Programming*, 68:241–265, 1995.
4. A. Eremin and M. Wallace. Hybrid benders decomposition algorithms in constraint logic programming. In *Procs. of the 7th Intern. Conference on Principles and Practice of Constraint Programming - CP 2001*, pages 1–15, Paphos, Cyprus, Nov. 2001. Springer.
5. I. E. Grossmann and V. Jain. Algorithms for hybrid milp/cp models for a class of optimization problems. *INFORMS Journal on Computing*, 13:258–276, 2001.
6. J. N. Hooker. A hybrid method for planning and scheduling. In *Procs. of the 10th Intern. Conference on Principles and Practice of Constraint Programming - CP 2004*, pages 305–316, Toronto, Canada, Sept. 2004. Springer.
7. J. N. Hooker. Planning and scheduling to minimize tardiness. In *Procs. of the 11th Intern. Conference on Principles and Practice of Constraint Programming - CP 2005*, pages 314–327, Sites, Spain, Sept. 2005. Springer.
8. J. N. Hooker and G. Ottosson. Logic-based benders decomposition. *Mathematical Programming*, 96:33–60, 2003.
9. P. Poletti, A. Poggiali, and P. Marchal. Flexible hardware/software support for message passing on a distributed shared memory architecture. In *2005 Design, Automation and Test in Europe Conference and Exposition DATE2005*, pages 736–741, 2005.
10. M. Ruggiero, A. Acquaviva, D. Bertozzi, and L. Benini. Application-specific power-aware workload allocation for voltage scalable mpsoc platforms. In *2005 International Conference on Computer Design*, pages 87–93, 2005.
11. M. Ruggiero, A. Guerri, D. Bertozzi, L. Benini, and M. Milano. Communication-aware allocation and scheduling framework for stream-oriented multi-processor systems-on-chip. In *2006 Design, Automation and Test in Europe Conference and Exposition DATE2006*, 2006.
12. E. S. Thorsteinsson. A hybrid framework integrating mixed integer programming and constraint programming. In *Procs. of the 7th International Conference on Principles and Practice of Constraint Programming - CP 2001*, pages 16–30, Paphos, Cyprus, Nov. 2001.

The Range Constraint: Algorithms and Implementation

Christian Bessiere[1], Emmanuel Hebrard[2], Brahim Hnich[3], Zeynep Kiziltan[4], and Toby Walsh[2]

[1] LIRMM, CNRS/University of Montpellier, France
bessiere@lirmm.fr
[2] NICTA and UNSW, Sydney, Australia
{ehebrard, tw}@cse.unsw.edu.au
[3] Izmir University of Economics, Izmir, Turkey
brahim.hnich@ieu.edu.tr
[4] University of Bologna, Italy
zkiziltan@deis.unibo.it

Abstract. We recently proposed a simple declarative language for specifying a wide range of counting and occurrence constraints. The language uses just two global primitives: the RANGE constraint, which computes the range of values used by a set of variables, and the ROOTS constraint, which computes the variables mapping onto particular values. In order for this specification language to be executable, propagation algorithms for the RANGE and ROOTS constraints should be developed. In this paper, we focus on the study of the RANGE constraint. We propose an efficient algorithm for propagating the RANGE constraint. We also show that decomposing global counting and occurrence constraints using RANGE is effective and efficient in practice.

1 Introduction

Constraints that put restrictions on the occurrence of particular values (*occurrence* constraints) or constraints that put restrictions on the number of values or variables meeting some conditions (*counting* constraints) are very useful in many real world problems, especially those involving resources. For instance, we may want to limit the number of distinct values assigned to a set of variables. Many of the global constraints proposed in the past are counting and occurrence constraints (see, for example, [14,4,15,2,5]). In [6], we show that many occurrence and counting constraints can be expressed by means of two new global constraints, RANGE and ROOTS, together with some classical elementary constraints. This language also provides us with a method to propagate counting and occurrence constraints. We just need to provide efficient propagation algorithms for the RANGE and ROOTS constraints. This paper focuses on the RANGE constraint. We give an efficient algorithm for propagating the RANGE constraint based on a flow algorithm. We propose an extension of the RANGE constraint where we have constraints on the cardinality of the set variables.

J.C. Beck and B.M. Smith (Eds.): CPAIOR 2006, LNCS 3990, pp. 59–73, 2006.
© Springer-Verlag Berlin Heidelberg 2006

We also show that decomposing occurence constraints and counting constraints using the RANGE constraint performs well in practice.

The rest of the paper is organised as follows. Section 2 gives the formal background. Section 3 shows how counting and occurrence constraints can be decomposed using the RANGE constraint. In Section 4, we propose a polynomial algorithm for the RANGE constraint and an extension to the case where the set variables are subject to constraints on their cardinality. Some experimental results are presented in Section 6. Finally, we conclude in Section 7.

2   Formal Background

A constraint satisfaction problem consists of a set of variables, each with a finite domain of values, and a set of constraints specifying allowed combinations of values for subsets of variables. We use capitals for variables (e.g. X, Y and S), and lower case for values (e.g. v and w). We write $D(X)$ for the domain of a variable X. A solution is an assignment of values to the variables satisfying the constraints. A variable is *ground* when it is assigned a value. We consider both *integer* and *set* variables. A set variable S is represented by its lower bound $lb(S)$ which contains the definite elements (that must belong to the set) and an upper bound $ub(S)$ which also contains the potential elements (that may or may not belong to the set).

Constraint solvers typically explore partial assignments enforcing a local consistency property using either specialized or general purpose propagation algorithms. Given a constraint C, a *bound support* on C is a tuple that assigns to each integer variable a value between its minimum and maximum, and to each set variable a set between its lower and upper bounds which satisfies C. A bound support in which each integer variable is assigned a value in its domain is called a *hybrid support*. If C involves only integer variables, a hybrid support is a *support*. A value (resp. set of values) for an integer variable (resp. set variable) is *bound* or *hybrid consistent with* C iff there exists a bound or hybrid support assigning this value (resp. set of values) to this variable. A constraint C is *bound consistent (BC)* iff for each integer variable X_i, its minimum and maximum values belong to a bound support, and for each set variable S_j, the values in $ub(S_j)$ belong to S_j in at least one bound support and the values in $lb(S_j)$ are those from $ub(S_j)$ that belong to S_j in all bound supports. A constraint C is *hybrid consistent (HC)* iff for each integer variable X_i, every value in $D(X_i)$ belongs to a hybrid support, and for each set variable S_j, the values in $ub(S_j)$ belong to S_j in at least one hybrid support, and the values in $lb(S_j)$ are those from $ub(S_j)$ that belong to S_j in all hybrid supports. A constraint C involving only integer variables is *generalized arc consistent (GAC)* iff for each variable X_i, every value in $D(X_i)$ belongs to a support. If all variables in C are integer variables, hybrid consistency reduces to generalized arc consistency, and if all variables in C are set variables, hybrid consistency reduces to bound consistency.

To illustrate these different concepts, consider the constraint $C(X_1, X_2, S)$ that holds iff the set variable S is assigned exactly the values used by the integer variables X_1 and X_2. Let $D(X_1) = \{1,3\}$, $D(X_2) = \{2,4\}$, $lb(S) = \{2\}$ and $ub(S) = \{1,2,3,4\}$. BC does not remove any value since all domains are already bound consistent (value 2 was considered as possible for X_1 because BC deals with bounds). On the other hand, HC removes 4 from $D(X_2)$ and from $ub(S)$ as there does not exist any tuple satisfying C in which X_2 does not take value 2.

A total function $\mathcal{F}$ from a set $\mathcal{S}$ into a set $\mathcal{T}$ is denoted by $\mathcal{F} : \mathcal{S} \longrightarrow \mathcal{T}$ where $\mathcal{S}$ is the domain of $\mathcal{F}$ and $\mathcal{T}$ is the range of $\mathcal{F}$. Throughout, we will view a set of variables, X_1 to X_n as a function $\mathcal{X} : \{1, .., n\} \longrightarrow \bigcup_{i=1}^{i=n} D(X_i)$. That is, $\mathcal{X}(i)$ is the value of X_i.

3   An Executable Language

One of the simplest ways to propagate a new constraint is to decompose it into existing primitive constraints. We can then use the propagation algorithms associated with these primitives. Of course, such decomposition may reduce the number of domain values pruned. In [6], we show that many global counting and occurrence constraints can be decomposed into two new global constraints, RANGE and ROOTS, together with simple non-global constraints over integer variables (like $X \leq m$) and simple non-global constraints over set variables (like $S_1 \subseteq S_2$ or $|S| = k$). Adding RANGE and ROOTS and their propagation algorithms to a constraint toolkit thus provides a simple executable language for specifying a wide range of counting and occurrence constraints.

We focus here on the RANGE constraint. Given the function $\mathcal{X}$ representing a set of variables X_1 to X_n, the RANGE constraint holds iff a set variable T is the range of this function, restricted to the indices belonging to a second set variable, S.

$$\text{RANGE}([X_1, .., X_n], S, T) \text{ iff } T = \bigcup_{i \in S} \mathcal{X}(i)$$

In [7], we present a catalog containing over 70 global constraints from [3] specified with this simple language containing RANGE and ROOTS constraints. We present here just a few examples of using RANGE to decompose a global constraint.

The NVALUE constraint is useful in a wide range of problems involving resources since it counts the number of distinct values used by a sequence of variables [11]. NVALUE$([X_1, .., X_n], N)$ holds iff $N = |\{X_i \mid 1 \leq i \leq n\}|$. This can be decomposed into a RANGE constraint:

$$\text{NVALUE}([X_1, .., X_n], N) \text{ iff } \text{RANGE}([X_1, .., X_n], \{1, .., n\}, T) \wedge |T| = N$$

Enforcing HC on the decomposition is weaker than GAC on the original NVALUE constraint. However, it is NP-hard to enforce GAC on a NVALUE constraint [8].

In [6], the USES constraint was introduced. USES is a variant of the USEDBY constraint [5]. USES$([X_1, .., X_n], [Y_1, .., Y_m])$ holds iff the set of values assigned to $Y_1, .., Y_m$ is a subset of the set of values assigned to $X_1, .., X_n$. This can be decomposed into a RANGE constraint:

$$\text{USES}([X_1, .., X_n], [Y_1, .., Y_m]) \text{ iff}$$

$$\text{RANGE}([X_1, .., X_n], \{1, .., n\}, T) \ \wedge \text{RANGE}([Y_1, .., Y_m], \{1, .., m\}, T') \ \wedge \ T' \subseteq T$$

Enforcing HC on the decomposition is weaker than GAC on the original USES constraint. However, it is NP-hard to enforce GAC on a USES constraint [6]. Thus, decomposition is a simple method to obtain a polynomial propagation algorithm.

The PERMUTATION constraint is an ALLDIFFERENT constraint where we additionally know $\Re$, the set of values to be taken. That is, the sequence of variables $[X_1, \ldots, X_n]$ is a permutation of the values in $\Re$ where $|\Re| = n$. In [6], the PERMUTATION constraint is decomposed using a single RANGE constraint:

$$\text{PERMUTATION}([X_1, \ldots, X_n], \Re) \ \text{iff} \ \text{RANGE}([X_1, \ldots, X_n], \{1, \ldots, n\}, \Re)$$

Enforcing HC on the decomposition is equivalent to GAC on the original PERMUTATION constraint [6].

4 Propagating the Range Constraint

Enforcing hybrid consistency on the RANGE constraint is polynomial. This can be done using a maximum network flow problem. In fact, the RANGE constraint can be decomposed using a global cardinality constraint (GCC) for which propagators based on flow problems already exist [15, 13]. This will be shown in Section 5. But the RANGE constraint does not need the whole power of maximum network flow problems, and thus HC can be enforced on it at a lower cost than that of calling a GCC propagator. In this section, we propose an efficient way to enforce HC on RANGE. To simplify the presentation, the use of the flow is limited to a constraint that performs only part of the work needed for enforcing HC on RANGE. This constraint, that we name $\text{OCCURS}([X_1, \ldots, X_n], T)$, ensures that all the values in the set variable T are used by the integer variables X_1 to X_n:

$$\text{OCCURS}([X_1, \ldots, X_n], T) \ \text{iff} \ T \subseteq \bigcup_{i \in 1..n} \mathcal{X}(i)$$

We first present an algorithm for achieving HC on OCCURS (Section 4.1), and then use this to propagate the RANGE constraint (Section 4.2).

4.1 Occurs Constraint

We achieve hybrid consistency on $\text{OCCURS}([X_1, \ldots, X_n], T)$ using a network flow. We use a unit capacity network [1] in which capacities between two nodes can only be 0 or 1. This is represented by a directed graph where an arc from node x to node y means that a maximum flow of 1 is allowed between x and y while the absence of an arc means that the maximum flow allowed is 0. The unit capacity network $G_C = (N, E)$ of the constraint $C = \text{OCCURS}([X_1, \ldots, X_n], T)$ is built in the following way. $N = \{s\} \cup N_1 \cup N_2 \cup \{t\}$, where s is a source node, t is a sink

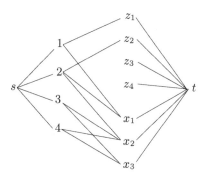

Fig. 1. Unit capacity network of the constraint $C = \text{OCCURS}([X_1, X_2, X_3], T)$ with $D(X_1) = \{1, 2\}$, $D(X_2) = \{2, 3, 4\}$, $D(X_3) = \{3, 4\}$, $lb(T) = \{3, 4\}$ and $ub(T) = \{1, 2, 3, 4\}$. Arcs are directed from left to right.

node, $N_1 = \{v \mid v \in \bigcup D(X_i)\}$ and $N_2 = \{z_v \mid v \in \bigcup D(X_i)\} \cup \{x_i \mid i \in [1..n]\}$. The set of arcs E is as follows:

$$E = (\{s\} \times N_1) \cup \{(v, z_v), \forall v \notin lb(T)\} \cup \{(v, x_i) \mid v \in D(X_i)\} \cup (N_2 \times \{t\})$$

G_C is quadripartite, i.e., $E \subseteq (\{s\} \times N_1) \cup (N_1 \times N_2) \cup (N_2 \times \{t\})$. In Fig. 1, we depict the network G_C of the constraint $C = \text{OCCURS}([X_1, X_2, X_3], T)$ with $D(X_1) = \{1, 2\}$, $D(X_2) = \{2, 3, 4\}$, $D(X_3) = \{3, 4\}$, $lb(T) = \{3, 4\}$ and $ub(T) = \{1, 2, 3, 4\}$. The intuition behind this graph is that when a flow uses an arc from a node v to a node x_i this means that X_i is assigned v, and when a flow uses the arc (v, z_v) this means that v is not necessarily used by the X_i's.[1] In Fig. 1 nodes 3 and 4 are linked only to nodes x_2 and x_3, which means that values 3 and 4 must necessarily be taken by one of the variables X_i (3 and 4 belong to $lb(T)$). On the contrary, nodes 1 and 2 are also linked to nodes z_1 and z_2 because values 1 and 2 do not have to be taken by a X_i (they are not in $lb(T)$).

In the particular case of unit capacity networks, a flow is any set $E' \subseteq E$: any arc in E' is assigned 1 and the arcs in $E \setminus E'$ are assigned 0. A *feasible* flow from s to t in G_C is a subset E_f of E such that $\forall n \in N \setminus \{s, t\}$ the number of arcs of E_f entering n is equal to the number of arcs of E_f going out of n, that is, $|\{(n', n) \in E_f\}| = |\{(n, n'') \in E_f\}|$. The value of the flow E_f from s to t, denoted $val(E_f, s, t)$, is $val(E_f, s, t) = |\{n \mid (s, n) \in E_f\}|$. A *maximum* flow from s to t in G_C is a feasible flow E_M such that there does not exist a feasible flow E_f, with $val(E_f, s, t) > val(E_M, s, t)$. A maximum flow for the network of Fig. 1 is given in Fig. 2. By construction a feasible flow cannot have a value greater than $|N_1|$. In addition, a feasible flow cannot contain two arcs entering a node x_i from N_2. Hence, we can define a function φ linking feasible flows and partial instantiations on the X_i's. Given any feasible flow E_f from s to t in G_C, $\varphi(E_f) = \{(X_i, v) \mid (v, x_i) \in E_f\}$. The maximum flow in Fig. 2 corresponds to

---

[1] Note that the edges go from the nodes representing the values to the nodes representing the variables. This is the opposite as the direction in the network flow problems used in the propagators of the ALLDIFF or GCC constraints [14, 15].

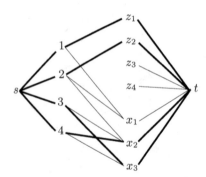

Fig. 2. A maximum flow for the network of Fig. 1. Bold arcs are those that belong to the flow. Arcs are directed from left to right.

the instantiation $X_2 = 4, X_3 = 3$. The way G_C is built induces the following theorem.

Theorem 1. *Let $G_C = (N, E)$ be the capacity network of a constraint $C = \text{OCCURS}([X_1, \ldots, X_n], T)$.*

1. *A value v in the domain $D(X_i)$ for some $i \in [1..n]$ is HC iff there exists a flow E_f from s to t in G_C with $val(E_f, s, t) = |N_1|$ and $(v, x_i) \in E_f$*
2. *If the X_i's are HC, T is HC iff $ub(T) \subseteq \bigcup_i D(X_i)$*

Proof. (1.$\Rightarrow$) Let I be a solution for C with $(X_i, v) \in I$. Build the following flow H: Put (v, x_i) in H; $\forall w \in I[T], w \neq v$, take a variable X_j such that $(X_j, w) \in I$ (we know there is at least one since I is solution), and put (w, x_j) in H; $\forall w' \notin I[T], w' \neq v$, add $(w', z_{w'})$ to H. Add to H the edges from s to N_1 and from N_2 to t so that we obtain a feasible flow. By construction, all $w \in N_1$ belong to an edge of H. So, $val(H, s, t) = |N_1|$ and H is a maximum flow with $(v, x_i) \in H$.

(1.$\Leftarrow$) Let E_M be a flow from s to t in G_C with $(v, x_i) \in E_M$ and $val(E_M, s, t) = |N_1|$. By construction of G_C, we are guaranteed that all nodes in N_1 belong to an arc in $E_M \cap (N_1 \times N_2)$, and that for every value $w \in lb(T)$, $\{n \mid (w, n) \in E\} \subseteq \{x_i \mid i \in [1..n]\}$. Thus, for each $w \in lb(T), \exists X_j \mid (X_j, w) \in \varphi(E_M)$. Hence, any extension of $\varphi(E_M)$ where each unassigned X_j takes any value in $D(X_j)$ and $T = lb(T)$ is a solution of C with $X_i = v$.

(2.$\Rightarrow$) If T is HC, all values in $ub(T)$ appear in at least one solution tuple. Since C ensures that $T \subseteq \bigcup_i \{X_i\}$, $ub(T)$ cannot contain a value appearing in none of the $D(X_i)$.

(2.$\Leftarrow$) Let $ub(T) \subseteq \bigcup_i D(X_i)$. Since all X_i's are HC, we know that each value v in $\bigcup_i D(X_i)$ is taken by some X_i in at least one solution tuple I. Build the tuple I' so that $I'[X_i] = I[X_i]$ for each $i \in [1..n]$ and $I'[T] = I[T] \cup \{v\}$. I' is still solution of C. So, $ub(T)$ is as tight as it can be wrt HC. In addition, since all X_i's are HC, this means that in every solution tuple I, for each $v \in lb(T)$ there exists i such that $I[X_i] = v$. So, $lb(T)$ is HC.   □

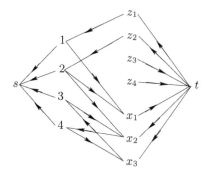

Fig. 3. Residual graph obtained from the network in Fig. 1 and the maximum flow in Fig. 2

Following Theorem 1, we need a way to check which edges belong to a maximum flow. *Residual graphs* are useful for this task. Given a unit capacity network G_C and a maximal flow E_M from s to t in G_C, the residual graph $R_{G_C}(E_M) = (N, E_R)$ is the directed graph obtained from G_C by reversing all arcs belonging to the maximum flow E_M; that is, $E_R = \{(x, y) \in E \setminus E_M\} \cup \{(y, x) \mid (x, y) \in E \cap E_M\}$. Given the network G_C of Fig. 1 and the maximum flow E_M of Fig. 2, $R_{G_C}(E_M)$ is depicted in Fig. 3. Given a maximum flow E_M from s to t in G_C, given $(x, y) \in N_1 \times N_2 \cap E \setminus E_M$, there exists a maximum flow containing (x, y) iff (x, y) belongs to a cycle in $R_{G_C}(E_M)$ [16]. Furthermore, finding all the arcs (x, y) that do not belong to a cycle in a graph can be performed by building the strongly connected components of the graph. We see in Fig. 3 that the arcs $(1, x_1)$ and $(2, x_1)$ belong to a cycle. So, they belong to some maximum flow and $(X_1, 1)$ and $(X_1, 2)$ are hybrid consistent. $(2, x_2)$ does not belong to any cycle. So, $(X_2, 2)$ is not HC.

HC on Occurs

We now have all the tools for achieving HC on any OCCURS constraint. We first build G_C. We compute a maximum flow E_M from s to t in G_C; if $val(E_M, s, t) < |N_1|$, we fail. Otherwise we compute $R_{G_C}(E_M)$, build the strongly connected components in $R_{G_C}(E_M)$, and remove from $D(X_i)$ any value v such that (v, x_i) belongs to neither E_M nor to a strongly connected component in $R_{G_C}(E_M)$. Finally, we set $ub(T)$ to $ub(T) \cap \bigcup_i D(X_i)$. Following Theorem 1 and properties of residual graphs, this algorithm enforces HC on OCCURS($[X_1, .., X_n], T$).

Complexity. Building G_C is in $O(nd)$. We need then to find a maximum flow E_M in G_C. This can be done in two sub-steps. First, we use the arc (v, z_v) for each $v \notin lb(T)$ (in $O(|\bigcup_i D(X_i)|)$). Afterwards, we compute a maximum flow on the subgraph composed of all paths traversing nodes w with $w \in lb(T)$ (because there is no arc (w, z_w) in G_C for such w). The complexity of finding a maximum flow in a unit capacity network is in $O(\sqrt{k} \cdot e)$ if k is the number of nodes and e the number of edges. This gives a complexity in $O(\sqrt{|lb(T)|} \cdot |lb(T)| \cdot n)$ for this second sub-step. Building the residual graph and computing the strongly

---

Algorithm 1. Hybrid consistency on RANGE

---

procedure $Propag\text{-}Range([X_1, \ldots, X_n], S, T)$;
1 Introduce the set of integer variables $Y = \{Y_i \mid i \in ub(S)\}$,
   with $D(Y_i) = D(X_i) \cup \{dummy\}$;
2 Achieve hybrid consistency on the constraint $\text{OCCURS}(Y, T)$;
3 Achieve hybrid consistency on the constraints $i \in S \leftrightarrow Y_i \in T$, for all $Y_i \in Y$;
4 Achieve GAC on the constraints $(Y_i = dummy) \vee (Y_i = X_i)$, for all $Y_i \in Y$;

---

connected components is in $O(nd)$. Extracting the HC domains for the X_i's is direct. There remains to compute BC on T, which takes $O(nd)$. Therefore, the total complexity is in $O(nd + n \cdot |lb(T)|^{3/2})$.

Incrementality. In constraint solvers, constraints are usually *maintained* in a locally consistent state after each modification (restriction) in the domains of the variables. It is thus interesting to ask about the total complexity of maintaining HC on OCCURS after an arbitrary number of restrictions on the domains (values removed from $D(X_i)$ and $ub(T)$, or added to $lb(T)$). Whereas some constraints are completely incremental (i.e., the total complexity after any number of restrictions is the same as the complexity of one propagation), this is not the case for constraints based on flow techniques like ALLDIFFERENT or GCC [14, 15]. They indeed potentially require the computation of a new maximum flow after each modification. Restoring a maximum flow from one that lost p edges is in $O(p \cdot e)$. If values are removed one by one (nd possible times), and if each removal affects the current maximum flow, the overall complexity over a sequence of restrictions on X_i's, S, T, is in $O(n^2 d^2)$.

4.2 Hybrid Consistency on Range

Enforcing HC on $\text{RANGE}([X_1, \ldots, X_n], S, T)$ can be done by decomposing it as an OCCURS constraint on new variables Y_i and some channelling constraints ([9]) linking T and the Y_i's to S and the X_i's. But the interesting point is that we do not need to *maintain* HC on the decomposition but we just need to propagate the constraints in *one pass*.

The algorithm *Propag-Range*, enforcing HC on the RANGE constraint, is presented in Algorithm 1. In line 1, a special encoding is built, where a Y_i is introduced for each X_i with index in $ub(S)$. The domain of a Y_i is the same as that of X_i plus a dummy value. The dummy value works as a flag. If OCCURS prunes it from $D(Y_i)$ this means that Y_i is necessary in OCCURS to cover $lb(T)$. Then, X_i is also necessary to cover $lb(T)$ in RANGE. In line 2, HC on OCCURS removes a value from a Y_i each time it contains other values that are necessary to cover $lb(T)$ in every solution tuple. HC also removes values from $ub(T)$ that cannot be covered by any Y_i in a solution. Line 3 updates the bounds of S and the domain of Y_i's. Finally, in line 4, the channelling constraints between Y_i and X_i propagate removals on X_i for each i which belongs to S in all solutions.

Theorem 2. *The algorithm Propag-Range is a correct algorithm for enforcing HC on* RANGE, *that runs in* $O(nd + n \cdot |lb(T)|^{3/2})$ *time, where d is the maximal size of* X_i *domains.*

Proof. Soundness. A value v is removed from $D(X_i)$ in line 4 if it is removed from Y_i together with *dummy* in lines 2 or 3. If a value v is removed from Y_i in line 2, this means that any tuple on variables in Y covering $lb(T)$ requires that Y_i takes a value from $D(Y_i)$ other than v. So, we cannot find a solution of RANGE in which $X_i = v$ since $lb(T)$ must be covered as well. A value v is removed from $D(Y_i)$ in line 3 if $i \in lb(S)$ and $v \notin ub(T)$. In this case, RANGE cannot be satisfied by a tuple where $X_i = v$. If a value v is removed from $ub(T)$ in line 2, none of the tuples of values for variables in Y covering $lb(T)$ can cover v as well. Since variables in Y duplicate variables X_i with index in $ub(S)$, there is no hope to satisfy RANGE if v is in T. Note that $ub(T)$ cannot be modified in line 3 since Y contains only variables Y_i for which i was in $ub(S)$. If a value v is added to $lb(T)$ in line 3, this is because there exists i in $lb(S)$ such that $D(Y_i) \cap ub(T) = \{v\}$. Hence, v is necessarily in T in all solutions of RANGE. An index i can be removed from $ub(S)$ only in line 3. This happens when the domain of Y_i does not intersect $ub(T)$. In such a case, this is evident that a tuple where $i \in S$ could not satisfy RANGE since X_i could not take a value in T. Finally, if an index i is added to $lb(S)$ in line 3, this is because $D(Y_i)$ is included in $lb(T)$, which means that the dummy value has been removed from $D(Y_i)$ in line 2. This means that Y_i takes a value from $lb(T)$ in all solutions of OCCURS. X_i also has to take a value from $lb(T)$ in all solutions of RANGE.

Completeness (Sketch). Suppose that a value v is not pruned from $D(X_i)$ after line 4 of *Propag-Range*. If $Y_i \in Y$, we know that after line 2 there was an instantiation I on Y and T, solution of OCCURS with $I[Y_i] = v$ or with $Y_i = dummy$ (thanks to the channelling constraints in line 4). We can build the tuple I' on $X_1, ..X_n, S, T$ where X_i takes value v, every X_j with $j \in ub(S)$ and $I[Y_j] \in I[T]$ takes $I[Y_j]$, and the remaining X_j's take any value in their domain. T is set to $I[T]$ plus the values taken by X_j's with $j \in lb(S)$. These values are in $ub(T)$ thanks to line 3. Finally, S is set to $lb(S)$ plus the indices of the Y_j's with $I[Y_j] \in I[T]$. These indices are in $ub(S)$ since the only j's removed from $ub(S)$ in line 3 are such that $D(Y_j) \cap ub(T) = \emptyset$, which prevents $I[Y_j]$ from taking a value in $I[T]$. Thus I' is a solution of RANGE with $I'[X_i] = v$. We have proved that the X_i's are hybrid consistent after *Propag-Range*.

Suppose a value $i \in ub(S)$ after line 4. Thanks to constraint in line 3 we know there exists v in $D(Y_i) \cap ub(T)$, and so, $v \in D(X_i) \cap ub(T)$. Now, X_i is hybrid consistent after line 4. Thus $X_i = v$ belongs to a solution of RANGE. If we modify this solution by putting i in S and v in T (if not already there), we keep a solution.

Completeness on $lb(S)$, $lb(T)$ and $ub(T)$ is proved in a similar way.

Complexity. The important thing to notice in *Propag-Range* is that constraints in lines 2–4 are propagated in sequence. Thus, OCCURS is propagated only once, for a complexity in $O(nd + n \cdot |lb(T)|^{3/2})$. Lines 1, 3, and 4 are in $O(nd)$. Thus, the complexity of *Propag-Range* is in $O(nd + n \cdot |lb(T)|^{3/2})$. This reduces to linear time complexity when $lb(T)$ is empty.

Incrementality. The overall complexity over a sequence of restrictions on X_i's, S and T is in $O(n^2 d^2)$. (See incrementality of OCCURS in Section 4.1.)   □

As we will show in the next section, the Range constraint can be decomposed using the GCC constraint. However, propagation on such a decomposition is in $O(n^2 d + n^{2.66})$ time complexity (see [13]). *Propag-Range* is thus significantly cheaper.

5   Range and Cardinality

Constraint toolkits like [10] additionally represent an interval on the cardinality of each set variable. This extra information is not taken into account by RANGE($[X_1, \ldots, X_n], S, T$) whereas it could improve propagation. We can easily extend the RANGE constraint to a constraint RANGE-CARD that involves this cardinality information. RANGE-CARD($[X_1, \ldots, X_n], S, M, T, N$) holds iff RANGE($[X_1, \ldots, X_n], S, T$) & $|S| = M$ & $|T| = N$. Unfortunately, enforcing HC on RANGE-CARD is NP-hard because it subsumes the NVALUE constraint (RANGE-CARD($[X_1, \ldots, X_n], \{1..n\}, n, T, N$) ≡ NVALUE($N, [X_1, \ldots, X_n]$)) and NVALUE is itself NP-hard to propagate [8]. However, we can partially take into account such cardinality information. RANGE-CARD($[X_1, \ldots, X_n], S, M, T, N$) can be decomposed using a GCC constraint:

$$\text{RANGE-CARD}([X_1, \ldots, X_n], S, M, T, N) \text{ iff}$$

$$
\begin{aligned}
&\text{GCC}([Y_1, \ldots, Y_n], [1, \ldots, m+1], [B_1, \ldots, B_{m+1}]) \wedge \\
\forall i \in [1..n] \quad &\quad i \in S \leftrightarrow Y_i \in T \quad\quad\quad\quad\quad\quad\quad\quad\quad\quad \wedge \\
\forall i \in [1..n] \quad &\quad (X_i = Y_i) \vee (Y_i = m+1) \quad\quad\quad\quad\quad\quad \wedge \\
\forall v \in [1..m+1] &\; v \in T \leftrightarrow B_v \neq 0 \quad\quad\quad\quad\quad\quad\quad\quad\quad \wedge \\
\forall v \in [1..m+1] &\; B_v \leq M - N + 1 \quad\quad\quad\quad\quad\quad\quad\quad\quad \wedge \\
&\quad\textstyle\sum_{v \in [1..m]} B_v = M
\end{aligned}
$$

where $m = |\bigcup_{i \in [1..n]}(D(X_i))|$, $m+1$ is a dummy value, and GCC($[X_1, \ldots, X_n]$, $[d_1, \ldots, d_m], [O_1, \ldots, O_m]$) holds iff the value d_i is used O_i times in $X_1, \ldots, X_n$, for all $i, 1 \leq i \leq m$. For sake of clarity we suppose that values are consecutive in the interval $[1..m]$ but this is not a restriction.

We have $\forall i \in [1..n], D(Y_i) = D(X_i) \cup \{m+1\}$ and $\forall v \in [1..m+1], D(B_v) = [0..n]$. We enforce GAC on the X's and Y's and BC on S, T and the B's. This algorithm has $O(n^2 d + n^{2.66})$ complexity (see [13]), which is typically worse than *Propag-Range* which ignores such cardinality information. It remains an open problem if we can extend *Propag-Range* to include some cardinality information, and if we can do so without changing its complexity.

6   Experimental Results

The purpose of this section is twofold. We demonstrate that decomposing global counting and occurrence constraints using RANGE is effective and efficient in practice. We show that propagating RANGE using the algorithm introduced in this paper is more effective than propagating it using the straightforward decomposition:

$$\text{RANGE}([X_1, \ldots, X_n], S, T) \quad \text{iff}$$

$$i \in S \rightarrow X_i \in T \wedge j \in T \rightarrow \exists i \in S.X_i = j \tag{1}$$

In order to isolate the effect of the RANGE constraint from other modelling issues, we used the following protocol: we randomly generated instances of binary CSPs and we added $\text{USES}([X_1, .., X_n], [Y_1, .., Y_n])$ constraints. Note that, it is NP-hard to achieve GAC on USES and there is no propagator available for this constraint in the literature. So, in all our experiments, we encode USES in three different ways:

[no-propag]: by putting the USES constraint in the model with no propagator but just a checker testing if it is satisfied or not,
[range]: by decomposing USES using RANGE as described in Section 3 and using the algorithm *Propag-Range* presented in Section 4,
[range-decomp]: by decomposing the RANGE constraints of the previous model using primitive constraints as in decomposition (1).

The problem instances are generated according to model B in [12], and can be described with the following parameters: the number of X and Y variables nx and ny in USES constraints, the total number of variables nz, the domain size d, the number of binary constraint m_1, the number of forbidden tuples t per binary constraint, and the number of USES constraints m_2. Note that the USES constraints can have overlapping or disjoint scopes of variables. We distinguish the two cases. All reported results are averages on 100 instances.

Our first experiment shows the effectiveness of decomposing USES with RANGE for propagation *alone* (not solving). We compared the number of values removed by propagation on the models obtained by representing USES constraints in the three different ways: no-propag, range, and range-decomp. (Note that in the no-propag model, the values are pruned only because of the binary constraints.) To simulate what happens inside a backtrack search, we randomly selected a subset of the variables and randomly assigned them values before propagation. Hence, in the experiments, the constraints are exposed to a wide range of different variable domains. We report the ratio of values removed by propagation on the following classes of problems:

class A : $\langle nx = 5, ny = 10, nz = 35, d = 20, m_1 = 70, t = 150, m_2 = 3 \ (overlap) \rangle$
class B : $\langle nx = 5, ny = 10, nz = 45, d = 20, m_1 = 90, t = 150, m_2 = 3 \ (disjoint) \rangle$

in which the number of assigned variables varies between 1 and 15. A failure of the propagation algorithm yields a ratio of 1 (all values are removed).

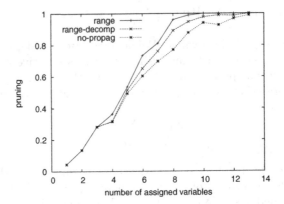

Fig. 4. Propagating random binary constraint satisfaction problems with three overlapping USES constraints (class A)

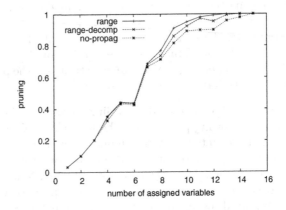

Fig. 5. Propagating random binary constraint satisfaction problems with three disjoint USES constraints (class B)

We observe in Figures 4 and 5 that domains can be reduced significantly using RANGE when propagating problems containing USES constraints. We also observe that propagating the RANGE constraint directly (**range** model) is more effective than propagating its decomposition (**range-decomp** model). The differences are greater when the USES constraints of the original problem overlap (Fig. 4) than when they are all disjoint (Fig. 5).

Our second experiment shows the efficiency of decomposing USES with RANGE when *solving* the problems. Our solver used the *smallest-domain-first* variable ordering heuristic with the lexicographical value ordering and a cutoff at 600 seconds. We compared the cost of solving the three types of models: **no-propag**, **range**, and **range-decomp**. We report the number of fails and the cpu-time needed to find the first solution on the following classes of problems:

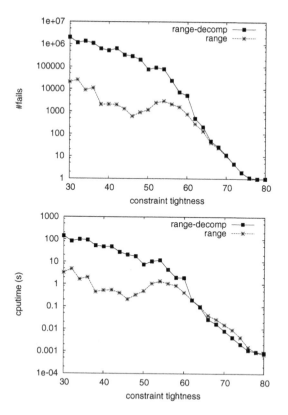

Fig. 6. Solving random binary constraint satisfaction problems with two overlapping USES constraints (class C)

$$class\ C : \langle nx = 5, ny = 10, nz = 25, d = 10, m_1 = 40, t, m_2 = 2 \rangle$$
$$class\ D : \langle nx = 5, ny = 10, nz = 30, d = 10, m_1 = 60, t, m_2 = 2 \rangle$$

in which t varies between 30 and 80.

We observe in Figures 6 and 7 that using the decomposition of RANGE (range-decomp model) is costly. This is due to the disjunction in the implementation of ∃. Note that the instances solved here (classes C and D) are much smaller than those used for propagation (classes A and B). Solving larger instances was impractical. Note also that we do not present the results where the RANGE constraint is not used (no-propag model) because they reached the cutoff in most of the instances not trivially over-constrained. So, this second experiment shows how efficiently RANGE can solve problems containing USES constraints. It also shows the clear benefit of using our algorithm in preference to the decomposition of RANGE over the under-constrained region. As the problems get over-constrained, the binary constraints dominate the pruning, and the algorithm gives a slight overhead in run-time, pruning equally with the decomposition of RANGE.

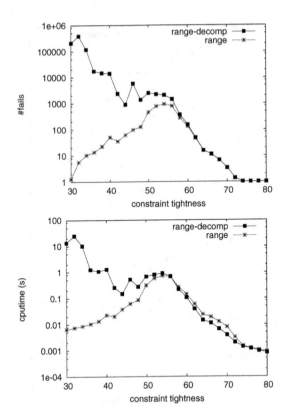

Fig. 7. Solving random binary constraint satisfaction problems with two disjoint USES constraints (class D)

7 Conclusion

RANGE and ROOTS are two global constraints that can express many other global constraints, such as occurrence and counting constraints [6]. We have presented a comprehensive study of the RANGE constraint. We proposed an algorithm for enforcing hybrid consistency on RANGE. We proposed a way to partially propagate RANGE-CARD, a constraint that combines RANGE with constraints on the cardinality of the set variables. Our experiments show the benefit we can obtain by incorporating the RANGE constraint in a constraint toolkit.

Acknowledgements

Hebrard and Walsh are supported by the National ICT Australia, which is funded through the Australian Government's *Backing Australias Ability* initiative, in part through the Australian Research Council. Hnich received support from Science Foundation Ireland (Grant 00/PI.1/C075).

References

1. R.K. Ahuja, T.L. Magnanti, and J.B. Orlin. *Network flows*. Prentice Hall, Upper Saddle River NJ, 1993.
2. N. Beldiceanu. Pruning for the minimum constraint family and for the number of distinct values constraint family. In *Proceedings CP'01*, pages 211–224, Paphos, Cyprus, 2001.
3. N. Beldiceanu, M. Carlsson, and J.X. Rampon. Global constraint catalog. Technical Report T2005:08, Swedish Institute of Computer Science, Kista, Sweden, May 2005.
4. N. Beldiceanu and E. Contejean. Introducing global constraints in chip. *Mathl. Comput. Modelling*, 20(12):97–123, 1994.
5. N. Beldiceanu, I. Katriel, and S. Thiel. Filtering algorithms for the *same* and *usedby* constraints. In *MPI Technical Report MPI-I-2004-1-001*, 2004.
6. C. Bessiere, E. Hebrard, B. Hnich, Z. Kiziltan, and T. Walsh. The Range and Roots constraints: Specifying counting and occurrence problems. In *Proceedings IJCAI'05*, pages 60–65, Edinburgh, Scotland, 2005.
7. C. Bessiere, E. Hebrard, B. Hnich, Z. Kiziltan, and T. Walsh. The Range and Roots constraints: some applications. Technical Report 2006-003, COMIC, January 2006.
8. C. Bessiere, E. Hebrard, B. Hnich, and T. Walsh. The complexity of global constraints. In *Proceedings AAAI'04*, pages 112–117, San Jose CA, 2004. to appear.
9. B.M.W. Cheng, K.M.F. Choi, J.H.M. Lee, and J.C.K. Wu. Increasing constraint propagation by redundant modeling: an experience report. *Constraints*, 4:167–192, 1999.
10. ILOG. *Reference and User Manual*. ILOG Solver 5.3, ILOG S.A., 2002.
11. F. Pachet and P. Roy. Automatic generation of music programs. In *Proceedings CP'99*, pages 331–345, Alexandria VA, 1999.
12. P. Prosser. An empirical study of phase transition in binary constraint satisfaction problems. *Artificial Intelligence*, 81:81–109, 1996.
13. C.G. Quimper, A. López-Ortiz, P. van Beek, and A. Golynski. Improved algorithms for the global cardinality constraint. In *Proceedings CP'04*, pages 542–556, Toronto, Canada, 2004.
14. J.C. Régin. A filtering algorithm for constraints of difference in CSPs. In *Proceedings AAAI'94*, pages 362–367, Seattle WA, 1994.
15. J.C. Régin. Generalized arc consistency for global cardinality constraint. In *Proceedings AAAI'96*, pages 209–215, Portland OR, 1996.
16. A. Schrijver. *Combinatorial Optimization - Polyhedra and Efficiency*. Springer-Verlag, Berlin, 2003.

On the Separability of Subproblems in Benders Decompositions

Marco Cadoli and Fabio Patrizi

Dipartimento di Informatica e Sistemistica
Università di Roma "La Sapienza", Italy
{cadoli, patrizi}@dis.uniroma1.it

Abstract. Benders decomposition is a well-known procedure for solving a combinatorial optimization problem by defining it in terms of a *master problem* and a *subproblem*. Its effectiveness relies on the possibility of synthethising *Benders cuts* (or nogoods) that rule out not only one, but a large class of trial values for the master problem. In turns, this depends on the possibility of *separating* the subproblem into several subproblems, i.e., problems exhibiting strong intra-relationships and weak inter-relationships. The notion of *separation* is typically given informally, or relying on syntactical aspects. This paper formally addresses the notion of separability of the subproblem by giving a semantical definition and exploring it from the computational point of view. Several examples of separable problems are provided, including some proving that a semantical notion of separability is much more helpful than a syntactic one. We show that separability can be formally characterized as equivalence of logical formulae, and prove the undecidability of the problem of checking separability.

1 Introduction and Motivations

Benders decomposition [1] is a well-known procedure for solving combinatorial optimization problems, which relies on the idea of distinguishing *primary* from *secondary* variables, defining a *master problem* over primary variables, and defining a *subproblem* over secondary variables given a trial value for primary variables. Every unsuccessful attempt to solve the subproblem is recorded as a *Benders cut* (or *nogood*) and added to the master problem, until an optimal solution is found, or the problem is proven to be unfeasible.

Two important factors that make the above procedure effective are: 1) the possibility of using different technologies for solving the master and the subproblem, e.g., ILP and CP, respectively, [6, 7], and 2) the possibility of synthethising Benders cuts that rule out not only one, but a large class of trial values for the master problem. In this paper we focus on the second factor, and specifically on the notion of *separability* of the subproblem, which intuitively means that it can be formulated using several subproblems exhibiting strong intra-relationships and weak inter-relationships. As a matter of fact, it has been noted in [6, 5, 2] that, if the subproblem is separable, then it is possible to design a Benders cut that excludes several instantiations of the primary variables, or, in other words, a nogood which is a partial, and not a total, assignment to the primary variables.

J.C. Beck and B.M. Smith (Eds.): CPAIOR 2006, LNCS 3990, pp. 74–88, 2006.

Therefore, the ability of recognizing separability of the subproblem is crucial for the efficiency of a Benders decomposition.

Let us introduce our running example, taken from [6, 5], which refers to a *machine scheduling* problem.

Example 1 (Machine Scheduling Problem [6, 5]). Machine Scheduling is the problem of finding an assignment of a set of jobs to a set of machines in such a way that 1) constraints on release and 2) due date are satisfied, 3) machines are single-task, and a cost function is minimized. Using the modelling language of the OPL system [?], one possible model is as follows:

```
// INPUT DESCRIPTION
{int+} Jobs=...;              //The set of jobs to be scheduled
int+ horizon=...;            //Max start time point for jobs
int+ n_machines=...;         //The number of machines range
Time[1..horizon];           //Range "Time" definition range
Machines[1..n_machines];    //Range "Machines" definition int+
ReleaseDate[Jobs]=...;       //Each job has a release date int+
DueDate[Jobs]=...;           //Each job has a due date int+
Cost[Jobs,Machines]=...;     //Machines incur different costs per
job int+ Duration[Jobs,Machines]=...;//Machines run at different
speeds per job
// SEARCH SPACE
var Machines Assignment[Jobs]; var Time StartTime[Jobs];
// OBJECTIVE FUNCTION
minimize
    sum (j in Jobs) Cost[j,Assignment[j]]
// CONSTRAINTS
subject to {
    forall (j in Jobs) // 1. RESPECT RELEASE DATE
        StartTime[j] >= ReleaseDate[j];
    forall (j in Jobs) // 2. RESPECT DUE DATE
        StartTime[j] + Duration[j,Assignment[j]] <= DueDate[j];
    forall (t in Time) // 3. MAX ONE JOB PER MACHINE AT EACH TIME POINT
        forall (m in Machines)
            sum (j in Jobs)
                (Assignment[j] = m &
                StartTime[j] <= t < (StartTime[j] + Duration[j,Assignment[j]])
                ) <= 1;
};
```

The Benders decomposition suggested in [6] selects **Assignment** and **StartTime** as primary and secondary variables, respectively. Moreover, it defines the master problem as the minimization problem on no constraints, nd the subproblem as the decision problem on constraint 1, 2, and 3 (for a given optimal instantiation of **Assignment** and ignoring the objective function). A given optimal instantiation **Assignment** which is unfeasible for the subproblem is called a Benders cut (or nogood), and the next iteration of the master problem includes the constraint **Assignment** ≠ **Assignment**, until a feasible instantiation is found, or the problem is proven to be unfeasible.

This is a "raw" version of the decomposition, which ignores the fact that the subproblem is *separable* wrt the machines. As an example, if we have three machines, we can consider three separate subproblems, one for each of them. If $\overline{\texttt{Assignment}}$ = [1,1,2,1,3,2] is optimal for the master problem, and no serial schedule of jobs 3,6 on the second machine exists, we can safely add the constraint Assignment[3] <> 2 \/ Assignment[6] <> 2. Constraints of the latter kind rule out a whole set of assignments (not just one) to the primary variables, and can be added for each unfeasible subproblem. This ultimately results in a more efficient decomposition. □

An informal notion of separability is typically used in the literature, but we claim that the importance of this concept calls for precise definitions and careful analysis.

Example 2 (Example 1, continued). Since constraint 3 is universally quantified wrt the machines "forall (m in Machines)", we can claim its separability just relying on an intuitive argument. The methodological problem is that this syntax-based argument is heavily dependent on the way the problem is formulated (cf. also forthcoming Example 6). To see this point, consider the following statement, equivalent to constraint 3.

```
//3'. NO TWO JOBS RUNNING ON THE SAME MACHINE AT EACH TIME POINT:
forall(t in Time)
  sum (i,j in Jobs: i <> j)
    ( Assignment[i] = Assignment[j] &
      StartTime[i] <= t < (StartTime[i] + Duration[i,Assignment[i]]) &
      StartTime[j] <= t < (StartTime[j] + Duration[j,Assignment[j]])
    ) = 0;
```

Since constraint 3' is not universally quantified wrt the machines, it is less clear that it separates. □

It is everyday experience that different formulations, all of them being intuitive, can be done for a problem, sometimes in the hope of having more performant models.

Example 3 (Example 1, continued). We can define a dependent array RunsOn storing for each time point and each job the machine that runs the job (or a negative number if the job is not running). In fact, in this way we can define the "single-task machines" constraint (3 or 3') by means of a global alldifferent constraint, just stating that running machines are all different at each time point. The alldifferent constraint often performs very well [10], especially in connection with "channelling constraints" [?].

```
range MachinesPlus[-card(Jobs)..n_machines];//negative numbers:
irrelevant var MachinesPlus RunsOn[Time,Jobs];
// DEFINITION OF RunsOn:
forall (t in Time)
  forall (j in Jobs){
    (StartTime[j] <= t < (StartTime[j] + Duration[j,Assignment[j]])
      => RunsOn[t,j] = Assignment[j])
    &
```

```
  (StartTime[j] > t \/ t >= (StartTime[j] + Duration[j,Assignment[j]])
    => RunsOn[t,j] = -j); // negative numbers are all different
};
// 3''. AT EACH TIME POINT RUNNING MACHINES ARE ALL DIFFERENT:
forall (t in Time)
  alldifferent (all (j in Jobs) RunsOn[t,j]);
```

Again, constraint 3" is not universally quantified wrt the machines, but it nevertheless separates. □

In this paper we investigate the possibility of automating the process of checking subproblem separability in the context of Benders decompositions. In particular, given a problem and applying to it a given Benders decomposition schema which leads to a constraint satisfaction subproblem [5], our goal is to state the conditions, if any, that make the subproblem separable. To this end, we first address the notion of subproblem separability by giving a semantical definition and then we explore it from the computational point of view, providing two theorems which show that i) separability can be formally characterized as equivalence of logical formulae and ii) the problem of checking separability is undecidable.

The exposition is structured as follows. In Section 2 we recall the definition of Benders decomposition, in Section 3 a formal definition of separation is given, while in Section 4 we show semantical and computational characterizations. Finally, Section 5 draws some conclusions.

2 Preliminaries

Given two arrays of variables $\boldsymbol{p} = (p_1, \ldots, p_n)$ (primary) and $\boldsymbol{s} = (s_1, \ldots, s_m)$ (secondary) which may take values, respectively, from sets $P = C_1^p \times \ldots \times C_n^p$ and $S = C_1^s \times \ldots \times C_m^s$, in this paper we consider problems of the form:

$$PB : \begin{cases} \min\{f(\boldsymbol{p})\} \ objective function \ (o.f.) \\ s.t. \\ \alpha(\boldsymbol{s}) & constraint \ 1 \ (c1) \\ \gamma(\boldsymbol{p}) & constraint \ 2 \ (c2) \\ \beta(\boldsymbol{p}, \boldsymbol{s}) & constraint \ 3 \ (c3) \\ \boldsymbol{p} \in P & primary \ variables \ domain \ (p.v.d.) \\ \boldsymbol{s} \in S & secondary \ variables \ domain \ (s.v.d.) \end{cases} \tag{1}$$

where α, γ and β are suitable representations of constraints in which, respectively, only $\boldsymbol{s}$ variables, only $\boldsymbol{p}$ variables, or both occur. In [6,2] generalizations of the above problem in which, e.g., variables from $\boldsymbol{s}$ may occur in the objective function, are studied. According to [6], such problems can be solved by applying a "logic-based" Benders Decomposition scheme that gives raise to the following problems:

$$MP^k : \begin{cases} \min\{f(\boldsymbol{p})\} & o.f. \\ s.t. \\ \gamma(\boldsymbol{p}) & c2 \\ CUT_{\boldsymbol{p}^i}(\boldsymbol{p}) \\ (i = 1, 2, \ldots k-1) \ Benders \ cuts \\ \boldsymbol{p} \in P & p.v.d. \end{cases} \qquad SP : \begin{cases} \alpha(\boldsymbol{s}) & c1 \\ \beta(\overline{\boldsymbol{p}}, \boldsymbol{s}) & c3 \\ \boldsymbol{s} \in S & s.v.d. \end{cases} \tag{2}$$

Master Problem (MP^k) is the problem of finding an assignment to $p \in P$ that minimizes the objective function $f(p)$ while satifying i) $\gamma(p)$ and ii) the Benders cuts $CUT_{p^i}(p)$ $(i = 1, \ldots, k - 1)$ generated at the previous $k - 1$ iterations. When $k = 1$, MP^1 contains no cut, and the decomposition is just a bipartition of the constraints of PB into i) those over variables involved in the objective function, put into MP^1, and ii) the remaining ones, belonging to SP.

Subproblem (SP) is the feasibility problem of checking whether there exists an assignment $\overline{s}$ that, along with a given assignment $\overline{p}$ obtained as solution of MP^k, satisfies the constraints $\alpha(s)$ and $\beta(\overline{p}, s)$. If such $\overline{s}$ exists then $(\overline{p}, \overline{s})$ is a solution to PB, otherwise, problem MP^{k+1} is generated by adding to MP^k a Benders cut $CUT_{p^k}(p)$.

Referring to Example 1, p is `Assignment`, s is `StartTime`, $\alpha(s)$ is constraint 1, $\beta(p, s)$ is the conjunction of constraints 2 and 3, and $\gamma(p)$ is a tautology.

One obvious desirable quality of Benders cuts is *soundness*, i.e., the guarantee that the above algorithm finds an optimal solution to PB for each instance. As an example, the constraint

$$CUT_{p^k}(p) \doteq (p \neq p^k), \tag{3}$$

where p^k is the solution to MP^k, is sound. The problem with (3) is that an unacceptably large number of cuts may be added to the Master problem, and this may reflect in inefficiency (cf. Example 1). In the next sections we look for conditions which may be helpful for having a significantly lower number of cuts.

3   Separation into Subproblems

Before formalizing the notion of separability introduced in Section 1, we need to clarify the role played by the selection of relevant input data. Referring to Example 1, every choice of the machine induces a selection of the release and due dates, costs, and durations. As an example, if $\overline{\texttt{Assignment}} = [1,1,2,1,3,2]$ and machine 2 is selected, then we only need the third and the sixth rows of input arrays `ReleaseDate` and `DueDate`. Analogously, we need only some entries of the `Cost` and `Duration` arrays.

In general, given a representation R of the instance, e.g., as a relational database over the schema $\mathcal{R}$, and an integer q representing the number of subproblems, we assume that there is a function $\sigma_1 : \mathcal{R} \times [1, q] \to \mathcal{R}$ that *selects* the input data relevant for the i-th subproblem $(1 \leq i \leq q)$.

Analogously, we need a way to select the variables relevant to the i-th subproblem. As an example, for the given $\overline{\texttt{Assignment}}$ and machine 2, we want to assign a `StartTime` just to jobs 3 and 6. In general, we assume that there is a function σ_2 that partitions the variables into q subsets, one for each subproblem. For the sake of simplicity, we assume that all the variables may take a value from the same set.

From now on, we represent problems with the following notation

$$\psi(R) = \exists F : D \to C \ s.t. \ \phi(R, F), \tag{4}$$

where R is a representation of the instance over the schema $\mathcal{R}$, F is the required assignment to the variables, D and C are the domain and the codomain of the assignment, respectively, and $\phi(R, F)$ is a representation of the constraints. We prefer the above notation over the notation as in (1) or (2) because it highlights the input, which is crucial for our purposes. Moreover it is worth reminding that, if C is finite and ϕ is a formula in first-order logic, then formulae of the kind (4) can represent every problem in the complexity class NP [?,?]. Finally, we note that there is a direct correspondence between the above notation and state-of-the-art modelling languages such as OPL. As an example, an array of variables in Example 1 corresponds to the existentially quantified function F in (4).

Definition 1 (Subproblems). *Given a problem ψ of the form (4), an integer $q \geq 1$ and two functions $\sigma_1 : \mathcal{R} \times [1, q] \rightarrow \mathcal{R}$ and $\sigma_2 : [1, q] \rightarrow 2^D$ such that $\{D_1, \ldots, D_q\}$ $(D_i = \sigma_2(i))$ is a partition of D, the following q problems are defined as the* subproblems *of ψ wrt σ_1 and σ_2:*

$$\psi_i(R) = \exists R_i, F_i : D_i \rightarrow C \ s.t. \ R_i = \sigma_1(R, i) \wedge \phi(R_i, F_i), \qquad (i = 1, \ldots, q).$$

Definition 1 can be used to obtain the subproblems in a syntactical way, by means of a symbolic manipulation of the problem. To see intuitively how the subproblems are obtained, we resort again to our running example.

Example 4 (Example 1, continued). Given an instance of the problem with q machines, and a value for `Assignment`, we consider the (sub)problem defined as the conjunction of constraints 1, 2, and 3, and no objective function. As mentioned before, σ_1 takes a machine i and the input, e.g., arrays `ReleaseDate`, `DueDate`, `Cost`, and `Duration`, and gives new arrays `ReleaseDate_i`, `DueDate_i`, `Cost_i`, and `Duration_i`. σ_1 can be represented by means of simple constraints, the following being an example for i = 1:

```
//INPUT:
{int+} Jobs=...; int+ horizon=...; int+ n_machines=...; range
Machines [1..n_machines]; int+ ReleaseDate[Jobs]=...; int+
DueDate[Jobs]=...; int+ Cost[Jobs,Machines]=...; int+
Duration[Jobs, Machines]=...;
//JOBS ASSIGNMENT:
Open Machines Assignment[Jobs];
//CONSTANTS DEFINITION:
int+ maxTime = max(j in Jobs)(DueDate[j]); int+ maxCost = max(j in
Jobs, m in Machines)(Cost[j,m]); int+ maxDuration = max(j in Jobs,
m in Machines)(Duration[j,m]);
//OUPUT:
{int+} Jobs_1={j | j in Jobs: Assignment[j]=1}; var Machines
n_machines_1 in 1..1; // n_machines_1 = 1 var int+ horizon_1 in
horizon..horizon; // horizon_1 = horizon var int+
ReleaseDate_1[Jobs_1] in 0..horizon; var int+ DueDate_1[Jobs_1] in
0..maxTime; var int+ Cost_1[Jobs_1,[1..1]] in 0..maxCost; var int+
Duration_1[Jobs_1,[1..1]] in 0..maxDuration;
//CONSTRAINTS:
solve{
```

```
forall(j in Jobs_1){
   ReleaseDate_1[j] = ReleaseDate[j];
   DueDate_1[j] = DueDate[j];
   Cost_1[j,1] = Cost[j,1];
   Duration_1[j,1] = Duration[j,1];
}
};
```

Note that, coherently with Definition 1 where the R_i are existentially quantified, all items of the form xxx_1, e.g. DueDate_1 and horizon_1, are variables that must be assigned, Jobs_1 being a syntactical exception, due to implementation reasons, that can be yet conceptually regarded as a variable.

The other function σ_2 takes a machine i and the variables, i.e., array StartTime, and gives a new array of variables StartTime_i. The representation of σ_2 is also simple, and is omitted for brevity.

Each subproblem can be simply represented by defining all constraints on the new symbols, e.g., by writing DueDate_1 instead of DueDate for the first subproblem. It is worth noting that this can be done for all versions of the machine scheduling problem, i.e., for Examples 1, 2, and 3.   □

Given an instance R of a problem of the form (4), we denote as $SOL(\psi(R))$ the set of solutions to $\psi(R)$, i.e., of the set of functions which satisfy the constraints. The following definition tells us how to integrate the solutions of the subproblems.

Definition 2 (Composition of solutions). *Given a problem $\psi(R)$ and its q subproblems $\psi_i(R)$ as in Definition 1, we define the* composition ($\bowtie$) *of the solutions $SOL(\psi_i(R))$ of the subproblems as follows:*

$$\bowtie_{i=1}^{q} SOL(\psi_i(R)) \doteq \{F : D \to C \ s.t. \ \forall i = 1, \ldots, q \ F|_{D_i} \in SOL(\psi_i(R))\},$$

where $F|_{D_i}$ denotes the selection of the assignments of F to the variables in D_i.

Now we need a way to relate a problem to its subproblems, which is *semantical*, i.e., based on the respective solutions. The following definition tells us that a problem is separated by (σ_1, σ_2) if its solutions can be obtained just by composing the solutions of its subproblems.

Definition 3 (Separation). *Given a problem of the form (4) and its q subproblems $\psi_i(R)$ as in Definition 1, ψ is (σ_1, σ_2)-separated into the q problems $\psi_1, \ldots, \psi_q$ iff*

$$\forall R \in \mathcal{R} \ \bowtie_{i=1}^{q} SOL(\psi_i(R)) = SOL(\psi(R)).$$

Referring again to the subproblem in the three versions of Examples 1, 2 and 3, it is possible to see that it is (σ_1, σ_2)-separated into q problems according to Definition 3, where q is the number of machines.

Of course, not all problems are separable, as shown by the next example.

Example 5. We add to the constraints of Example 1 a further constraint which avoids more than 2 machines running at the same time, useful, e.g., to reduce noise or energy consumption.

```
forall (t in Time) // 4. MAX TWO JOBS RUNNING AT EACH TIME POINT
    sum (j in Jobs)(
      StartTime[j] <= t < (StartTime[j] + Duration[j,Assignment[j]]
    ) <= 2;
```

The latter constraint is added to the subproblem, and the Master problem is unchanged. With functions σ_1 and σ_2 defined as in Example 4, it is possible to see that the current version of the subproblem is not (σ_1, σ_2)-separated. We do that by 1) exhibiting an instance, 2) solving separately the subproblems obtained applying Definition 1, 3) composing their solutions according to Definition 2, and 4) showing that a solution which does not satisfy the original problem arises.

The instance is as follows:

```
Jobs = {1,2,3,4,5,6}; n_machines = 3; horizon = 15;
ReleaseDate[Jobs] = [1,10,2,4,9,8]; DueDate[Jobs] =
[4,18,10,14,14,18]; Cost[Jobs,Machines] =[    //  m1  m2  m3
                      [ 2 , 3 , 6 ], //j1
                      [ 7 , 8 , 11], //j2
                      [ 6 , 5 , 7 ], //j3
                      [ 10, 12, 12], //j4
                      [ 7 , 7 , 6 ], //j5
                      [ 12, 5 , 6 ], //j6
                   ];

Duration[Jobs,Machines] =[    //  m1  m2  m3
                      [ 3 , 2 , 4 ], //j1
                      [ 6 , 4 , 5 ], //j2
                      [ 7 , 7 , 6 ], //j3
                      [ 5 , 8 , 7 ], //j4
                      [ 3 , 5 , 4 ], //j5
                      [ 5 , 6 , 5 ], //j6
                   ];
```

We assume that solving the Master Problem led to the assignment $\overline{\text{Assignment}}$ = [1,1,2,1,3,2]. Now, applying σ_1 as in Example 4 to select the relevant data for jobs assigned to, e.g., machine 2, we obtain the following data set:

```
n_jobs_2 = 2; n_machines_2 = 1; horizon_2 = 15;
ReleaseDate_2[Jobs_2] = [2,8]; DueDate_2[Jobs_2] = [10,18];
Cost_2[Jobs_2,[2..2]] = [ //  m2
                      [ 5 ], //j3
                      [ 5 ]  //j6
                   ];
Duration_2[Jobs_2,[2..2]] = [ //  m2
                      [ 7 ], //j3
                      [ 6 ]  //j6
                   ];
```

which represents the input to the 2nd separated subproblem. The input to the other subproblems can be obtained in the same way.

A solution to the three subproblems is as follows:

```
// Machine 1, jobs 1, 2, 4
StartTime_1 = [1,10,4];
// Machine 2, jobs 3, 6
StartTime_2 = [2,9];
// Machine 3, job 5
StartTime_3 = [9];
```

which does not satisfy the fourth constraint. As an example, in time points 10, 11 and 12, all three machines are running.   □

One may argue whether the semantical criterion for checking problem separation as defined by Definition 3 is really necessary or not, and in particular whether simpler criteria based on syntactic aspects are equally effective. As an example, we could build the primal constraint graph [4] of the subproblems –as defined previously– of Example 1, i.e., a graph with a node for each variable and an edge between any pair of variables syntactically occurring in the same constraint. A weaker notion of separability could be based on the fact that the graph we obtain has one component for each machine. Anyway, as shown by the next example, a problem with *redundant* constraints arises.

Example 6 (Example 1, continued). Let the following constraint be added to the machine scheduling problem specification:

```
/* 4. If machines are less than jobs, then at least two jobs start
at different time points. (card() returns the cardinality of a
set)*/ n_machines < card(Jobs) =>
          sum(i,j in Jobs:i<>j)(StartTime[i]<>StartTime[j])>=2;
```

Note that constraint 3 logically implies constraint 4, hence any solution satisfying 1-3 also satisfies 4. Note also that constraint 4 involves all the secondary variables, hence its primal constraint graph is a complete graph with card(Jobs) nodes, one for any StartTime component, representing the fact that all the secondary variables are somehow mutually constrained. As a consequence, a syntactic definition based on the constraint graph would fail to recognize separability, while Definition 3 does not.   □

Definition 3 is clarified also by the following example.

Example 7 (Protein Folding). [8] This problem specification models a simplified version of an important problem in computational biology which consists in finding the spatial conformation of a protein (i.e., a sequence of amino-acids) with minimal energy.

The simplifications with respect to the real problem are twofold: firstly, the 20-letter alphabet of amino-acids is reduced to a two-letter alphabet, namely H and P. H represents *hydrophobic* amino-acids, whereas P represents polar or *hydrophilic* amino-acids. Secondly, the conformation of the protein is limited to a bi-dimensional discrete space. Nonetheless, these limitations have been proven to be very useful for attacking the whole protein conformation prediction protein, which is known to be NP-complete [3] and very hard to solve in practice.

In this formulation, given the sequence (of length n) of amino-acids of the protein (the so called primary structure of the protein), i.e., a sequence of length n with elements in $\{H,P\}$, we aim to find a connected shape of this sequence on a bi-dimensional grid (whose points have coordinates in the integral range $[-(n-1), (n-1)]$, the sequence starting at $(0,0)$), which is not overlapping, and maximizes the number of "contacts", i.e., the number of non-sequential pairs of H amino-acids for which the Euclidean distance of the positions is 1 (the overall energy of the protein is defined as the opposite of the number of contacts).

Protein Folding can be modeled as a planning problem, whose input is a representation of the protein and the output is a sequence of moves that maximizes the number of contacts by folding it.

More specifically, a protein is described by a sequence $s \in \{0,1\}^n$ (1 and 0 representing, respectively, H and P), that can be arranged on the grid by means of four moves: up, $d$own, $l$eft, $r$ight, each of which sets the position of an amino-acid wrt its predecessor. The first element is always placed in the center of the grid and the sequence cannot cross itself. Figure 1 shows a 1-contacts configuration for a sequence $s =< 1,0,1,0,0,1 >$ of length 6, obtained by applying, in the reported order, the moves $u\ r\ r\ u\ l$.

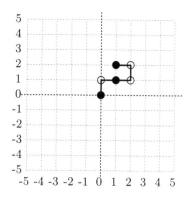

Fig. 1. An instance of the Protein folding problem

More formally, given:

- an array $\bar{s} \in \{0,1\}^n$, representing the protein;
- a set $T_{move} = [1, n-1] \in \mathbb{N}$;
- a set $T_{pos} = [1, n] \in \mathbb{N}$;
- a set $B = [-n+1, n-1] \in \mathbb{Z}$;
- a set $M = \{u =< 0,1 >, d =< 0, -1 >, l =< -1,0 >, r =< 1,0 >\} \in \mathbb{Z}^2$;
- the set of variables $V = \{m_1, \ldots, m_{n-1}, x_1, \ldots, x_n, y_1, \ldots, y_n\}$, $m_i \in M, \forall i \in T_{move}$ and $< x_i, y_i >\in B^2, \forall i \in T_{pos}$;

and defined:

- $\boldsymbol{moves} = (m_1, \ldots, m_{n-1}) \in M^{n-1}$, the array representing the sequence of moves;

- $pos = (< x_1, y_1 >, \ldots, < x_n, y_n >) \in (B^2)^n$, the array whose i-th component represents the position of the i-th amino-acid;
- the function

$$Hits(pos, \overline{s}) = \sum_{\substack{t,t' \in T_{move} \ s.t. \\ t' > t+1 \ \wedge \ \overline{pos_t pos_{t'}} = 1}} \overline{s_t} \cdot \overline{s_{t'}}$$

that counts the number of contacts for a particular pos;

The Protein Folding Problem can be stated as follows:

$$PF : \begin{cases} \max\{Hits(pos, \overline{s})\} & o.f. \\ s.t. \\ pos_1 = < 0,0 > & start\ condition \\ \forall t \in T_{move}\ move_t = pos_{t+1} - pos_t & channelling\ constraints \\ \forall(t,t') \in T^2_{pos} \mid t > t'\ pos_t \neq pos_{t'} & no\ crossing\ constraints \end{cases}$$

In such form, it can be decomposed as in (2) by considering pos as the array of primary variables and $moves$ that of secondary, obtaining the Master Problems MP^k_{PF} and the Subproblem SP_{PF}:

$$MP^k_{PF} : \begin{cases} \max\{Hits(pos, \overline{s})\} & o.f. \\ s.t. \\ pos_1 = < 0,0 > & start\ condition \\ \forall(t,t') \in T^2_{pos} \mid t > t'\ pos_t \neq pos_{t'} & no\ crossing\ constraints \\ CUT_{pos^i}(pos) \\ (i = 1, \ldots k - 1) & Benders\ cuts \end{cases}$$

$$SP_{PF} : \{ \forall t \in T_{move}\ move_t = \overline{pos_{t+1}} - \overline{pos_t}\quad channelling\ constraints$$

The Subproblem is to find a sequence of moves such that the given positions obtained by solving the Master Problem are feasible.
Now, apply Definition 3:

- define σ_2 as the function selecting, for each $i = 1, \ldots, n - 1$, the set of secondary variables $\{m_i\}$;
- define σ_1 as the function selecting, for each $i = 1, \ldots, n - 1$, the pair (pos_i, pos_{i+1}) from pos.

Such functions define the $n - 1$ problems

$$SP^i_{PF} : \{ move_i = \overline{pos_{i+1}} - \overline{pos_i}\ , i = 1, \ldots, n - 1$$

of checking whether there exists a move such that the $(i + 1)$-th amino-acid is in a position reachable from that of the i-th.
To see that SP_{PF} is (σ_1, σ_2)-separated, observe that the generic array $moves$ is a solution to SP_{PF} for a given assignment $\overline{pos}$ iff

$$moves = (\overline{pos}_2 - \overline{pos}_1, \ldots, \overline{pos}_n - \overline{pos}_{n-1})$$

and that each of its components is a solution to SP^i_{PF}, $(i = 1, \ldots, n)$. Conversely, given $(n-1)$ solutions $moves_i$ to SP^i_{PF}, $(i = 1, \ldots, n)$, the array

$$\boldsymbol{moves} = (moves_1, \ldots, moves_n)$$

is a solution to SP_{PF}, as assignment $\overline{pos}_{i+1}$ of SP^i_{PF} is, by construction, the same as $\overline{pos}_i$ of problem SP^{i+1}_{PF} forall $i = 1, \ldots, n-1$. $\square$

4   Characterization of Separation

Definition 3 gives a semantical notion of separation of a problem in subproblems. A practical difficulty is that it is not obvious how to use it for proving separation, since we would have to consider all possible instances, solve the problem and the candidate subproblems, and check that their solutions coincide.

The following theorem shows that in principle it is not necessary to do that, and reduces the problem of checking separation to the problem of equivalence of two logical formulae.

Theorem 1. *Given a problem $\psi(\boldsymbol{R})$, an integer q, two functions σ_1, σ_2, and q problems $\psi_1, \ldots, \psi_q$ as in Definition 3, ψ is (σ_1, σ_2)-separated into $\psi_1, \ldots, \psi_q$ iff the following formula is a tautology*

$$\psi \equiv \bigwedge_{i=1}^{q} \psi_i. \tag{5}$$

Proof. (Only if part.) By hypothesis, the following q problems (σ_1, σ_2)-separate ψ:

$$\psi_i(\boldsymbol{R}) = \exists \boldsymbol{R}_i, F_i : \sigma_2(D, i) \to C \ s.t. \ \boldsymbol{R}_i = \sigma_1(\boldsymbol{R}, i) \wedge \phi(\boldsymbol{R}_i, F_i), \quad (i = 1, \ldots, q). \tag{6}$$

Now, given an instance $\boldsymbol{R} \in \mathcal{R}$, if F is a solution to $\psi(\boldsymbol{R})$ then any restriction $F|_{D_i}$ to $D_i = \sigma_2(D, i)$ is a solution to $\psi_i(\boldsymbol{R})$ for each $i = 1, \ldots, q$. In fact, since separation holds, it holds that

$$SOL(\psi(\boldsymbol{R})) = \{F : D \to C \ s.t. \ \forall i = 1, \ldots, q \ F|_{D_i} \in SOL(\psi_i(\boldsymbol{R}))\}.$$

Hence, for any instance $\boldsymbol{R}$, if F solves $\psi(\boldsymbol{R})$ then it solves the q problems $\psi_i(\boldsymbol{R})$ or, equivalently, F is a solution to the problem

$$\bigwedge_{i=1}^{q} \psi_i(\boldsymbol{R})$$

and then

$$\forall \boldsymbol{R} \ SOL(\psi(\boldsymbol{R})) \subseteq SOL(\bigwedge_{i=1}^{q} \psi_i(\boldsymbol{R})).$$

In order to prove that (5) is a tautology, we must show that also the inverse containment holds. To this end, consider a solution $G : D \to C$ to the problem

$\bigwedge_{i=1}^{q} \psi_i(\boldsymbol{R})$ *for a generic instance* $\boldsymbol{R}$. *By definition,* G *solves all of the* $\psi_i(\boldsymbol{R})$ *problems and, recalling the form (6) of* ψ_i, *it is straightforward that* $G|_{\sigma_2(D,i)}$ *solves* $\psi_i(\boldsymbol{R})$. *In other words,* G *is such that*

$$\forall i = 1, 2, \ldots, q \ \ G|_{\sigma_2(D,i)} \in SOL(\psi_i(\boldsymbol{R})).$$

But, due to separability, it yields $G \in SOL(\psi(\boldsymbol{R}))$ *and hence*

$$\forall \boldsymbol{R} \ \ SOL(\psi(\boldsymbol{R})) \supseteq SOL(\bigwedge_{i=1}^{q} \psi_i(\boldsymbol{R})).$$

(If part.) Assuming (5) is a tautology, $F : D \to C$ *is a solution to* $\psi(\boldsymbol{R})$, *for any* $\boldsymbol{R}$, *if and only if* F *solves the problem* $\bigwedge_{i=1}^{q} \psi_i(\boldsymbol{R})$ *or, equivalently, the* q *problems* $\psi_i(\boldsymbol{R})$ *(i = 1,...,q). Consequently, any solution* F *to* $\psi(\boldsymbol{R})$ *is such that* $\forall i = 1, \ldots, q \ \ F|_{D_i} \in SOL(\psi_i(\boldsymbol{R}))$ *and viceversa. Hence, for any* $\boldsymbol{R}$,

$$SOL(\psi(\boldsymbol{R})) = \{F : D \to C \ s.t. \ \forall i = 1, \ldots, q \ F|_{D_i} \in SOL(\psi_i(\boldsymbol{R}))\}. \qquad \square$$

Theorem 1 calls for an equivalence check among logical formulae. This task is undecidable even for first-order formulae [?], and actually this lower bound applies also to this case, as shown by the next theorem.

Theorem 2. *Given a problem* $\psi(\boldsymbol{R})$, *an integer* q, *two functions* σ_1, σ_2, *and* q *problems* $\psi_1, \ldots, \psi_q$ *as in Definition 3, it is not decidable to check whether* ψ *is* (σ_1, σ_2)*-separated or not.*

Proof. (sketch) Consider a problem $\psi(\boldsymbol{R})$ *of the form:*

$$\exists F : D \to C \ s.t. \ (\forall m \in M \ \eta(\boldsymbol{R}, F, m)) \wedge (\xi(\boldsymbol{R}) \to \exists m \in M \ \pi(\boldsymbol{R}, F, m)), \quad (7)$$

where M *is a finite domain such that* $q = |M|$ *and* $\xi(\boldsymbol{R})$ *is a first-order formula which represents a "filter" on input data. Assume that the problem*

$$\exists F : D \to C \ s.t. \ \forall m \in M \ \eta(\boldsymbol{R}, F, m), \qquad (8)$$

is (σ_1, σ_2)*-separated, and that the problem*

$$\exists F : D \to C \ s.t. \ \exists m \in M \ \pi(\boldsymbol{R}, F, m),$$

is not (σ_1, σ_2)*-separated. As an example for the former problem just take the third constraint from Example 1. As an example of the latter problem, just take the fourth constraint from Example 5.*

   Of course the problem

$$\exists F : D \to C \ s.t. \ \forall m \in M \ \eta(\boldsymbol{R}, F, m) \quad \wedge \quad \exists m \in M \ \pi(\boldsymbol{R}, F, m)$$

is not (σ_1, σ_2)*-separated, essentially being a conjunction of constraints, the former being* (σ_1, σ_2)*-separated and the latter being not* (σ_1, σ_2)*-separated.*

   Now note the role played by formula $\xi(\boldsymbol{R})$ *in problem (7). If* $\xi(\boldsymbol{R})$ *is identically false, then problem (7) coincides with problem (8), and it is* (σ_1, σ_2)*-separated. If* $\xi(\boldsymbol{R})$ *is not identically false, then we can find an instance showing that problem (7) is not* (σ_1, σ_2)*-separated. Summing up, problem (7) is* (σ_1, σ_2)*-separated iff first-order formula* $\xi(\boldsymbol{R})$ *is identically false, which is not decidable [?].* $\qquad \square$

The undecidability of the problem of checking separation puts severe restrictions on the possibility of mechanizing the process of finding, or at least validating, Benders decompositions. Nevertheless, it has been shown in [?] that current Automated Theorem Provers technology can be effectively used for checking properties, such as existence of symmetries or dependence among arrays of variables, similar to separation. It is the purpose of future research to investigate on the applicability of the methodology of [?] to the separation problem.

5   Conclusions

In this paper we have analyzed the notion of separation of problems. This is a concept interesting *per se*, and finds an immediate application in the context of Benders decompositions. In fact, it is well known that such decompositions are effective only if the subproblem is formulated using several subproblems exhibithing strong intra-relationships and weak inter-relationships.

In the literature, informal notions of separation of subproblems are typically used, but in this paper we have shown that it is not easy at all to come up with a clear syntactical definition of separability. Examples 1-4 show that formulations of a problem which look similar from the syntactical point of view may or may not be separable. A precise, semantical definition of separation has been provided, which has been characterized both from the logical and from the computational points of view.

In particular, we have shown that separation can be reduced to checking equivalence of second-order logic formulae, and that the problem of checking whether a given selection of input data corresponds to a separation or not is not decidable.

We are currently working on finding other computational results, e.g., special cases of Theorem 1 which call for first-order, instead of second-order, equivalence. Moreover, since the notion of separation into subproblems seems to be related to the concept of *database integration*, especially in the context of different information sources, cf. e.g., [9], we plan to extend our definitions in the traditional database context.

Acknowledgements

The authors would like to thank the anonymous reviewers for their useful comments, which have been very helpful for improving readability of the paper.

References

1. J.F. Benders. Partitioning procedures for solving mixed-variables programming problems. *Numerische Mathematik*, 4:238–252, 1962.
2. H. Cambazard and N. Jussien. Integrating Benders decomposition within constraint programming. Number 3709, pages 752–756. Springer-Verlag, 2005.
3. Pierluigi Crescenzi, Deborah Goldman, Christos H. Papadimitriou, Antonio Piccolboni, and Mihalis Yannakakis. On the complexity of protein folding. *J. of Comp. Biology*, 5(3):423–466, 1998.

4. Rina Dechter. Constraint Networks. In Stuart C. Shapiro, editor, *Encyclopedia of Artificial Intelligence*, volume 1. Addison-Wesley Publishing Company, 1992.

5. J. Hooker. *Logic-based methods for optimization: combining optimization and constraint satisfaction.*, chapter 19, pages 389–422. Wiley and Sons, 2000.

6. J.N. Hooker and G. Ottosson. Logic-based Benders decomposition. *Mathematical Programming*, 96:33–60, 2003.

7. V. Jain and I.E. Grossmann. Algorithms for hybrid MILP/CP models for a class of optimization problems. *INFORMS Journal on Computing*, 13:258–276, 2001.

8. Kit Fun Lau and Ken A. Dill. A lattice statistical mechanics model of the conformational and sequence spaces of proteins. *Macromolecules*, 22:3986–3997, 1989.

9. P. S. Medcraft, U. Schiel, and C.S. Baptista. Database integration using mobile agents. Number 2872, pages 160–167. Springer-Verlag, 2003.

10. J.F. Puget. A fast algorithm for the bound consistency of alldiff constraints. In *AAAI/IAAI*, pages 359–366, 1998.

A Hybrid Column Generation and Constraint Programming Optimizer for the Tail Assignment Problem

Sami Gabteni[1] and Mattias Grönkvist[1,2]

[1] Carmen Systems AB,
Odinsgatan 9,
S-411 03 Göteborg, Sweden
{sami.gabteni, mattias.gronkvist}@carmensystems.com
[2] Department of Computer Science and Engineering,
Chalmers University of Technology,
Eklandagatan 86, S-412 96 Göteborg, Sweden

Abstract. Tail Assignment is the problem of assigning flight legs to aircraft while satisfying all operational constraints, and optimizing some objective function. In this article, we present a hybrid column generation and constraint programming solution approach. This approach can be used to quickly produce solutions for operations management, and also to produce close-to-optimal solutions for long and mid term planning scenarios. We present computational results which illustrate the practical usefulness of the approach.

1 Introduction

The airline planning process is usually divided into timetable creation, fleet planning, and crew planning. These steps are often handled sequentially, without feedback. Most often, the fleet planning process itself consists of two stages – *Fleet Assignment* and *Aircraft Routing*. Fleet Assignment is the process of assigning aircraft types to the flights in the schedule [16], while maximizing the revenue. Aircraft Routing handles the construction of anonymous maintenance feasible flight sequences for each fleet or subfleet.

Once the long-term planning is done, the schedules are handed over to operations management, which handles operational issues such as sick crew, replanning due to changes in forecast, delays, airport closures, etc. Operations management typically lasts from a few days from flight departure up to the actual departure time. In the long-term planning stage careful optimization is possible, but in the operational stage, disruptions must be solved as quickly as possible.

Usually, the Tail Assignment problem is solved a few days before operations, as an extension to Aircraft Routing, to adjust the planned routes to various operational constraints and assign them to identified aircraft. However, we propose

J.C. Beck and B.M. Smith (Eds.): CPAIOR 2006, LNCS 3990, pp. 89–103, 2006.

that Tail Assignment should be solved earlier in the planning process. Considering individual aircraft early in the planning process makes it easier to implement planned optimized solutions without degrading the quality.

In [13] and [14] we have shown how constraint programming can be used as a standalone solution method, and as a preprocessing method, for Tail Assignment. This article presents the complete solution approach, where constraint programming is combined with column generation and local search. The result is an approach that can meet both the running time requirements of operations management and the quality requirements of long and mid-term planning.

Section 2 describes the Tail Assignment problem in detail and gives a brief survey of the relevant literature. Section 3 presents a column generation solution approach, and Section 4 briefly summarizes the constraint model presented in [14]. Section 5 discusses the complete hybrid approach, combining column generation, constraint programming and local search, and Section 6 shows computational results. In Section 7 we conclude.

2   The Tail Assignment Problem

The Tail Assignment problem is the problem of assigning flight legs to aircraft, identified by their *tail numbers*. As a result, each aircraft is assigned a *route* consisting of a sequence of legs which obeys all operational constraints, making the solution to the problem a fully operational aircraft deployment.

The operational constraints can be categorized in two groups: general constraints applying to subsets of aircraft, such as fleets and sub-fleets, and tail constraints, applying specifically to particular tail numbers. The latter can relate to any technical aspect, such as limited fuel capacity, noise level, or in-flight entertainment system functionalities. Pre-assigned activities, such as heavy maintenance, is a particular type of tail constraint. These aspects make each aircraft unique in the way it can be operated. The tail constraints are the fundamental difference between Tail Assignment and Aircraft Routing.

The simplest form of constraint is the *leg based* constraint, which only depends on a single flight leg. These constraints are often called *flight restrictions*. Other constraints depend on two connecting flights, and are hence *connection based*. The minimum buffer time between the arrival and departure of an aircraft is a good example of a connection based constraint, which might or might not depend on the specific place or time of the connection. More complex constraints involve longer sequences of flight legs. A typical example is the maintenance checks which need to be performed with regular intervals. The intervals are usually defined in terms of aircraft activity, such as airborne hours or landings.

The Tail Assignment process is, at most airlines, only seen as a feasibility problem, as the goal is to find a feasible assignment of aircraft to flight legs. However, it can in some cases be driven by operational quality objectives. For example, maximizing the number of long connections increases aircraft availability to handle incidents on the day of operations. Long connections at appropriate airports can also offer alternative slots for minor maintenance activities. Later in this article, we will present planning scenarios which optimize financial aspects of the Tail Assignment problem.

The most intuitive data representation for the Tail Assignment data is a *flight network*, where each node represents a flight leg, or some other activity such as a planned maintenance activity for a specific aircraft, and each arc represents a possible connection between two activities.[1] Modeling individual operational constraints requires a dated model, which accurately considers the start locations and maintenance histories of each aircraft.

2.1 Literature Review

The assumption that all aircraft within a fleet are equivalent has put most research focus on Aircraft Routing rather than Tail Assignment. So, rather than a specific Tail Assignment survey, we here provide an overview on how the related problems have been addressed. For a more complete survey, we refer to [15].

Maintenance routing the problem of creating maintenance feasible aircraft routes. It is a feasibility problem rather than an optimization problem. Gopalan and Talluri [12] describe a system for maintenance routing implemented at US-Air. The maintenance requirements are simplified to a restriction on each aircraft to return to a maintenance base every three days, and it is assumed that the lines of flying during daytime (LOFs) are fixed. This problem can be solved in polynomial time, but Talluri [23] has shown that the three-day problem is a special case, and that the general N-day problem is NP-hard.

Through Assignment is a financially driven problem: A through flight is a two leg flight between two locations, via the hub, that does not require an aircraft change for passengers going from the departure airport of the first leg to the arrival airport of the second leg. Through assignment is the problem of deciding which leg-to-leg connections are to be through flights. In a number of references, for example [1, 18], Through Assignment modeling is combined with simplified operational constraints, such as maintenance requirements.

Most Aircraft Routing approaches combine the maintenance routing aspect with a cost function, which is often through value based, but can also capture other aspects. Kabbani and Patty [19] model the aircraft routing problem for American Airlines as a set partitioning problem, where each column represents a week-long aircraft route. This makes it possible to handle general maintenance constraints, but the drawback is long running times. In [6], Clarke et al. solve an aircraft rotation problem for Delta Air Lines, building maintenance feasible routes while maximizing through values. They require all aircraft to fly the same cyclic route (rotation). They formulate the problem as a TSP with side constraints, and solve it with Lagrangian relaxation.

In [3], Barnhart et al. solve a combined fleet assignment/aircraft routing problem by an approach based on maintenance feasible *strings* of activities, that are combined to create feasible routes, within a branch-and-price framework. Short-haul instances with up to 190 flights are solved successfully. Elf et al. [10] propose

---

[1] From now on, the term *activity* will be used to denote either a flight leg, a sequence of hard-locked flight legs, or a planned maintenance activity.

an aircraft rotation planning model for minimizing delay risk. In their model, a 'delay risk' is either individual connections being too short, or consecutive visits to certain airports. Maintenance is not considered in their model, and a solution method based on Lagrangian relaxation is proposed. The existing literature does not consider individual aircraft requirements, and is typically based on cyclic rather than dated models.

Many references on integrating constraint programming and column generation take the approach of solving the master problem with standard LP techniques, and use constraint programming to solve the often complex pricing problem. For example, Fahle et al. [11] on the Airline Crew Rostering problem, and Rousseau et al. [22] on the Vehicle Routing Problem (VRP). The latter shows promising results on some of the well-known Solomon instances. In [5], Caprara et al. describe a combined CP/OR Crew Rostering application at the Italian State Railway. Their main solution method is constraint programming, and they use a Lagrangian relaxation to obtain lower bounds.

3   A Column Generation Solution Approach

Let us start by describing a mathematical model for the Tail Assignment problem. Let F represent the set of all flights, R the set of all possible routes satisfying individual route constraints, and let x_r be binary decision variables that are 1 if route $r \in R$ is used and 0 otherwise. c_r is the cost of using route r, and a_{fr} is 1 if activity $f \in F$ is covered by route r and 0 otherwise. Now, we can formulate a path-based Tail Assignment model (PATH-TAS):

$$\min \quad \sum_{r \in R} c_r x_r \tag{1}$$

$$\sum_{r \in R} a_{fr} x_r = 1 \quad \forall f \in F \tag{2}$$

$$x_r \in \{0, 1\} \tag{3}$$

Here, constraints (2) make sure that all activities are covered exactly once. To handle unassigned activities, the set R contains columns covering single activities, at a high penalty cost. This can also be seen as adding slack variables and including them in constraint (2). While PATH-TAS can be formulated very compactly, it contains an exponential number of variables hidden in the definition of R. However, PATH-TAS has the nice property that it separates the generation of routes from the selection. This makes it easier to capture various complicated constraints, such as maintenance constraints, as we will demonstrate.

Column generation [7] is the process of solving a linear program with only a subset of the variables available explicitly. It uses a dual solution to generate new primal variables that can improve the current solution, until no such variables can be found. This is very useful to address problems involving too many variables, like PATH-TAS, and in fact most real-life large-scale transportation scheduling problems. Column generation can be seen as a generalization of the revised simplex method. The problem of finding negative reduced cost variables, which can improve the current solution, is called the *pricing problem*. The linear relaxation of the variable-restricted master linear program is called the *restricted master problem* (RMP).

For model PATH-TAS, the RMP is a relaxed set partitioning problem, for which several solution methods can be used. We have chosen to use the primal simplex or barrier LP solvers of Ilog CPLEX [17] or Dash Xpress [8].

3.1   The Pricing Problem

Assuming that the cost of route r is a sum of costs on activities (or connections), the reduced cost of route r is $\bar{c}_r = \sum_{f \in F_r}(c_f - \pi_f)$, where F_r is the set of activities covered by route r, π_f is the dual value of the row corresponding to activity f, and c_f is the cost of activity f. To handle maintenance constraints, we must introduce resource constraints in the pricing problem, which can be reset along a route to model the maintenance checks. The introduction of resource constraints turns the pricing problem from a normal shortest path problem into an NP-hard *resource constrained shortest path problem* (RCSPP).

The resource constrained shortest path problem is solved using a standard label-setting algorithm [9]. The possibility to reset the resource consumptions does not require any substantial change to the algorithm. The activity start times give a total ordering of the nodes in the network, and by pulling labels order of increasing start time, only a single label pulling sweep is required. *Dominance* is used to restrict saved labels to the efficient routes. However, an exact approach based on a pure dominance scheme does not prevent an explosion of the number of labels created. Instead, a heuristic implementation [15] of the algorithm is used, which imposes a limit on the number of labels created at each node. If λ is the maximum number of labels for any node, the complexity of the resulting heuristic algorithm is $\lambda|F|$.

The resource consumption updates and resets are controlled via Carmen System's *Rave* rule system [2]. Rave is a stand-alone module, making it possible to quickly add any resource constraints by defining the update and reset procedures in Rave.

3.2   Accelerating the Column Generator

While implementing the standard column generation algorithm is rather straight-forward, much effort must be put into making it perform well in practice. We refer to [20] for a fairly recent survey on common techniques to accelerate column generation. Here, we will briefly describe two of the acceleration techniques we have used in our implementation.

- **The maximum number of labels**
  For each node, two limits for the number of labels are used — one soft limit and one hard limit. As long as the number of labels is less than the soft limit,

we insert new labels if they are *not dominated*, and remove dominated labels after insertion. If the number of labels is above the soft limit, but below the hard limit, we insert new labels if they are *lexicographically smaller* than the lexicographically largest label already present. Our labeling algorithm is thus heuristic in that it stores only a limited number of labels at each node, and by using a lexicographical ordering of the labels. Observe that in order to make our pricing routine exact, in the sense that at least one negative reduced cost route is found if one exists, we might need to dynamically increase the node label limits in case no route is found, and the label limit is reached for some node.

– **Dual re-evaluation**
The pricing problem is solved by decomposition into aircraft specific pricing problems, so that columns are generated for all aircraft using the same dual values. Using the same dual values for all aircraft has the effect that activities with attractive dual values will be covered by routes for several aircraft within the same pricing iteration. To avoid similar columns from being generated for all aircraft, we use a dual re-evaluation scheme which modifies the dual values when columns are selected for insertion into the RMP: Between each aircraft-specific pricing problem we re-evaluate the dual values according to $\pi_f^{new} = \pi_f^{old} + (\bar{c}_r)/(\xi \times |F_r|)$, where $|F_r|$ is the set of activities covered by route r, and ξ is a smoothing parameter.

Observe that when $\xi = 1.0$, the reduced cost of the column will be evenly spread over the activities it covers. The new reduced cost thus becomes 0, and simulates the introduction of the column in the RMP basis. By adjusting ξ, it is possible to tune the behavior of the re-evaluation. Computational tests [14] have shown that a using $\xi = 10.0$ in general has a dramatically positive effect on column generation convergence.

4 A Constraint Programming Solution Approach

In [14], we present a constraint programming model for the Tail Assignment problem. We will here just summarize the model, to make it easier to understand the following sections. The constraint model focuses only on feasibility, i.e. finding a feasible solution with all activities assigned, regardless of its quality.

Firstly, `successor` variables are introduced for all activities containing the possible successors of the activity. Since the rules related to activity-to-activity connections are already modeled in the `successor` variables, no more constraints concerning these rules need to be explicitly added to the model. Secondly, `vehicle` variables are introduced for all activities, the domains of which initially contain all aircraft which are allowed to operate the activity. These variables are used to model preassigned activities as well as flight restriction rules. Since all activities must have unique successors to form disjoint routes through the flight network, all `successor` variables must take unique values, which means that an `all_different` [21] constraint over all `successors` is added. Since the `successor` and `vehicle` are mutually redundant, it is crucial to keep the domains of these variables consistent. This is done with a specialized *tunneling constraint*.

The constraints described so far are sufficient to model the Tail Assignment feasibility problem, assuming the maintenance constraints are relaxed. However, in order to improve the propagation, and thus the computational performance, redundant constraints and variables are added: `predecessor` variables are connected to the `successor` variables via an `inverse` constraint, `all_different` constraints are added for the `vehicle` variables of overlapping activities, and a special propagation algorithm for flight restrictions is added.

Finally, we re-use the column generation pricing solver to handle the maintenance constraints. As the pricing solver is based on labeling, it is easy to check whether each activity is reachable, i.e. there exists at least one partial route arriving at each activity, and there exists at least one legal route per aircraft. In the case of a partially instantiated flight network, e.g. during the search when some follow-ons are fixed, this feasibility check cannot guarantee to find all infeasibilities. However, for a fully fixed network, it will always find all infeasibilities, which is enough to make it usable for our purposes. We will refer to the full constraint model as CSP-TAS, and to a relaxed model lacking the maintenance constraint handling and the extra flight restriction propagation as CSP-TAS^{relax}.

5   A Hybrid Solution Approach

The column generation solution approach described in Section 3 is well suited for finding optimal solutions to the Tail Assignment problem. Unfortunately, column generation has a tendency to converge slowly, which means that using a pure column generation algorithm it often takes a long time to obtain even a feasible solution, let alone an optimal solution. The constraint programming approach described in Section 4, on the other hand, is especially designed to quickly finding feasible solutions, and ignores the optimality aspect.

In this section, we will present a solution approach that combines the column generation and constraint programming approaches, giving a method that can quickly produce initial solutions, as well as close-to-optimal solutions. We will start this section by showing how the constraint programming model can improve column generation performance by means of powerful preprocessing and as a feasibility analyzer in a heuristic branching algorithm. We will then describe the full hybrid approach.

5.1   Preprocessing Using the Constraint Model

In [13] we describe how a constraint programming model can be used for preprocessing the flight network input. The basic idea is simply to take the constraint model described above, and *only* apply constraint propagation, without any search. This will remove activity-to-activity connections which can never be used in a solution from the flight network. We show that this kind of preprocessing is more powerful than traditional preprocessing methods, and has a huge impact on the size of the flight network, and consequently on the overall column generation performance.

5.2   The Heuristic Branching Algorithm

Section 3 only discusses the first phase of the column generation process, finding a good LP relaxation. Once a sufficiently good relaxed solution has been found by the column generation algorithm, the next stage of the solution process consists of an integer programming approach to find a sufficiently good integer solution.

data

PROPAGATE: A function that propagates the constraint model. Returns a set of removed connections, and new restriction sets.

COLUMNGENERATION(i): Solve problem using i column generation iterations, return a relaxed solution, and if possible, lower and upper bounds

σ: Number of column generation iterations to perform between fixing iterations

ϵ: Relative optimality tolerance

z_u: Known upper bound on integer objective, or ∞

δ: Number of restrictions to undo when backtracking

Recall that F_r is the set of activities covered by route r.

```
1:  procedure LOOK-AHEADINTEGERHEURISTIC
2:      z ← network flow lower bound
3:      (x̄, z, z̄) ← COLUMNGENERATION(100)              ▷ 100 iter., as an example
4:      loop
5:          Restrict the solution space by removing connections C
6:          for all connections (f, f′) ∈ C do
7:              POST successorf ≠ successorf′ in CSP-TASrelax
8:          end for
9:          (C′, R′) ← PROPAGATE(CSP-TASrelax)               ▷ R′ = ⋃t∈T R′t
10:         conflict ← false
11:         if some domain in CSP-TASrelax empty then
12:             Restore CSP-TASrelax
13:             conflict ← true
14:         end if
15:         if ¬conflict then
16:             Remove from RMP all columns k s.t.
                    |(Fk ∩ f) ∪ (Fk ∩ f′)| = 2 for some (f, f′) ∈ (C ∪ C′)
17:             Remove from pricing network all connections in C ∪ C′
18:             Rt ← R′t  ∀t ∈ T        ▷ Rt: activities which aircraft t cannot operate
19:             z′ ← network flow lower bound
20:             (x̄, z′, z̄) ← COLUMNGENERATION(σ)
21:         end if
22:         if ¬conflict and (x̄) integral and z̄ < zu
                  and 100 × (z̄−z)/z < ε then
23:             return (x̄, z, z̄)
24:         else if conflict or x̄ integral or z̄′ ≥ zu then
25:             Undo the last δ solution space restrictions
26:         end if
27:     end loop
28: end procedure
```

Fig. 1. The general integer fixing heuristic with look-ahead

For the integer phase, we have chosen to implement an integer heuristic inspired by branch-and-price [4]. The heuristic consists of iteratively restricting the solution space with variable or connection fixes, and re-generating columns. This process is repeated until an integer solution is found. When a conflict is detected, or the objective gets too heavily deteriorated by the fixing process, some of the fixes are revised, and other parts of the search tree are explored. To avoid the situation where the fixing process fixes something which can never lead to a solution, the constraint model is used to check feasibility of fixing decisions, and to propagate the effects of the fixing decisions once done.

Figure 1 shows the heuristic fixing process with constraint programming look-ahead. The decision in step 5 about which connections to fix is taken by looking at the current LP solution, and either fixing variables taking a value close to 1, or by fixing connections which seem good from an integer point of view [15]. Observe that we use constraint model CSP-TAS^{relax} rather than the full model CSP-TAS for the look-ahead. The reason is simply that CSP-TAS takes longer to propagate, and adds very little in terms of initial propagation. Once we have made sure that the constraint model is not inconsistent, we remove the fixed connections from the flight network and the conflicting columns from the RMP, update our network flow lower bound, and re-generate columns. If the solution quality is acceptable and the current solution is integral, we stop. If the solution quality has deteriorated we backtrack, and otherwise we continue fixing. Since there always exists at least one connection to fix as long as the relaxed solution $\bar{x}$ is non-integral [15], we know that the algorithm will find an integer solution, even if it might be one with activities unassigned.

However, even using this fixing algorithm, it might sometimes take too long to find an initial solution. To make it possible to find solutions even faster, more aggressive fixing must be used, which in turn means that even more look-ahead must be used to avoid conflicts. In this case, instead of just propagating CSP-TAS^{relax} before each fixing decision, we actually solve it to check that a solution exists before proceeding with the fixing. If we also check the maintenance constraint and the flight restriction propagation algorithm from CSP-TAS at the root node, the resulting look-ahead is powerful enough to allow us to fix very aggressively, only re-generate very few columns, and still find a solution with all activities assigned.

5.3   A Local Search Improvement Approach

The typical solution process using our hybrid solution approach is to first find an initial solution, regardless of quality. This proves that a solution exists, and gives a hint about how the final solution might look. Initial solutions are found either using CSP-TAS directly, or using the aggressive fixing algorithm described in the previous section. In this step, a global network flow lower bound is also calculated.

Once an initial solution has been found, a local search improvement process is started. This is done by selecting a subset of aircraft, and a time interval within the planning period, for which the tail assignments are re-optimized. All

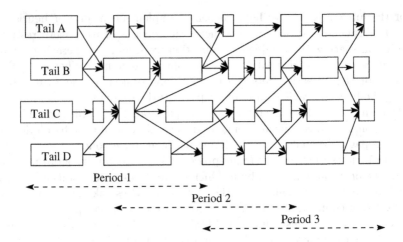

Fig. 2. The sliding time window improvement strategy

activities outside the interval, or which are assigned to unselected aircraft, are excluded from the re-optimization. The restricted problem is optimized using the hybrid approach presented in Section 5.2, but with less aggressive settings so as to obtain close-to-optimal solutions for the restricted problems. This is repeated over a sequence of problem restrictions, until a sufficiently good global solution has been found. Using the improvement process on reasonably sized subproblem, the method scales well to long planning periods or instances with a large number of aircraft.

The most common strategy for selecting the problem restriction, is to use a *sliding time window*. After solving a problem restriction based on a certain time interval, a slightly later interval is used to define the following problem restriction, as figure 2 illustrates. This is repeated several times over the planning period, with variable time interval lengths. Aircraft subsets are only used for instances with a very large number of aircraft. Improving over the entire planning period is possible, but typically leads to very long running times. Also, since most of the rules and costs in Tail Assignment only depend on single connections, and not entire routes, the quality loss when using small time intervals is limited.

5.4 Controlling Solution Quality

Obviously, the solution process which we propose is heuristic, in the sense that it does not guarantee that an optimal solution is found. To improve running times, a number of heuristic algorithmic steps have been added to various parts of our process:

- the lexicographical ordering and the limit on the number of labels at each node make the pricing algorithm heuristic;
- stopping the column generation algorithm prior to LP optimality means we do not always obtain an LP lower bound;

- the heuristic branching algorithm is inherently heuristic;
- the time window improvement method might get stuck in a local optimum.

A few of these shortcomings are possible to avoid by using alternative techniques, which are typically more time consuming. However, for our purposes the approximations are acceptable. To control the solution quality, we continually monitor and update the network flow lower bound obtained by ignoring all maintenance and tail-dependent constraints.

6 Computational Results

While the Tail Assignment approach we have presented can be used to model various types of planning scenarios, ranging from fleet planning to operations management, we will here illustate its usefulness by two planning scenarios, one fairly standard scenario, and one more experimental scenario. First, we will show how operational robustness can be increased by our Tail Assignment approach, and then we will show how commercial aircraft leasing costs can be decreased.

6.1   Minimizing the Number of Used Medium-Length Connections

Operational robustness is a measure of how robust a planned Tail Assignment solution is with respect to events occurring close to the day of operations. Robustness is difficult to measure, short of actually observing the real-world behavior. Here, we will identify criteria that are considered to improve or deteriorate robustness, from experience, and optimize using these criteria.

Short connections, slightly above the minimum connection time, are desirable since they give high utilization of the aircraft. Connections longer than some limit, e.g. three hours, are also desirable, because they allow the aircraft to be used as a standby aircraft, in case of operational disruptions. Medium-length connections, on the other hand, are considered bad for robustness. Medium-length connections sometimes force aircraft to occupy gates during a long periods, and make it impossible to use aircraft for standby duty.

To prevent medium-length connections, we use an objective function penalizing $2 - 6$ hour connections, with extra penalty for $3 - 5$ hour connections. Connections longer than 6 hours and shorter than 2 hours are penalized by a small factor which decreases with increasing connection time. Table 1 shows the running times and the number of 2-6 and 3-5 hour connections, when using the CSP-TAS model, the initial method based on aggressive fixing, and when using the time window improvement method starting from the CSP-TAS solution. The aggressive initial method is included mainly as a comparison in terms of running times and solution quality for the other methods. The improvement strategy is the following: Start with a period length of 24 hours and a step length of 12 hours, and perform a sweep over the entire planning period. Repeat at most 20 such sweeps, increasing the period by six hours in every sweep. The step is always half the period length. All running times are reported as *minutes:seconds*.

Table 1. Running times and number of 2-6 and 3-5 hour connections used in the initial and improved solutions.

Instance	CSP-TAS			Aggr. fixing			Improved CSP-TAS		
	2-6 h	3-5 h	Time	2-6 h	3-5 h	Time	2-6 h	3-5 h	Time
1A	151	30	00:05	144	30	00:11	111	16	02:50
1B	60	17	00:12	50	11	00:24	0	0	01:55
1C	319	107	01:01	298	98	02:47	169	61	23:13
1D	347	100	01:51	309	80	05:49	141	38	28:25
1E	443	103	01:29	412	98	10:03	214	56	31:15

Table 2. The test instances. The columns show the number of activities, aircraft and maintenance constraints.

Instance	#Activities	#Aircraft	#Constraints
1A	1338	9	2
1B	2378	17	1
1C	4932	33	1
1D	5816	31	1
1E	5571	31	1

Using the CSP-TAS initial method followed by an improvement run, we find initial solutions within 2 minutes. The aggressive initial method only gives slightly better solution quality than CSP-TAS, and takes longer time. Within around 30 minutes we obtain solutions with substantially fewer medium-length connections. For one of the instances we can even get rid of all medium-length connections. Using the flow lower bound we can show that the final results are optimal for all instances.

Information about the test instances is shown in Table 2. Instance 1A includes an 'A check' maintenance constraints, forcing all aircraft to be maintained every 450 flying hours. For all instances, there is also one generic resource constraint, forcing the aircraft to return to their home base every six days. There are also a number of flight restrictions present.

6.2   Commercial Planning: Minimizing the Utilization of Leased Aircraft

Airlines often lease parts of their fleet, to make their capacity more flexible. They will then avoid the high costs associated with owning the aircraft, but on the other hand have to pay substantial fees when actually using the aircraft. The fee of using a leased aircraft of course varies depending on the type of aircraft, and the leasing agreement, but € 820 (about $1000) per flying hour is not an unrealistic estimate.

Table 3. Running times and total flying time for the leased aircraft in the initial and improved solutions

| | | Initial method | | | | Improvement method | |
| | | CSP-TAS | | Aggr. fixing | | Improved CSP-TAS | |
Instance	#Leased Aircraft	Lease time	Time	Lease time	Time	Lease time	Time
3A	1	458:00	00:03	488:10	00:25	275:45	04:54
3B	2	534:50	00:12	541:55	00:21	348:50	12:40
3C	4	1676:12	00:57	1689:57	03:06	1234:30	93:46
3D	4	1556:13	01:43	1625:52	03:53	1029:55	228:39
3E	4	1777:06	01:29	1709:33	04:21	1144:35	255:02

Since using leased aircraft is so expensive, in a fleet consisting of mixed leased and owned aircraft it is desirable to utilize the leased aircraft as little as possible, but still cover all activities. To show how our tail assignment optimizer can minimize the utilization of leased aircraft, we have re-used the instances from Section 6.1, and simulated that a few randomly selected aircraft in each instance are leased. The cost function is the same as that in Section 6.1, except that the leased aircraft are given a penalty of 50 cost units per flying minute. Observe that adding costs which depend on the individual aircraft is not a problem in model PATH-TAS. The maintenance and flight restriction constraints are also the same as in Section 6.1. Observe that the cost function does not model the true lease cost, but is a mix between two optimization criteria.

Table 3 shows the total flying time for the leased aircraft, and the optimization running time for the five instances. In the improved solutions the leased aircraft are used substantially less than in the initial solutions. On average for the five instances, the flying time per leased aircraft in the improved solutions is decreased by 131 hours and 15 minutes, compared to the initial solutions. With an estimated hourly lease cost of € 820, this means monthly savings of about € 820 ×131.25 = € 107625 per leased aircraft. Observe that since we only compare initial solutions to improved solutions, and the leased aircraft are randomly selected, our results should be interpreted as indicative of possible savings rather than actual savings. However, what we demonstrate is that our improvement optimizer is able to produce much better solutions than solution approaches which do not care about these costs. And such approaches are not uncommon, many airlines for example do not plan their tail assignments until very late in the planning process, at which time it is difficult to consider lease costs.

7   Conclusions

We have presented the Tail Assignment problem as a way to obtain operationally feasible aircraft assignments. Unlike existing approaches for this problem, the Tail Assignment problem considers all individual operational constraints, including maintenance, flight restrictions and preassigned maintenance activities. The Tail Assignment problem is solved for a fixed time period, to make it possible to consider specific activities and irregular schedules. Also, the Tail Assignment problem uses a fairly general objective function, which makes it possible to model many kinds of planning scenarios.

We have presented mathematical and constraint models for the Tail Assignment problem, and shown how the constraint model can be used to accelerate and strengthen a column generation solution approach for the mathematical model. The hybrid column generation and constraint programming solution approach uses constraint programming to quickly produce initial solutions, as well as check feasibility during a column generation-based heuristic branching algorithm. Further, the entire process is wrapped into local search, resulting in a solution algorithm which scales well to large instances. The solution algorithm is able to quickly produce initial solutions respecting all operational rules, as well as close to optimal solutions.

We have demonstrated the usefulness of the method on a set of real-world instances. We showed that the optimizer can help reduce the number of medium-length connections, which increases aircraft usage, and to reduce aircraft leasing costs. Initial solutions where obtained within a few minutes, while solutions of high quality where obtained given longer running times.

References

1. R. K. Ahuja, J. Liu, J. Goodstein, A. Mukherjee, J. B. Orlin, and D. Sharma. Solving Multi-Criteria Combined Through Fleet Assignment Models. In T. A. Ciriani, G. Fasano, S. Gliozzi, and R. Tadei, editors, *Operations Research in Space and Air*, pages 233–256. Kluwer Academic Publishers, 2003.
2. E. Andersson, A. Forsman, S. E. Karisch, N. Kohl, and A. Sørensson. Problem Solving in Airline Operations. Carmen Research and Technology Report CRTR-0404, Carmen Systems AB, Gothenburg, Sweden, June 2004.
3. C. Barnhart, N. L. Boland, L. W. Clarke, E. L. Johnson, G. L. Nemhauser, and R. G. Shenoi. Flight String Models for Aircraft Fleeting and Routing. *Transportation Science*, 32(3):208–220, August 1998.
4. C. Barnhart, E. L. Johnson, G. L. Nemhauser, M. W. P. Savelsbergh, and P. H. Vance. Branch-and-Price: Column Generation for Solving Huge Integer Programs. *Operations Research*, 46(3):316–329, May-June 1998.
5. A. Caprara, F. Focacci, E. Lamma, P. Mello, M. Milano, P. Toth, and D. Vigo. Integrating Constraint Logic Programming and Operations Research Techniques for the Crew Rostering Problem. *Software − Practice and Experience*, 28(1):49–76, January 1998.
6. L. W. Clarke, E. L. Johnson, G. L. Nemhauser, and Z. Zhu. The Aircraft Rotation Problem. *Annals of Operations Research*, 69:33–46, 1997.
7. G. B. Dantzig and P. Wolfe. Decomposition Principle for Linear Programs. *Operations Research*, 8:101–111, 1960.
8. Dash Optimization Ltd. *Xpress-Optimizer Reference Manual, release 14*, 2002.
9. M. Desrochers and F. Soumis. A generalized permanent labelling algorithm for the shortest path problem with time windows. *INFOR*, 26(3):191–212, 1988.

10. M. Elf, M. Jünger, and V. Kaibel. Rotation Planning for the Continental Service of a European Airline. In W. Jager and H.-J. Krebs, editors, *Mathematics – Key Technologies for the Future. Joint Projects between Universities and Industry*, pages 675–689. Springer Verlag, 2003.

11. T. Fahle, U. Junker, S. E. Karisch, N. Kohl, M. Sellmann, and B. Vaaben. Constraint Programming Based Column Generation for Crew Assignment. *Journal of Heuristics*, 8(1):59–81, 2002.

12. R. Gopalan and K. T. Talluri. The Aircraft Maintenance Routing Problem. *Operations Research*, 46(2):260–271, March–April 1998.

13. M. Grönkvist. Using Constraint Propagation to Accelerate Column Generation in Aircraft Scheduling. In *Proceedings of CPAIOR'03*, May 2003.

14. M. Grönkvist. A Constraint programming Model for Tail Assignment. In *Proceedings of CPAIOR'04, vol. 3011 of Lecture Notes in Computer Science*, pages 142–156. Springer-Verlag, April 2004.

15. M. Grönkvist. *The Tail Assignment Problem*. PhD thesis, Department of Computing Science, Chalmers University of Technology, Gothenburg, Sweden, 2005.

16. C. A. Hane, C. Barnhart, E. L. Johnson, R. E. Marsten, G. L. Nemhauser, and G. Sigismondi. The fleet assignment problem: solving a large-scale integer program. *Mathematical Programming*, 70:211–232, 1995.

17. ILOG Inc. *ILOG CPLEX 7.5 Reference Manual*, 2001.

18. A. I. Jarrah and J. C. Strehler. An optimization model for assigning through flights. *IIE Transactions*, 32(3):237–244, March 2000.

19. N. M. Kabbani and B. W. Patty. Aircraft Routing at American Airlines. In *Proceedings of the Thirty-Second Annual Symposium of AGIFORS*, 1992.

20. M. E. Lübbecke and J. Desrosiers. Selected Topics in Column Generation. Les Cahiers du GERAD G-2002-64, Department of Mathematical Optimization, Braunschweig University of Technology, and GERAD, 2002. Submitted to *Operations Research*.

21. J.-C. Régin. A filtering algorithm for constraints of difference in CSPs. In *Proceedings of AAAI-94*, pages 362–367, 1994.

22. L.-M. Rousseau, M. Gendreau, and G. Pesant. Solving small VRPTWs with Constraint Programming Based Column Generation. In *Proceedings of CPAIOR'02*, March 2002.

23. K. T. Talluri. The Four-Day Aircraft Maintenance Problem. *Transportation Science*, 32:43–53, 1998.

The Power of Semidefinite Programming Relaxations for MAX-SAT

Carla P. Gomes, Willem-Jan van Hoeve, and Lucian Leahu

Dpt. of Computer Science, Cornell University, Ithaca, NY 14853, USA
{gomes, vanhoeve, lleahu}@cs.cornell.edu

Abstract. Recently, Linear Programming (LP)-based relaxations have been shown promising in boosting the performance of exact MAX-SAT solvers. We compare Semidefinite Programming (SDP) based relaxations with LP relaxations for MAX-2-SAT. We will show how SDP relaxations are surprisingly powerful, providing much tighter bounds than LP relaxations, across different constrainedness regions. SDP relaxations can also be computed very efficiently, thus quickly providing tight lower and upper bounds on the optimal solution. We also show the effectiveness of SDP relaxations in providing heuristic guidance for iterative variable setting, significantly more accurate than the guidance based on LP relaxations. SDP allows us to set up to around 80% of the variables without degrading the optimal solution, while setting a single variable based on the LP relaxation generally degrades the global optimal solution in the overconstrained area. Our results therefore show that SDP relaxations may further boost exact MAX-SAT solvers.

1 Introduction

In recent years, we have witnessed a tremendous progress in the state-of-the-art of encodings and algorithms for Boolean Satisfiability (SAT). For example, in areas such as planning and finite model-checking, we are now able to solve large SAT problems with up to a million variables and five million constraints. More generally, SAT encodings have been shown to be very powerful in several practical domains, such as electronic design automation, AI planning, and hardware and software verification. The key algorithmic improvements that have been incorporated into state-of-the-art SAT solvers have been largely based on artificial intelligence (AI) and constraint programming (CP) techniques. For example, for complete solvers, the underlying backtrack search strategy has been enhanced by a series of increasingly sophisticated techniques, such as non-chronological backtracking, fast pruning and propagation methods, nogood (or clause) learning, and more recently randomization and restarts. While we have recently seen an increasing dialogue between the artificial intelligence (AI) and constraint programming (CP) community and the Operations Research community concerning the study and design of algorithms for SAT and variants, it has been surprisingly difficult to integrate OR based relaxations into practical approaches for SAT. For example, despite a significant amount of beautiful Linear Programming (LP) results for SAT (see e.g., [1, 2]), practical state-of-the-art solvers do

J.C. Beck and B.M. Smith (Eds.): CPAIOR 2006, LNCS 3990, pp. 104–118, 2006.

not incorporate LP relaxation techniques. The main reason seems to be the fact that, in the case of SAT, the inference performed by LP is basically equivalent to the inference performed by unit propagation, which is considerably less expensive than LP.[1] Nevertheless, when it comes to MAX-SAT, the optimization counterpart of SAT in which the objective is to assign values to boolean variables maximizing the number of satisfied clauses, there seems to be a more clear role for hybrid approaches that combine AI and OR based techniques. In fact, recently Xing and Zhang [3] made an interesting contribution in the area of hybrid approaches for MAX-SAT, showing how one can use the information provided by linear programming to effectively compute lookahead lower bounds on the number of clauses unsatisfiable. Joy et al ([4]) have also shown how LP-based relaxations can be effective for MAX-2-SAT.

Another area that has received considerable attention in combinatorial optimization is *Semidefinite Programming* (SDP). In a semidefinite programming formulation a linear function of a symmetric matrix is optimized, subject to linear equality constraints and the constraint that the matrix be positive semidefinite. Semidefinite programming is a special case of convex programming and to some extent is similar to linear programming. In particular, the simplex, ellipsoid, and interior point methods developed for LP can be generalized to solve SDP programs. Furthermore, the rich set of LP results in dual theory, a powerful tool for sensitivity analysis and for computing bounds on the objective function have also been generalized to SDP [5, 6]. Moreover, SDP has gained considerable importance in the context of combinatorial optimization since it has been shown that it leads to tighter relaxations than those based on LP for several combinatorial problems. For example, SDP has been shown to provide very good approximations for several combinatorial problems, in particular for the stable set problem [7], for the maximum cut problem and for MAX-SAT [8]. Approximation algorithms are procedures that provide a feasible solution in polynomial time (see e.g. [9]). A key aspect that characterizes approximation algorithms is the fact that they provide some guarantee on the quality of the solution. The quality of an approximation algorithm is the maximum "distance" between its solutions and the optimal solutions, evaluated over all the possible instances of the problem. In a seminal paper, Goemans and Williamson [8] used SDP to obtain improved approximations for the Max-Cut and the MAX-2-SAT problem. In this work they present a randomized approximation algorithm for MAX-2-SAT that produces solutions of expected value at least .87856 times the optimal value. Subsequently, Feige and Goemans [10] extended this work, developing an .931-approximation algorithm for MAX-2-SAT. In our experiments we apply the "classical" SDP relaxation of Goemans and Williamson [8].

In this paper we study the quality of SDP based relaxations for MAX-SAT. In particular, we are interested in comparing the power of SDP relaxations against LP based relaxations, since LP relaxations have been shown to be useful in speeding up practical solvers [3]. In the work reported in this paper we address

---

[1] Unit propagation recursively sets the literals corresponding to unit clauses to true, eliminating all the clauses in which the literal appear, until a fix-point is reached.

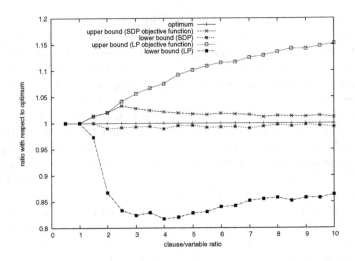

Fig. 1. Lower and upper bounds based on LP and SDP relaxations

the following research questions: (1) *How do the LP and SDP solutions compare as upper bounds on the optimal solution?* This is a critical question since we can use the relaxations as admissible heuristics to prune the search space. As we will see, in the overconstrained area, the upper bound based on the SDP relaxation is considerably tighter than the one provided by the LP relaxation: the upper bound provided by LP for MAX-SAT instances is always equal to the total number of clauses, therefore not informative in the overconstrained area; the upper bound given by the SDP relaxation is surprisingly close to the optimal solution (within less than 3% of the real optimal value, when varying the ratio of clauses to variables up to 10) (2) *How do the assignments based on the LP and SDP relaxations compare as lower bounds on the optimal solution?* Once again we see that the SDP relaxation outperforms the LP relaxation considerably, especially in the overconstrained area. In fact while the SDP lower bound is always within 1% of the optimal solution, the lower bound provided by the LP relaxation can be as far as 18% from optimal (see figure 1). We note that the LP solver runs faster than the SDP solver. Nevertheless, the runtimes for the SDP solver are very good, a little over 1 second per instance, on average, for an 80 variable problem, independently of the number of clauses. We also compared the quality of the solutions obtained from the SDP relaxation as a lower bound on the optimal solution against Walksat, one of the best performing local search methods for MAX-SAT. We gave Walksat about 5 minutes per instance (note that the SDP solver takes less than 2 seconds per instance). Interestingly, as the instances become more and more overconstrained, the SDP solutions become better than those provided by Walksat. Furthermore, because Walksat is a local search solver, it does not provide an upper bound on the optimal solution, a key aspect of the SDP relaxation. (3) *To what extent the relaxations provide a global perspective of the search space and therefore to what extent they can be used as heuristics to guide a complete solver?* In order to address this issue we performed the following experiment: set the X highest values suggested by the LP/SDP relaxation; check if the optimal value of the resulting instance is still the same as the original optimal value. Once again the SDP relaxation clearly outperforms the LP relaxation. In fact, in the overconstrained area, the setting of a single value dictated by the LP relaxation generally results in a value for the optimal solution lower than the original value. The SDP relaxation on the other hand is much more robust: We can set up to an average of 84% of variables based on the SDP suggestions, without changing the value of the optimal solution. This result suggests that the SDP relaxation can be a very valuable heuristic for setting variable values in a backtrack search strategy.

2   Preliminaries

The Boolean satisfiability problem (SAT) is a decision problem at the core of complexity theory, artificial intelligence, logic and hardware design and verification. We consider the problem in conjunctive normal form (CNF). A formula F in CNF is a conjunction of clauses, where each clause is a disjunction of literals. Each literal is a logical variable (x) or its negation ($\bar{x}$). The SAT problem is to determine whether there exists a variable assignment that makes the formula true (i.e., each clause is true). k-SAT represents the satisfiable problem where the clauses are constrained to have the length equal to k.

MAX-SAT is the optimization version of SAT. Given a formula we want to maximize the number of simultaneously satisfied clauses. Given an algorithm for MAX-SAT we can solve SAT, but not viceversa, therefore MAX-SAT is more complex than SAT. The distinction becomes obvious when considering the case when the clauses are restricted to two literals per clause (2-SAT): 2-SAT is solvable in linear time, while MAX-2-SAT is NP-hard [11].

Given the importance of the problem, the complexity of SAT has received much attention. Previous results show easy-hard-easy patterns in terms of the problem hardness, as a function of the clause/variable ratio (C/V). Furthermore, phase-transition phenomena have been reported with respect to satisfiability (i.e., a sudden change from many satisfiable instances to none). For instance, for 2-SAT this phase transition has been proven to occur when C/V is 1 [12].

Recently there have been promising results when using LP for MAX-SAT, after several unsuccessful efforts to apply integer LP to MAX-SAT (e.g.,[1,4]). Xing and Zhang developed MaxSolver ([3]), an efficient exact algorithm for (weighted) MAX-SAT. Their solver uses a DPLL-based branch and bound algorithm and it successfully uses a lookahead LP lower bound. This lower bound is only applied to the nodes that have unit clauses, to avoid fractional values equal to 1/2 (in the case of MAX-2-SAT). MaxSolver also incorporates existing and novel unit propagation rules, a binary-clause first rule and a dynamic-weighting variable ordering rule.

SDP relaxations have also been deployed for the MAX-2-SAT problem. Following the randomized polynomial time algorithm of Goemans and Williamson [8] which had an approximation ratio of 0.87856, there has been a series of theoretical results improving this approximation ratio. The most recent improvement is a 0.940-approximation algorithm [13]. A thorough survey of SDP-based approximation algorithms for the MAX-SAT problem is presented in [14]. See [15] for interesting results on the SDP for the so-called 2+p-SAT problem.

3 LP and SDP Formulations for MAX-SAT

3.1 LP Formulation

We consider the following ILP formulation for the MAX-SAT problem from [16]. With each clause C_j we associate a variable $z_j \in \{0, 1\}$. 1 corresponds to the clause being satisfied and 0 to the clause not being satisfied. For each variable x_i we associate a corresponding variable y_i in the ILP. y_i can take the values 0 and 1, corresponding to x_i being false or true, respectively. Let C_j^+ be the set of indices of positive literals that appear in clause C_j, and C_j^- be the set of indices of negative literals (i.e., complemented variables) that appear in clause C_j. The problem can be formally stated as follows:

$$\max \sum_{j=1}^{m} z_j$$

subject to

$$\sum_{i \in C_j^+} y_i + \sum_{i \in C_j^-} (1 - y_i) \geq z_j, \quad \forall j$$

where

$$y_i, z_j \in \{0, 1\}, \quad \forall i, j.$$

The formulation ensures that a clause is true only if at least one of the variables that appear in the clause has the value 1. Since, we want to maximize $\sum_{j=1}^{m} z_j$ and z_j can be set to 1 only when clause C_j is satisfied, it follows that the objective function counts the number of satisfied clauses. By relaxing the integrality constraint, we obtain an LP relaxation for the MAX-SAT problem. This ILP formulation is equivalent to the ILP used in [3] to compute the lower bound and to the ILP solved at each node by the MAX-SAT branch and cut algorithm in [4].

It is interesting to note that there exists a trivial way to satisfy all the clauses: setting each variable y_i to 0.5. Using this assignment, the sum of literals for each clause is exactly 1, hence the clause can be satisfied and the objective function is equal to the number of clauses. The value 0.5 is not at all informative, lying half way between 0 and 1, it gives no information whether the corresponding Boolean variable should be set to true or false. As the problem becomes more constrained (i.e., the number of clauses increases) the corresponding 2-SAT problem is very likely to be unsatisfiable, hence any variable assignment different than 0.5 would lead to a less than optimal objective value. Naturally, the LP solver finds the highest possible objective value (i.e., the number of clauses) when setting all variables to 0.5.

3.2   Semidefinite Programming

In this section we briefly introduce semidefinite programming. A large number of references to papers concerning semidefinite programming are on the web pages of Helmberg[2] and Alizadeh[3]. A general introduction to semidefinite programming applied to combinatorial optimization is given in e.g. [6].

Semidefinite programming makes use of positive semidefinite matrices of variables. A matrix $X \in \mathbb{R}^{n \times n}$ is said to be positive semidefinite (denoted by $X \succeq 0$) when $y^\mathsf{T} X y \geq 0$ for all vectors $y \in \mathbb{R}^n$. Semidefinite programs have the form

$$
\begin{aligned}
&\max \operatorname{tr}(WX) \\
&\text{s.t. } \operatorname{tr}(A_j X) \leq b_j \ (j = 1, \ldots, m) \\
&\qquad X \succeq 0.
\end{aligned}
\tag{1}
$$

Here $\operatorname{tr}(X)$ denotes the trace of X, which is the sum of its diagonal elements, i.e. $\operatorname{tr}(X) = \sum_{i=1}^{n} X_{ii}$. The matrix X, the cost matrix $W \in \mathbb{R}^{n \times n}$ and the constraint matrices $A_j \in \mathbb{R}^{n \times n}$ are supposed to be symmetric. The m reals b_j and the m matrices A_j define m constraints.

We can view semidefinite programming as an extension of linear programming. In particular, when the matrices W and A_j $(j = 1, \ldots, m)$ are all diagonal matrices[4], the resulting semidefinite program is equal to a linear program, where the matrix X is replaced by a non-negative vector of variables $x \in \mathbb{R}^n$. In particular, then a semidefinite programming constraint $\operatorname{tr}(A_j X) \leq b_j$ corresponds to a linear programming constraint $a_j^\mathsf{T} x \leq b_j$, where a_j represents the diagonal of A_j.

Theoretically, semidefinite programs have been proved to be polynomially solvable to any fixed precision using the so-called ellipsoid method (see for instance [7]). In practice, nowadays fast 'interior point' methods are being used for this purpose (see [5] for an overview).

3.3   Semidefinite Relaxation for MAX-2-SAT

We applied the semidefinite relaxation of MAX-2-SAT proposed by Goemans and Williamson [8]. The relaxation follows from a quadratic programming formulation of MAX-2-SAT. We first introduce this integer quadratic program.

Let the MAX-2-SAT problem consist of boolean variables $x_1, x_2, \ldots, x_n$ and a set of clauses C on these variables. To each variable x_i $(i = 1, \ldots, n)$, we associate a variable $y_i \in \{-1, 1\}$. Moreover, we introduce a variable $y_0 \in \{-1, 1\}$. We define x_i to be true if and only if $y_i = y_0$, and false otherwise.

Next, we express the truth value of a boolean formula in terms of its variables. Given a formula c, we define its *value*, denoted by $v(c)$, to be 1 if the formula is true, and 0 otherwise. Hence,

$$
v(x_i) = \frac{1 + y_0 y_i}{2}
$$

---

[2] http://www-user.tu-chemnitz.de/~helmberg/semidef.html
[3] http://new-rutcor.rutgers.edu/~alizadeh/sdp.html
[4] A diagonal matrix is a matrix whose non-diagonal entries are zero.

gives the value of a boolean variable x_i as defined above. Similarly,

$$v(\overline{x}_i) = 1 - v(x_i) = \frac{1 - y_0 y_i}{2}.$$

Hence, the value of the formula $x_i \vee x_j$ can be expressed as

$$v(x_i \vee x_j) = 1 - v(\overline{x}_i \wedge \overline{x}_j) = 1 - v(\overline{x}_i)v(\overline{x}_j) = 1 - \frac{1 - y_0 y_i}{2}\frac{1 - y_0 y_j}{2}$$
$$= \frac{1 + y_0 y_i}{4} + \frac{1 + y_0 y_j}{4} + \frac{1 - y_i y_j}{4}.$$

The value of other clauses can be expressed similarly. If a variable x_i is negated in a clause, then we replace y_i by $-y_i$ in the above expression.

Now we are ready to state the integer quadratic program for MAX-2-SAT:

$$\max \sum_{c \in C} v(c) \tag{2}$$
$$\text{s.t.} \quad y_i \in \{-1, 1\} \; \forall i \in \{0, 1, \ldots, n\}.$$

It is convenient to rewrite this program as follows. We introduce an $(n+1) \times (n+1)$ matrix Y, such that entry Y_{ij} represents $y_i y_j$ (we index the rows and columns of Y from 0 to n). Then program (2) can be rewritten as

$$\max \operatorname{tr}(WY) \tag{3}$$
$$\text{s.t.} \quad Y_{ij} \in \{-1, 1\} \; \forall i, j \in \{0, 1, \ldots, n\}, i \neq j,$$

where W is an $(n + 1) \times (n + 1)$ matrix representing the coefficients in the objective function of (2). For example, if the coefficient of $y_i y_j$ is w_{ij}, then $W_{ij} = W_{ji} = \frac{1}{2}w_{ij}$.

The final step consist in relaxing the conditions $Y_{ij} \in \{-1, 1\}$ by demanding that Y should be positive semidefinite and $Y_{ii} = 1 \; \forall i \in \{0, 1, \ldots, n\}$. Hence, the semidefinite relaxation of MAX-2-SAT is given by the following program

$$\max \operatorname{tr}(WY)$$
$$\text{s.t.} \quad Y_{ii} = 1 \; \forall i \in \{0, 1, \ldots, n\}, \tag{4}$$
$$Y \succeq 0.$$

Program (4) provides an upper bound on the solution to MAX-2-SAT problems. Furthermore, the values Y_{0i}, representing $y_0 y_i$, correspond to the original boolean variables x_i $(i = 1, \ldots, n)$. Namely, if Y_{0i} is close to 1, variable x_i is "close to true". Similarly, if Y_{0i} is close to -1, variable x_i is "close to false".

Example 1. Consider the MAX-2-SAT problem on the variables x_1 and x_2, with one clause $x_1 \vee \overline{x}_2$. The semidefinite relaxation is

$$\max \tfrac{3}{4} + \tfrac{1}{4}Y_{01} - \tfrac{1}{4}Y_{02} + \tfrac{1}{4}Y_{12}$$
$$\text{s.t.} \quad Y_{ii} = 1 \; (i = 0, 1, 2)$$
$$Y \succeq 0.$$

An optimal solution is

$$Y = \begin{bmatrix} 1.0 & 0.5 & -0.5 \\ 0.5 & 1.0 & 0.5 \\ -0.5 & 0.5 & 1.0 \end{bmatrix}$$

with objective value 1.125, which is larger than 1, the number of clauses.

The suggestion made by this relaxation is $Y_{01} = 0.5$ and $Y_{01} = -0.5$. This corresponds to "x_1 close to true" and "x_2 close to false". Indeed, this leads to an optimal solution for the MAX-2-SAT problem.

Example 1 shows that the solution to the semidefinite relaxation may overestimate the actual solution value. In this particular case it is even higher than the number of clauses. Moreover, the fractional solution values are quite far from integrality in this example. In practice however, we will see that the semidefinite relaxation provides surprisingly tight bounds and near-integral values for the variables.

An interesting aspect of program (4) is that the problem instance is entirely encapsulated in the objective function. Hence, the solution process is likely to be independent of the number of clauses, because the model size remains constant for a given number of variables. This is an important property when such relaxations need to be applied in practice.

4   Experimental Setup and Results

We have used random MAX-2-SAT instances generated by Selman's MWFF package [17]. For our experiments, we have solved the LP relaxation using ILOG CPLEX libraries. For the semidefinite relaxation we have used the solver CSDP, version 5.0 [18]. To compute the optimum value for the MAX-2-SAT instances, we used MaxSolver, the complete solver from [3].

4.1   Quality and Fractionality of Relaxations

We begin by examining the objective value of the LP and SDP relaxations across different constraindness regions of the problem. We examine MAX-2-SAT instances for 80 variables, varying the C/V ratio from 0.5 to 10. We also solve the instances with a complete solver [3]. (The LP and SDP relaxation can be computed for much higher numbers of variables. However the need for an exact solution limits us to around 80 variables.) The left plot in Figure 2 depicts the median objective value returned by the LP and SDP relaxation versus the median value returned by the MAX-2-SAT solver (bottom curve), as a function of the C/V ratio. Unless otherwise noted, each data point in all the plots corresponds to the median of 100 instances. Since both the LP and the SDP solve a relaxed version of the problem, the objective value is overestimated, and therefore the relaxations provide upper bounds on the optimal solution. We have shown in section 3.1 that we can always find an assignment which makes the objective value of the LP relaxation equal to the number of clauses, hence in the plot the fraction of satisfiable clauses is always 1. While the LP relaxation provides no information about the true solution to the problem, the SDP relaxation follows closely the behavior of the curve corresponding to the maximum fraction of satisfiable instances.[5] Hence, the SDP relaxation is able to adapt to the difficulty of the instances and provides a meaningful upper bound on the maximum number of satisfiable clauses (especially as C/V increases).

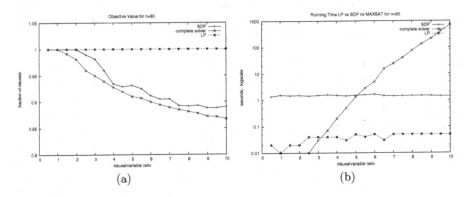

Fig. 2. (a) Objective value of the LP and SDP relaxations vs. optimum solution for different C/V values. (b) Runtime of the LP and SDP relaxations vs. the MAX-2-SAT solver in [3].

The plot on the right depicts the runtime in seconds of the two relaxations versus the MAX-2-SAT solver. The point of this plot is to observe whether the relaxations are affected by the C/V ratio. We first note that the runtime of the two relaxations is hardly affected by the constraindness of the problem, while the runtime of the MAX-2-SAT solver grows exponentially in the C/V ratio (the plot's y axis is a logarithmic scale). Naturally, the complete solver requires much more time in the over-constrained region, as it has to prove optimality of the found solution. The LP relaxation is computationally the least expensive across the board, except for the under-constrained region, where the problem is easy and the complete solver is able to quickly examine the search space.

In order to understand what enables the SDP relaxation to be more informed than the LP relaxation, we continue by studying the fractionality of the values returned by the two relaxations. Figure 3 plots the distribution of these values, in intervals of length 0.1 from 0 to 1 for the LP relaxation and from -1 to 1 for the SDP relaxation averaged over all instances (C/V ratio varying from 0.5 to 10). The ends of the two intervals correspond to the Boolean values false and true, respectively. The closeness of a value to one of the ends can be interpreted as the "confidence" of the relaxation that the corresponding Boolean variable should be set to true/false. We observe that the LP relaxation only sets variables to three values: 0, 1 and 0.5. 0 and 1 correspond to false and true, respectively, however

---

[5] Note that SDP can provide an objective value greater than the number of clauses; in those cases, for obvious reasons, we consider the number of clauses in the formula as the upper bound on the optimal value.

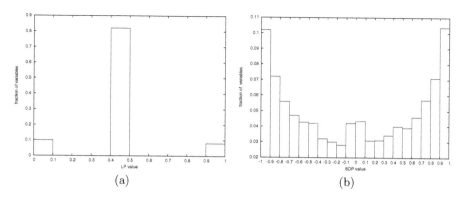

Fig. 3. Distribution of the values returned by the (a) LP and (b) SDP relaxation (averaged over instances with C/V ranging from 0.5 to 10)

0.5 gives us no information as to what should be assigned to the corresponding Boolean value. In contrast, the majority of values returned by the SDP relaxation are concentrated towards the ends of the interval $[-1, 1]$, thus suggesting more informed guidance about the way the variables should to be set.

To further describe the fractionality of the relaxations we examine them across different constraindness regions. Figure 4 a) plots the fraction of variables that are equal to 0.5 after solving the LP relaxation. In the under-constrained region (i.e., low clause/variable ratio) the fraction is 0, then as we pass the $C/V = 1$ point, the percentage goes up, and it approaches 1 (i.e., all variables) as the problem becomes over-constrained.

Similarly, figure 4a) plots the fraction of SDP variables whose values lie in the neighborhood of 0. We represent the fraction of variables equal to 0 and also those variables whose absolute value lies in the interval $(0, 0.1]$. In the under-constrained region the fraction of variables set to 0 is very high (close to 0.8). As we add more clauses this fraction sharply decreases and stabilizes at 0. These variables correspond to boolean variables that do not appear in the formula, hence the high fraction of such variables in the under-constrained region.

Figure 4b) plots the fraction of high confidence variables. For LP (variables having the value 0 or 1) we see a significant change around $C/V = 1$ and then the fraction decreases all the way to 0. (in fact this curve is the complement of the curve representing the variables set to 0.5 by LP). For SDP we plot the fraction of variables that are greater than 0.7 in absolute value. This fraction is low in the under-constrained region (as most of the variables are set to 0 as explained above) and it goes up to roughly half the variables as the problem becomes more constrained.

The plots demonstrate that the two relaxations examined behave very differently across the constraindness regions. As the problem gets harder (i.e., more constrained) the LP relaxation "defaults" to an uninformative assignment (all variables are assigned 0.5). In contrast, the SDP relaxation provides a good upper bound on the optimal solution (see Figure 2 and 1), with runtimes below 2

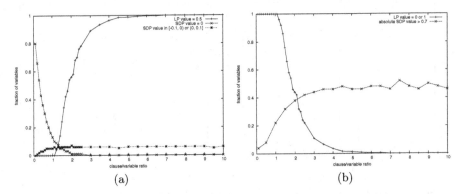

Fig. 4. a) Fraction of variables having the value a) 0.5 computed by LP and 0 or smaller than 0.1 in absolute value computed by SDP and b) 0 or 1 computed by the LP relaxation and above 0.7 in absolute value by the SDP relaxation

seconds per run. Furthermore, the SDP values assigned to the Boolean variables are less fractional than those delivered by LP (Figure 3) and therefore can be used more effectively as heuristics, as we will see in the next section.

4.2   SDP and LP as a Backtrack Search Heuristic

Related to the fractionality of the SDP relaxation is the question of whether the SDP relaxation provides a good global perspective of the search space and therefore could be used as a heuristic by MAX-SAT solvers. To test the heuristic power of SDP we used the following method:

Algorithm 1.   SDP-based Heuristic
Input: a MAX-2-SAT instance and x (the number of variables to be set using the SDP relaxation).
    Step 1: solve the SDP relaxation.
    Step 2: set x variables to the value suggested by the SDP relaxation[6].
    Step 3: given the (partial) assignment of variables from step 2, compute the MAX-2-SAT for the original instance s.t. the maximizing assignment extends this partial assignment.
Output: the maximum number of sat clauses.

To put into perspective the heuristic power of the SDP relaxation, we compare it to that of the LP relaxation. The following method was used:

Algorithm 2.   LP-based Heuristic
Input: a MAX-2-SAT instance and x (the number of variables to be set using the LP relaxation).
    Step 1: solve the LP relaxation.

---
[6] We consider variables in decreasing order of their absolute value.

Step 2: set x variables to the value suggested by the LP relaxation[7].

Step 3: given the (partial) assignment of variables from step 2, compute the MAX-2-SAT for the original instance s.t. the maximizing assignment extends this partial assignment.

Output: the maximum number of sat clauses[8].

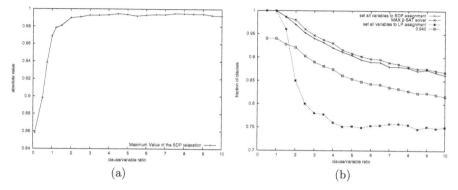

(a)   (b)

Fig. 5. a) The maximum value returned by the SDP relaxation and b) the heuristic power of SDP vs LP

Figure 5 a) plots the maximum absolute variable value returned by the SDP relaxation as we increase the constraidness of the problem. Given the fact that this value is very close to 1 and the high density of variables having high "confidence" level, the following experiments are justified. In the first experiment conducted, we set all the variables ($x = n$) to the value suggested by the relaxations and then we report the number of satisfied clauses. Figure 5 b) shows the results for the two relaxations versus the MAX-SAT value. Each data point corresponds to the median of 100 instances. The SDP performance is quite impressive, as it stays very close to the optimum across the board. When using the LP relaxation the performance degrades significantly as the problem becomes more constrained. The SDP's heuristic power does not seem to be affected by the constraidness of the problem. We have also included in figure 5 b) the curve corresponding to the best theoretical guarantee for an SDP based approximation algorithm (i.e., 0.940-approximation [13]). The results in figure 5, show that on random instances an algorithm using our SDP heuristic comes on average within 0.99 of the optimum. Thus, for most cases, when an estimate of the MAX-2-SAT solution is needed, simply setting the values suggested by the SDP relaxation should be a good heuristic.

---

[7] First consider the variables that were set to 0 or 1 by the LP relaxation and set the corresponding Boolean variables to false or true, respectively. If we exhaust all such variables, we consider variables that are 0.5 and randomly set the corresponding variable to true or false. Note that LP only assigns 0, 1, and 0.5 values to variables. Since the procedure is randomized we perform it many times for one instance.

[8] Because we perform the variable setting several times for every input instance, we return the median.

We also varied the number of variables set (x) between 0 and n and studied the effect on the maximum number of satisfiable clauses. We discovered that the median maximum number of satisfiable clauses when using SDP remains the same as the optimum, when we set up to 84% of the variables (i.e., 42 variables for n=50). It is at this point that we observe a slight change in performance – see figure 6 a).

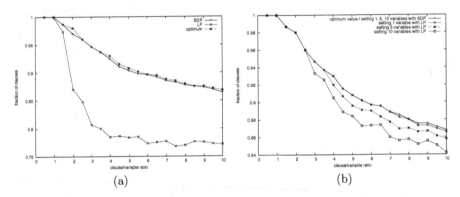

Fig. 6. Change in the number of satisfied clauses as we set a) 84% of the variables and b) 1, 5 and 10 variables using the LP and SDP

Once again, we compared SDP with LP and observed that when using LP the performance starts degrading even after setting just 1 variable and it continues to drop as we increase x (figure 6 b)). In contrast, SDP makes no mistakes when setting 1 and 2 variables and we have found just one instance (in over 2000 random instances) for $x = 3$, where the maximum number of satisfied clauses decreased by 1. For $x = 10$, the number of instances where the SDP suggestion is sub-optimal is four. These results show that the SDP relaxation is very informed and by following the SDP suggestion we remain very close to optimal performance.

In section 4.1 we showed that the objective function of the SDP relaxation provides a good upper bound for MAX-2-SAT. We test the potential of the SDP relaxation to provide a lower bound, by using the SDP relaxation to set all variables and we compare it to Walksat [19]. We ran Walksat with a cutoff of 10^8 flips. The average Walksat runtime was approximately five minutes, while the SDP relaxation was computed in approximately 1 second. The results are presented in figure 7. The plot on the left side depicts the upper bound, lower bound, optimum value and Walksat value for the C/V ratio ranging from 0.5 to 10. We note that the curves are very close to each other, but Walksat still provides a better lower bound. In the plot on the right side, as we increase the C/V ratio up to 100, we see that SDP outperforms Walksat, namely it provides a better bound, using considerably less time (recall from section 4.1 that the runtime for the SDP relaxation does not depend on the C/V ratio).

Given these very promising results for the guiding power of the SDP relaxation, we believe that the performance of MAX-SAT solvers could be further

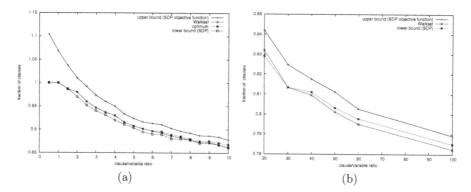

Fig. 7. SDP vs Walksat as a lower bound.

improved by incorporating the information provided by the SDP relaxation, for instances in the over-constrained region, where the time required to compute the SDP relaxation is considerably smaller than the time needed by the current MAX-SAT solvers (see figure 2 b)).

5   Conclusions and Future Work

In this paper we show how SDP relaxations are surprisingly powerful, providing much tighter bounds than LP relaxations, across different constrainedness regions. SDP relaxations can also be computed very efficiently, thus quickly providing tight lower and upper bounds on the optimal solution. We also show that heuristic guidance based on the SDP relaxation for iterative variable setting is significantly more accurate than the guidance based on the LP relaxation. SDP allows us to set up to around 84% of the variables without degrading the optimal solution, while setting a single variable based on the LP relaxation generally degrades the global optimal solution in the overconstrained area. We also compared SDP against Walksat: Interestingly, as the instances become more and more constrained, the lower bound provided by the SDP relaxation outperforms Walksat. The SDP relaxation runs much faster than Walksat. (We allocated 5 minutes per run for Walksat while SDP took less than 2 seconds per run.) Furthermore, Walksat has the limitation of not providing upperbounds. In our experiments, the SDP upper bound is always less than 3% above the optimal solution. Our results therefore show that SDP relaxations may further boost exact MAX-SAT solvers.

References

1. Hooker, J.N., Fedjiki, C.: Branch-and-cut solution of inference problems in propositional logic. Annals of Math. and Artificial Intelligence **1** (1990)
2. Warners, J.: Nonlinear approaches to satisfiability problems. PhD thesis, Technische Universiteit Eindhoven (1999)

3. Xing, Z., Zhang, W.: Maxsolver: An efficient exact algorithm for (weighted) maximum satisfiability. Artificial Intelligence **164**(1-2) (2005) 47–80
4. Joy, S., Mitchell, J., Borchers, B.: A branch and cut algorithm for MAX-SAT and weighted MAX-SAT. Satisfiability Problem: Theory and Applications **35** (1997) 519–536
5. Alizadeh, F.: Interior point methods in semidefinite programming with applications to combinatorial optimization. SIAM Journal on Optimization **5**(1) (1995) 13–51
6. Goemans, M., Rendl, F.: Combinatorial Optimization. In Wolkowicz, H., Saigal, R., Vandenberghe, L., eds.: Handbook of Semidefinite Programming. Kluwer (2000) 343–360
7. Grötschel, M., Lovász, L., Schrijver, A.: Geometric Algorithms and Combinatorial Optimization. John Wiley & Sons (1988)
8. Goemans, M., Williamson, D.: Improved Approximation Algorithms for Maximum Cut and Satisfiability Problems Using Semidefinite Programming. Journal of the ACM **42**(6) (1995) 1115–1145
9. Hochbaum, D.S.: (Editor) Approximation algorithms for NP-Hard problems. PWS Publishing Company (1997)
10. U. Feige and M. Goemans: Approximating the value of two prover proof systems, with applications to max2sat and max dicut. In: Proceedings of the 3rd Israel Symposium on Theory of Computing and Systems. (1995)
11. Garey, M., Johnson, D.: Computers and Intractibility. Freeman (1979)
12. Chvátal, V., Reed, B.: Mike gets some (the odds are on his side). In: 33th Annual Symposium of Foundations of Computer Science. (1992) 620–627
13. Lewin, M., Livnat, D., Zwick, U.: Improved rounding techniques for the MAX 2-SAT and MAX DI-CUT problems. In: IPCO. (2002) 67–82
14. Anjos, M.: Semidefinite optimization approaches for satisfiability and maximum-satisfiability problems. Journal on Satisfiability, Boolean Modeling and Computation **1** (2005) 1–47
15. de Klerk, E., van Maaren, H.: On semidefinite programming relaxation of 2+p-sat. Annals of Math. and Artificial Intelligence **37** (2003)
16. Motwani, R., Raghavan, P.: Randomized Algorithms. Cambridge University Press (1995)
17. Selman, B.: Mwff - a program for generating random MAX k-SAT instances. (1993)
18. Borchers, B.: A C Library for Semidefinite Programming. Optimization Methods and Software **11**(1) (1999) 613–623
    http://www.nmt.edu/~borchers/csdp.html.
19. Selman, B., Kautz, H., Cohen, B.: Local search strategies for satisfiability testing (1996)

Expected-Case Analysis for Delayed Filtering

Irit Katriel*

BRICS**, University of Aarhus, Århus, Denmark
irit@daimi.au.dk

Abstract. One way to address the tradeoff between the efficiency and the effectiveness of filtering algorithms for global constraints is as follows: Instead of compromising on the level of consistency, compromise on the frequency at which arc consistency is enforced during the search. In this paper, a method is suggested to determine a reasonable filtering frequency for a given constraint.

For dense instances of *AllDifferent* and its generalization, the *Global Cardinality Constraint*, let n and m be, respectively, the number of nodes and edges in the variable-value graph. Under the assumption that propagation is random (i.e., each edge removed from the variable-value graph is selected at random), it is shown that recomputing arc consistency only after $\Theta(m/n)$ edges were removed results in a speedup while, in the expected sense, filtering effectiveness is comparable to that of enforcing arc consistency at each search step.

1 Introduction

At the heart of the propagation-search technique in constraint programming is a tradeoff between the amount of resources (space, time) spent on propagation and the amount of pruning that it achieves. One type of propagation is filtering, i.e., the task of identifying useless values in variable domains and removing them. Global constraints play an important role in this game by enabling a larger number of useless values to be identified compared to the decomposition into a semantically equivalent set of simpler constraints. A "good" global constraint is a constraint that allows a significant increase in the amount of filtering with a low computational overhead.

An important question is that of *incremental* filtering, i.e., the development of fast algorithms that re-filter a constraint after a small change has occurred, such as the removal of a value from a variable domain due to filtering with respect to another constraint. For *AllDifferent* and its generalization, the *Global Cardinality Constraint* (GCC), incremental arc consistency can be performed faster than computing arc consistency from scratch. The arc consistency algorithm first finds a solution, i.e., a flow in the variable-value graph of the constraint (which is augmented into a flow network by adding a source and a sink), and then computes the strongly connected components (SCCs) of the residual graph w.r.t. this flow [4, 5]. Let n and m be, respectively, the number of nodes and edges in the variable-value graph. Then the running time is dominated by the time to find a single solution. Hopcroft and Karp's algorithm [1] for maximum bipartite

---

* Supported by the Danish Research Agency (grant # 272-05-0081).
** Basic Research in Computer Science, funded by the Danish National Research Foundation.

matching in $O(m\sqrt{n})$ time applies directly to *AllDifferent*, and as recently shown by Quimper et al. [3], can also be used for *GCC*.

An incremental arc consistency algorithm for *AllDifferent* and *GCC* receives an arc consistent flow network, a flow in it, and an edge that should be removed. If the removed edge is a flow edge, a new flow can be constructed after finding an augmenting path between the edge's endpoints, which takes linear time. Recomputing SCCs also takes time which is linear in the size of the graph. Although linear-time is much better than $O(m\sqrt{n})$, it is still a pity when it is wasted, i.e., when the algorithm does not discover any inconsistent edges.

We say that an edge in an arc consistent constraint's flow network is *important* if its removal would render at least one other edge "inconsistent". It is an interesting question whether there are fast methods to determine whether a given edge is important, or even to estimate *how* important it is, i.e., how many edges become inconsistent once it is removed?

In this paper we address a different aspect of this issue. We show that the number of important edges cannot be very large for the flow-based constraints mentioned above. More precisely, it is at most linear in the number of nodes in the variable-value graph. The implication of this is as follows. Assume that the filtering is random. That is, every edge in the variable-value graph of a constraint is equally likely to be removed. Then the expected number of edges that are removed until an important edge is removed, is $\Theta(m/n)$. Now, assume that instead of computing arc consistency incrementally in $O(m+n)$ time after every edge deletion, we compute arc consistency from scratch in $O(m\sqrt{n})$ time only after $\Theta(m/n)$ edges were deleted. Then the total time spent on filtering during these deletions drops from $O(m^2/n)$ to $O(m\sqrt{n})$. The filtering effectiveness, in the expected sense, is comparable. If the constraint is dense, i.e., $m = \Omega(n\sqrt{n})$, we have obtained a speedup.

The rest of the paper is organized as follows. In Section 2 we prove the main technical result the paper, which is an upper bound on the number of important edges in a flow network. In Section 3 we apply this bound to the specific flow networks of *AllDifferent* and *GCC* and conclude the result mentioned above on the filtering frequency. In Section 4 we describe how a solver can maintain the information it needs to perform delayed filering. Finally, in Section 5 we point out some directions for further research.

2   The Important Edges of a Flow Network

The constraints we address can be modeled by a flow network such that there is a one-to-one correspondence between the integral feasible flows in the network and the solutions to the constraint. Computing arc consistency then amounts to identifying edges that do not belong to any integral feasible flow, and flow theory tells us that given a flow f, the inconsistent edges are exactly the edges that do not belong to f and which connect two distinct SCCs in the residual graph corresponding to f [4, 5].

In the following we abuse terminology and speak of an edge in a flow network as being consistent if it belongs to at least one integral feasible flow. Furthermore, we say that the network is arc consistent if all of its edges are consistent and that an edge is important if its removal from the network would render at least one other edge inconsistent.

In the rest of this section we show an upper bound on the number of important edges in a flow network:

Theorem 1. *Let G be a flow network and let F be an integral feasible flow in G. The number of important edges in G is bounded from above by $|F| + 2n$, where $|F|$ is the maximum number of edges participating in a feasible flow (which can be bounded from above by the value of the flow because it is an integral flow).*

The proof is by showing that there exists a set S of cardinality less than $|F| + 2n$ such that S contains all important edges: We include all $|F|$ flow edges in S and then add less than $2n$ non-flow edges.

A non-flow edge is important if and only if its removal increases the number of SCCs in the residual graph of the network with respect to the given flow. This means that an edge between two SCCs cannot be important and we can consider one SCC at a time. For each SCC, we construct a DFS tree starting at an arbitrary node. The tree partitions the edges within the SCC into four sets: tree edges, forward-edges (non-tree edges from a node to its descendent), back-edges (non-tree edges from a node to its ancestor) and cross-edges (the remaining edges, which go from a node to a non-ancestor which was visited earlier by the DFS traversal). Clearly, a forward-edge cannot be important because the tree path between its endpoints implies that its removal does not change the reachability relation in the graph. We include all $n - 1$ tree edges in S and show in the next three lemmas that at most n back- and cross-edges can be important. In the following, we assume that each node is an ancestor of itself.

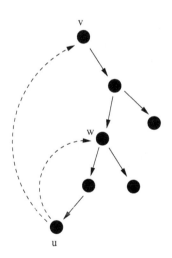

Fig. 1. Illustration for the proof of Lemma 1. Edges of the DFS tree are solid.

Lemma 1. *Every node can have at most one outgoing important back-edge.*

Proof. Assume that there are two important back-edges (u,v) and (u,w) outgoing from a node u. Since v and w are both ancestors of u in the tree, one of them must be an

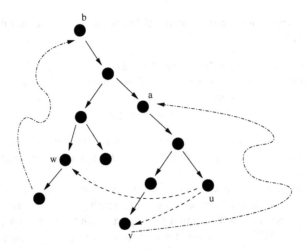

Fig. 2. Illustration for the proof of Lemma 2. Edges of the DFS tree are solid.

ancestor of the other. Assume, w.l.o.g., that v is an ancestor of w (see Figure 1). After removing the edge (u, w), the graph still has a path from u to w, namely the concatenation of the edge (u, v) and the path $T(v, w)$ (the unique path from v to w in the DFS tree). So w is still reachable from u, and hence any two nodes in the SCC are still reachable from each other. This contradicts the assumption that the edge (u, w) is important.   □

Lemma 2. *Every node can have at most one outgoing important cross-edge.*

Proof. Assume that there are two important cross-edges (u, v) and (u, w) outgoing from a node u. Since the graph is strongly connected, there is a path from v to u. Since u has a cross edge to v, it was visited by the DFS traversal after v, so the path from v to u must visit a common ancestor a of v and u before it reaches u. Similarly, there is a path from w to u which reaches a common ancestor b of w and u before it reaches u. Since a and b are both ancestors of u, one of them is an ancestor of the other. Assume, w.l.o.g., that b is an ancestor of a and hence also of v (see Figure 2). After removing the edge (u, v), there still is a path from from u to v, namely the edge (u, w) followed by the path from w to b, followed by the tree path $T(b, v)$ from b to v. Hence, any two nodes in the SCC are still reachable from each other. This contradicts the assumption that the edge (u, v) is important.   □

Lemma 3. *A node cannot simultaneously have an important outgoing back-edge and an improtant outgoing cross-edge.*

Proof. Assume that there is a cross-edge (u, v) and a back-edge (u, w) outgoing from u. Then w is not an ancestor of v, because otherwise when the edge (u, v) is removed, the edge (u, w) followed by the path $T(w, v)$ forms a path from u to v, and this contradicts our assumption that the edge (u, v) is important.

This implies that w was visited by the DFS search after the search has backtracked from v. Since the graph is strongly connected, there is a path from v to u and it must visit

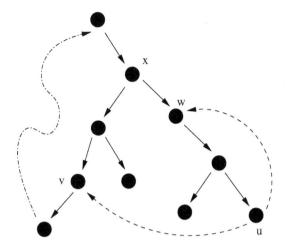

Fig. 3. Illustration for the proof of Lemma 3. Edges of the DFS tree are solid.

a common ancestor x of v and u before it reaches u. So the path from v to x, followed by the path $T[x, w]$ forms a path from v to w that does not visit u. Together with the edge (u, v), this forms an alternative path from u to w, contradicting our assumption that the edge (u, w) is important.   □

Thus, we have proved Theorem 1.

3   Important Edges for *AllDifferent* and *GCC*

The flow network corresponding to the *AllDifferent* and *GCC* constraints contains edges that do not correspond to variable-value pairs. Therefore, a refined upper bound is due. The proof of the following theorem is similar to the proof of Theorem 1, while counting only edges between nodes of the network that represent variables of the constraint or values in their domains, and noticing that $|F| \leq n$.

Theorem 2. *Let C be a constraint defined on a set X of n domain variables and let G be a flow model for C. The number of important variable-value pairs in C is at most 3n.*

Perhaps a more careful analysis could yield a tighter upper bound, but we wish to point out that the upper bound in Theorem 2 is within a constant factor of optimal. Figure 4 shows the variable-value graph of an *AllDifferent* constraint with $2n$ important edges: There are two solutions, and each edge belongs to exactly one of them. Removing any edge would destroy one of the solutions and turn the other edges that belong to it into inconsistent edges.

Assume that when a value is removed from the domain of one of the variables of a constraint, it is selected at random among all possibilities. In other words, a random edge is removed from the variable-value graph. Selection without replacement corresponds to the *hypergeometric distribution*. If we have a bin with M balls, of which

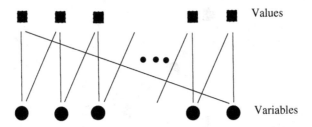

Fig. 4. An *AllDifferent* variable-value graph with $2n$ important edges

N are black and the remaining are white, and we randomly select x balls (without replacement), the expected number of selected black balls is xN/M. In other words, after selecting M/N balls, we are expected to have selected one black ball. Returning to our setting, we get that on expectation, an important edge is lost after $m/3n$ edges were removed from the variable-value graph of an *AllDifferent* or *GCC* constraint.

If we re-filter with respect to the constraint only after losing $m/3n$ edges, we need to recompute arc consistency from scratch, which takes $O(m\sqrt{n})$ time. On the other hand, if we re-filter every time an edge is lost, we spend a linear amount of time for each edge, for a total of $O(m^2/n)$. Whenever $m = \omega(n\sqrt{n})$, the delayed re-filtering is asymptotically faster.

Corollary 1. *For a dense instance of AllDifferent or GCC, re-filtering after $\Theta(m/n)$ edges are lost is more efficient and its effectiveness, in the expected sense, is comparable to that of re-filtering every time an edge is lost.*

4   Implementation

The simplest way to support delayed filtering in a solver appears to be as follows. For each constraint, the solver should have two counters. The first keeps track of the number of edges in the variable-value graph of this constraint, and the second indicates how many edges need to be lost before the constraint is to be re-filtered.

The number of edges in the variable-value graph is computed once when the constraint is posted: It is equal to the sum of the cardinalities of the domains of the variables. From that point on, it is decremented every time an edge is lost. This requires that we associate with every variable a list of the constraints that are defined on it. When a value is removed from its domain, this means that each of these constraints lost an edge from its variable-value graph so we traverse this list and decrement the edge count of each of the constraints.

The number of edges that should be lost before a re-filtering occurs is calculated once at the beginning and again every time re-filtering is performed. It is calculated using the edge-count and the number of variables as well as the specific frequency function for the constraint. In the case of *AllDifferent* and *GCC*, we have shown that the function $f(m,n) = m/3n$ makes sense (under the assumption that propagation is random).

5   Conclusion and Directions for Further Research

In the design of filtering algorithms for global constraints we should take into account the context in which they are used. For example, it is usually more important to find a fast incremental algorithm than to fine-tune the algorithm that filters a constraint from scratch. Another example is the analysis of propagation patterns (e.g., [2]), which deals with the question of how many times a constraint can be re-filtered before a fixpoint is reached. In this paper we suggested an additional way to look at propagation in the context of the search.

We assumed in our analysis that filtering is random in the sense that whenever an edge is removed from the variable-value graph, it is selected at random among all possibilities. Is this assumption reasonable in practice? Even if it is, the model we used disregards the propagation of filtering between the constraints. In other words, when we delay the removal of a useless edge, we also delay the propagation of this removal. Furthermore, if the CSP contains only complex constraints and the solver delays filtering for all of them, it could take many search steps before *any* propagation is performed, and this is clearly not what we intended. In summary, it is not immediately clear how to apply the analysis technique proposed in this paper to delayed filtering. An evaluation, and possibly a refinement of our model of the propagation-search process is necessary.

Another assumption we made is that the desired filtering frequency for a constraint depends on the constraint, and not on the CSP that it is in. It could be that the same constraint can exhibit very different filtering behaviors in different CSPs, and a filtering function associated with a constraint is meaninless if it does not take the context of the constraint into account. On the other hand, it could be that the filtering behavior depends both on the constraint and on the context, in which case it may work better if the sovler initially uses our pre-calculated frequency function, and then constantly *adjusts* the filtering frequency as it proceeds: Whenever a filtering step was useless, the frequency is reduced and whenever it is successful, the frequency is increased.

Acknowledgements. I wish to thank the anonymous referees for numerous helpful suggestions, many of which were incorporated into the discussion in Section 5.

References

1. J.E. Hopcroft and R.M. Karp. An $n^{5/2}$ algorithm for maximum matching in bipartite graphs. *SIAM J. Computing*, 2(4):225–231, 1973.
2. L. Mercier and P. Van Hentenryck. Strong polynomiality of resource constraint propagation, 2005.
3. C.-G. Quimper, A. López-Ortiz, P. van Beek, and A. Golynski. Improved algorithms for the *global cardinality* constraint. In *CP 2004*, volume 3258 of *LNCS*, pages 542–556. Springer-Verlag, 2004.
4. J.-C. Régin. A filtering algorithm for constraints of difference in CSPs. In *AAAI-94*, pages 362–367, 1994.
5. J.-C. Régin. Generalized Arc-Consistency for Global Cardinality Constraint. In *Proceedings of the 13th National Conference on Artificial Intelligence (AAAI-96)*, pages 209–215, 1996.

Plan B: Uncertainty/Time Trade-Offs for Linear and Integer Programming

Claire Kenyon and Meinolf Sellmann

Brown University, Department of Computer Science
115 Waterman St, Providence, RI 02912, U.S.A.
{claire, sello}@cs.brown.edu

Abstract. We address the following dilemma: When making decisions in real life, we often face the problem that, while we have time to contemplate about a problem, we are not entirely sure what the exact parameters of our problem will be. And, on the other hand, as soon as the real world is revealed to us, we need to act quickly and have no more time to rethink our actions extensively.

We suggest an approach that allows to trade uncertainty for time and marginal quality loss and discuss its applicability to combinatorial optimization problems that can be formulated as linear and integer linear programs. The core idea consists in solving a polynomial number of problems in the extensive time period before the day of operation, so that, as soon as complete information is available, a feasible near-optimal solution to the problem can be found in sublinear time.

1 Introduction

When solving real-world optimization problems, we face a trade-off between time and uncertainty. Often, there is enough time to consider a problem before a solution needs to be realized, but the information available about the problem at that stage is associated with a lot of uncertainty; complete information may be available only immediately before the first decision needs to be made, when almost no time is left anymore to optimize. Consider, for instance, a delivery company that, every night, needs to solve a vehicle routing problem with time windows to schedule the deliveries for the next day in a network with some notoriously unreliable links. In the morning, when the trucks are about to be loaded, it may be much clearer what the delay on the uncertain links is going to be, but waiting 20 minutes to reschedule before loading can begin is just not an option.

Related Work: There has been conducted extensive research that tries to tackle uncertainty. It is impossible to cover this very active research area entirely here, so we need to focus on the work that appears most relevant with respect to our contribution. One of the first big areas to consider unknown future events was online optimization (for an introduction see [4]) where decisions need to be taken already before future events are visible. The worst-case character of adversarial competitive analysis is often perceived as too pessimistic for real world applications, though, since we may want to optimize our expected rewards rather than to shelter ourselves against the worst case.

J.C. Beck and B.M. Smith (Eds.): CPAIOR 2006, LNCS 3990, pp. 126–138, 2006.

This motivation combined with the assumption that stochastic information about future events may be available beforehand has lead to the paradigm of stochastic optimization (for an overview on stochastic programming see for example [3]). The standard scenario considered here is a two-stage process where actions can be taken in stage one at a known cost that must then be augmented to a feasible plan by taking additional actions in stage two for costs that were unknown in stage one but that are drawn from a known probability distribution.

A paradigm that treats uncertainty where all decisions need to be taken before the real parameters are revealed to us is robust optimization (see for example [2]). In robust optimization, a solution needs to be computed that is required to be feasible with respect to all possible future scenarios. It is of course appealing to have one solution that will work no matter what, and in some critical applications this appears necessary. However, the loss in solution quality when comparing against the quality that could have been achieved in the scenario that we really face eventually can be enormous.

Therefore, it has been suggested to drop the requirement that just one solution must satisfy all future scenarios. This approach is very natural: In critical circumstances, humans often do not only make one plan for the future, but instead prepare one or several back-up plans that can kick in when the original one turns out to be infeasible. So we may also decide to compute more than one solution in order to be prepared for changing future constraints.

The classic example for this approach is the k-shortest path problem where the k best solutions to the problem are computed (see [7] for an overview of literature). The problem is of course interesting in and by itself. With respect to robustness, however, it is easy to construct examples where this approach fails miserably when just one edge is not available when it comes to executing the shortest path, because all k best solutions relied on the existence of that one edge.

Consequently, some approaches try to compute solutions that differ a lot, in the hope that different solutions are less likely to fail for the same change in the requirements. One example for this approach is the so-called penalty method [12] where an additional cost is introduced that penalizes similarity to an existing solution (whereby it is unclear how to reasonably quantify the trade-off between solution quality and similarity). When restricting ourselves to pure feasibility problems, another example is the search for so-called (a, b)-super solutions in constraint programming [9]. Here, solutions are sought-after that are robust in the sense that, if a variables lose their values, then the solution can be repaired by reassigning the critical a variables plus at most b others. This notion of robustness implicitly assumes that there is a limited amount of time available to re-schedule if the scenario which was planned for changes. Then, super solutions have the additional benefit that the original plan only needs to be adapted slightly to be legal. A modification of this idea is presented in [6] where the notion of multi-consistency is introduced. The idea of computing a set of back-up solutions is taken to extremes in [10] where linear programming sensitivity analysis is used to compute a potentially exponentially large set of optimal solutions for linear programs with changing objectives.

Contribution: In this paper, we assume that there is extensive time available before the day of operation when our solution will need to be realized, but that during this period the problem is associated with uncertainty. Then, when the exact problem is revealed to us, there is only very little time left to compute a high-quality feasible solution. Under which conditions, and how, can we efficiently prepare ourselves in that situation? In more formal terms, we address the question whether we can trade time for uncertainty by precomputing a set of solutions (our "plan Bs") such that, for every possible scenario, there exists a feasible solution in our set that is near optimal. Applied to the example of the delivery company, we investigate whether a set of optimization problems could be solved overnight so that we are guaranteed to have a near-optimal schedule waiting to be executed for the scenario that we are facing in the morning.

We do not assume that stochastic information is available. Like earlier work on uncertain constraint satisfaction problems (see, for example, [15] and [16]), we consider linear and integer linear programming (LP and IP) formulations. LPs and IPs are known to be capable of modelling a large variety of combinatorial optimization problems, and can be used efficiently in problems such as AI planning [14]. However, we focus on LPs and IPs for a different reason, namely because there exists an established standard form that, as we shall see, we can exploit both to define uncertainties as well as to determine what scenarios should be considered to prepare ourselves effectively for the future.

We make two important remarks: First, the uncertainties that we consider are defined in terms of linear programming formulations while, a priori, it is unclear how uncertainties in problem parameters translate into uncertain parameters of a linear programming formulation. Note that, in principle, it is possible that one uncertainty translates into many correlated uncertain parameters in the corresponding linear programming model. Extra care must be taken to decide in which scenarios the following work applies. Second, we use (integer) linear programs strictly as modelling entities, while the solution method for the problems that we consider can always be chosen freely.

The class of uncertain linear programs that we consider represents a subset of so-called linear programs with interval coefficients (LPICs). LPICs were introduced in [5] where an exponential-time algorithm was presented that allows to assess the range in which the optimal objective function value can lie depending on what values the coefficients in their respective interval will take. This gap-analysis is not trivial because LPICs allow unsigned variables and equality constraints. Excluding these two cases, we develop a method that provides worst case quality guarantees for uncertain LPs and IPs. We show that, even with signed variables and in the absence of equality constraints, no polynomial number of precomputed solutions suffices to prepare for all possible future scenarios. Despite this negative result, for a variety of more restricted cases, we can devise polynomial time algorithms that allow to find a near-optimal feasible solution in sublinear time as soon as all uncertainty has vanished. The restrictions guarantee, in particular, that infeasibility will not occur.

2   Uncertain Linear Programs

Definition 1 (ϵ-approximation). *Consider the problem $L = (P, f)$ of maximizing a function f over a set P into the non-negative rationals, and let $OPT(L)$ denote the optimal value $\max_P f$, if it exists. Given $\epsilon > 0$, an ϵ-approximation is an element $x \in P$ such that $(1 + \epsilon)f(x) \geq f(y)$ for all $y \in P$.*

We use the following paradigm. Given partial knowledge on the optimization problem, precompute a set of candidate solutions such that, once, at a later time, (P, f) is known exactly, then it is possible to use that prearrangement so as to find very fast (in sublinear time) a near-optimal solution to the problem. In this section we show how and under what circumstances this can be achieved for LPs.

For any two matrices or vectors, we write $A \geq B$ when this inequality holds for every component. Henceforth we assume that we are given two m by n matrices $\underline{A} \leq \overline{A}$, two vectors $\underline{b} \leq \overline{b} \in \mathbb{Q}^m$, two vectors $\underline{p} \leq \overline{p} \in \mathbb{Q}^n$, a desired approximation guarantee $\epsilon > 0$, and a constant $k \in \mathbb{N}$.

Definition 2 (Uncertain Linear Program and Prearrangement). *An uncertain linear program (ULP) S_{LP} is given as $(\underline{A}, \overline{A}, \underline{b}, \overline{b}, \underline{p}, \overline{p})$ and consists of the set of all LPs of the form*

$$\max p \cdot x$$

$$s.t. \begin{cases} Ax \leq b \\ x \geq 0, \end{cases}$$

such that $\underline{A} \leq A \leq \overline{A}$, $\underline{b} \leq b \leq \overline{b}$, and $\underline{p} \leq p \leq \overline{p}$.

A prearrangement of S_{LP} with parameter ϵ is a set X of rational vectors such that for every linear program $L = (A, b, p)$ in S_{LP}, there exists an ϵ-approximation in X.

ULPs, as defined above, represent a subset of linear programs with interval coefficients (LPICs) that were introduced in [5]. Note that, in ULPs, equality constraints are not allowed, and neither are unsigned variables. In order to avoid infeasibility issues we further restrict the ranges to positive intervals:

Definition 3 (Range-positive). *Given a ULP, we say that it is range-positive wrt A, iff (1) all coefficients of $\underline{A}$ are non-negative: $\underline{A} \geq 0$, and moreover, (2) uncertain coefficients must be positive: $\overline{a}_{ij} \neq \underline{a}_{ij} \Rightarrow \underline{a}_{ij} > 0$. Similarly, we say that it is range-positive wrt b (resp. p) iff (1) all coefficients of $\underline{b}$ (resp. $\underline{p}$) are non-negative, and moreover, (2) uncertain coefficients must be positive.*

We will also need to restrict the number of parameters that can vary more than marginally. We introduce the following notation:

Given two vectors (or matrices) v, w, and given $\delta \geq 0$ we write $v \geq_k^\delta w$, iff $v \geq w$ and for all but k coordinates, it holds that $w_i \leq v_i \leq w_i + \delta |w_i|$.

Definition 4 (Bounded Uncertainties). *Given a ULP S_{LP}, $k \in \mathbb{N}$, and $\delta > 0$, a Linear Program with k-δ-Bounded Uncertainties is the subset $S_{BLP} \subset S_{LP}$ consisting of the LPs of S_{LP} such that $A \geq_k^\delta \underline{A}$, $b \geq_k^\delta \underline{b}$ and $p \geq_k^\delta \underline{p}$.*

To ease the presentation, we use the lower end of each interval as the default value of the coefficient, but this is by no means necessary: the default value for each coefficient could also be explicitly specified and lie anywhere in the given range.

Note that Definition 4 does not require that we know *which* k parameters in A, b, and p will change by more than a small factor from their default value. We will discuss in a moment how knowledge about which k parameters could change more than marginally can be exploited to yield to more efficient prearrangement algorithms.

When starting to think about the problem, at first, even when only one parameter in an LP can change over a larger range, it is not clear that there exists a polynomial-size prearrangement. For this, we need to prove that, when a range-positive parameter in the LP is changed slightly, then the objective function can only change marginally, too. From this, it follows that, for range-positive ULPs with bounded uncertainties, we can efficiently compute a prearrangement that will allow us to lookup a solution later in sublinear time.

Theorem 1. *Given $\epsilon > 0$, let S_{BLP} denote a ULP with $k\text{-}\epsilon/4$-bounded uncertainties that is range-positive wrt A, b, and p. Let C denote the maximum of all ratios $\overline{p}_i/\underline{p}_i$, $\overline{b}_j/\underline{b}_j$, and $\overline{a}_{ij}/\underline{a}_{ij}$ (where by convention $0/0 = 1$). We can compute a prearrangement X of size polynomial in m, n, $\log C$, and $1/\epsilon$. The computation time of X equals $|X|$ times the time to solve an LP of S_{BLP}.*

Moreover, once X is computed, given the coefficients of the LP which deviate more than marginally from their anticipated value, one can compute an ϵ-approximation of the modified problem P in time sublinear in the size of the ULP.

Proof: Here is the algorithm.

Prearrangement: Let $\delta := \epsilon/4$. For each subset I_p of $[1, n]$ of size k, each subset I_b of $[1, m]$ of size k, and each subset I_A of $[1, m] \times [1, n]$ of size k, let $n_p^i = \log_{1+\delta}(\overline{p}_i/\underline{p}_i)$ for $i \in I_p$, $n_b^i = \log_{1+\delta}(\overline{b}_i/\underline{b}_i)$ for $i \in I_b$, and $n_A^{ij} = \log_{1+\delta}(\overline{a}_{ij}/\underline{a}_{ij})$ for $(i, j) \in I_A$.

For each list of integers $((k_i)_{i \in I_p}, (\ell_i)_{i \in I_b}, (m_{ij})_{(i,j) \in I_A})$ such that $0 \leq k_i \leq n_p^i$, $0 \leq \ell_i \leq n_b^i$, and $0 \leq m_{ij} \leq n_A^{ij}$, consider the LP defined by

$$p_i = \begin{cases} \underline{p}_i & \text{if } i \notin I_p \\ \underline{p}_i(1+\delta)^{k_i} & \text{if } i \in I_p \end{cases},$$

$$b_i = \begin{cases} \underline{b}_i & \text{if } i \notin I_b \\ \underline{b}_i(1+\delta)^{\ell_i} & \text{if } i \in I_b \end{cases}, \text{ and}$$

$$a_{ij} = \begin{cases} \underline{a}_{ij} & \text{if } \overline{a}_{ij} = \underline{a}_{ij}, \text{ else} \\ \underline{a}_{ij}(1+\delta) & \text{if } (i, j) \notin I_A \\ \underline{a}_{ij}(1+\delta)^{m_{ij}} & \text{if } i \in I_A \end{cases}.$$

Notice that, since all coefficients a_{ij} and b_i are non-negative, 0 is always a feasible solution of this LP, hence there always is a (possibly infinite) optimal solution. Compute an optimal solution, and add the LP and its solution to X. We store the elements of X represented as $((k_i)_{i \in I_p}, (\ell_i)_{i \in I_b}, (m_{ij})_{(i,j) \in I_A})$ in a data structure enabling fast search, say, a balanced tree for example (whereby we use the lexicographic order on the representing tuples to define an order on the LPs).

Solution Lookup: Given X, we can gather an ϵ-approximation for any given LP $P = (A, b, p) \in S_{BLP}$ in sublinear time. When we are given a list of at most k coefficients and their new values, we compute $((k_i)_{i \in I_p}, (\ell_i)_{i \in I_b}, (m_{ij})_{(i,j) \in I_A})$ such that:

- $\forall i \in I_p :\ \underline{p}_i(1+\delta)^{k_i-1} < p_i \leq \underline{p}_i(1+\delta)^{k_i}$.
- $\forall i \in I_b :\ \underline{b}_i(1+\delta)^{\ell_i} \leq b_i < \underline{b}_i(1+\delta)^{\ell_i+1}$.
- $\forall (i, j) \in I_A :\ \underline{a}_{ij}(1+\delta)^{m_{ij}-1} < a_{ij} \leq \underline{a}_{ij}(1+\delta)^{m_{ij}}$.

Then, we search and output the optimal solution of the corresponding LP P' in X.

Correctness: We first claim that any feasible solution x' of P' is also a feasible solution of program P. Indeed, by definition $x' \geq 0$ and $A'x' \leq b'$. Thus:

$$Ax' \leq A'x' \leq b' \leq b. \tag{1}$$

Second, let us analyze the quality of the solution that we return. Let x denote an optimal solution for P. We claim that $x/(1+\delta)^2$ is a feasible solution for P':

$$A'x/(1+\delta)^2 \leq Ax/(1+\delta) \leq b/(1+\delta) \leq b' \tag{2}$$

and the scaled solution remains non-negative. Since x' is an optimal solution for P', $p' \geq p$, and $\delta = \epsilon/4$, we have

$$p \cdot x' \geq p' \cdot x'/(1+\delta) \geq p' \cdot x/(1+\delta)^3 \geq p \cdot x/(1+\epsilon). \tag{3}$$

Complexity: The set X has size at most $4\binom{n}{k}\binom{m}{k}\binom{mn}{k}\log^{3k}(C)/\epsilon$. Note that this is polynomial in the description size of the instance of the problem for constant k. Constructing X amounts to solving $|X|$ LPs from S_{BLP}, which can thus be done in polynomial time.

Regarding the lookup of a solution, we see that the computation of the representing tuple can then be performed in time $O(3k \log \log C)$. Finding the corresponding solution then can be done in time $O(\log|X|)$. Since $|X|$ is polynomial in the size of the given ULP and $1/\epsilon$, the total lookup time is in $O(\log(1/\epsilon) + \log(|S_{BLP}|) + k)$ where $|S_{BLP}|$ denotes the input size of the ULP. Consequently, the lookup can be performed in time sublinear in the size of the ULP. □

Note that Theorem 1 considers the most general case where all parameters can change, k parameters can change more than just marginally, and we have absolutely no idea which ones. If we knew those k parameters in advance, the theorem implies that a pre-arrangement of size $O(\log^{3k}(C)/\epsilon)$ is sufficient, which is very manageable and quite encouraging. Between the two extremes, whenever the size of the set from which the k non-marginal parameters are drawn from can be limited, this knowledge can be leveraged to make much smaller prearrangements.

We now prove that the assumption of bounded uncertainties of Theorem 1 is not arbitrary but necessary.

Theorem 2. *There exists a ULP S_{LP} which is range-positive wrt A, b, and p, and such that every prearrangement X for S_{LP} has size at least exponential in the size of the description of S_{LP}.*

Proof: Consider the following ULP, where $m = n$. $\underline{A} = \overline{A}$ is the identity matrix, $\underline{p} = \overline{p}$ is the all-ones vector, $\underline{b}$ is the all-ones vector, and $\overline{b}$ is the vector whose entries are all equal to 2. In other words, S_{LP} is the set of LPs of the form:

$$\max \sum_i x_i$$

$$s.t. \begin{cases} \forall i, & x_i \leq b_i \\ \forall i, & x_i \geq 0, \end{cases}$$

where for each i we have $1 \leq b_i \leq 2$.

Any prearrangement of S_{LP} must also be a prearrangement of the subset S' consisting of those LPs such that exactly half of the b_i's are equal to 1 and the other half are equal to 2. Any such LP has value $(n/2) + 2(n/2) = 3n/2$. Let $x \in X$ be in a prearrangement of S'. Without loss of generality, we can assume that every coordinate x_i is either 1 or 2. Setting $\xi := \epsilon/(1 + \epsilon)$, we must have: $\sum_i x_i \geq (1 - \xi)3n/2$, so the number k of ones in x satisfies

$$k + 2(n - k) \geq (1 - \xi)3n/2,$$

hence $k \leq n/2 + 3\xi n/2$. But x can only be a feasible solution for at most $\binom{k}{k-n/2}$ elements of S', which is at most $\binom{n/2+3\xi n/2}{3\xi n/2}$. Since S' has cardinality $\binom{n}{n/2}$, any prearrangement of S' must have cardinality at least

$$\frac{\binom{n}{n/2}}{\binom{n/2+3\xi n/2}{3\xi n/2}},$$

which is exponential in n for $\xi < 1/3$ or $\epsilon < 1/2$. By scaling the right hand side, we can generalize the result for all $\xi < 1$ and $\epsilon > 0$.   □

At this stage, the significance of our positive result is rather limited since we can really only handle LPs with non-negative coefficients and $\leq$ constraints. We find that analogous results hold for LPs of the form: $\min p \cdot x$ such that $Ax \geq b, x \geq 0$ (in other words, we can tackle both "covering" and "packing" linear programs). However, one may wonder how necessary is it to assume that the coefficients are all non-negative. We now prove that the range-positive assumptions of Theorem 1 can be somewhat relaxed, although at the cost of restricting the uncertainty.

Theorem 3. *Let S_{BLP} denote a ULP with bounded uncertainties. Assume that at least one of the following sets of conditions hold.*

- *$A = \overline{A}$ and S_{BLP} is range-positive wrt b and p.*
- *$b = \overline{b}$ and S_{BLP} is range-positive wrt A and p.*
- *$b = \overline{b}$, $p = \overline{p}$, and S_{BLP} is range-positive wrt A.*
- *$A = \overline{A}$, $b = \overline{b}$, and S_{BLP} is range-positive wrt p.*

Then the conclusion of Theorem 1 still holds.

Proof: The algorithm is identical to the one used in the proof of Theorem 1 except for a slightly improved setting of δ. Then, only minor differences occur in the analysis of correctness.

- Under the first set of conditions, we set $\delta := \epsilon/3$. In the analysis, we then replace Equation 2 by:

$$A'x/(1 + \delta) = Ax/(1 + \delta) \leq b/(1 + \delta) \leq b',$$

thus proving that $x/(1+\delta)$ is feasible for P'. We can then replace Equation 3 by:

$$p \cdot x' \geq p' \cdot x'/(1+\delta) \geq p' \cdot x/(1+\delta)^2 \geq p \cdot x/(1+\epsilon).$$

Notice that it may now happen that P is infeasible, but that this is detected by the algorithm, since P is infeasible iff P' is also infeasible.

- Under the second set of conditions, either $\underline{b} \geq 0$ in which case the problem is range-positive wrt b and we can apply Theorem 1, or there exists no feasible solution to any LP in S_{BLP}.
- Under the third set of conditions, we set $\delta := \epsilon$ and replace Equation 2 by:

$$A'x/(1+\delta) \leq Ax \leq b = b',$$

thus proving that $x/(1+\delta)$ is feasible for P'. We can then replace Equation 3 by:

$$p \cdot x' = p' \cdot x' \geq p' \cdot x/(1+\delta) = p \cdot x/(1+\delta) = p \cdot x/(1+\epsilon).$$

- Under the last set of conditions, we set $\delta := \epsilon$. We notice that x is trivially feasible for P'. Therefore, we can replace Equation 3 by:

$$p \cdot x' \geq p' \cdot x'/(1+\delta) \geq p' \cdot x/(1+\delta) \geq p \cdot x/(1+\epsilon). \qquad \square$$

By slight adaptation of our algorithm and its analysis, it is easy to show that analogous results still hold when we consider ULPs of the form $\min p \cdot x$ such that $Ax \geq b$ and $x \geq 0$. As a consequence, by arguing about linear programming duality, we can further extend Theorem 1 as follows.

Corollary 1. *The conclusion of Theorem 1 is still valid when one of the following holds:*

- $\underline{p} = \overline{p}$ *and* S_{BLP} *is range-positive wrt A and b.*
- $\underline{A} = \overline{A}$, $\underline{p} = \overline{p}$, *and* S_{BLP} *is range-positive wrt b.*

What we have achieved now is that we can handle arbitrary ULPs with bounded uncertainties where only the part (constraint matrix, objective or right hand side) that is subject to potential changes is required to be range-positive. We believe that this requirement is mild enough to allow to tackle practical applications where it is often the case that objective and/or right hand side entries all have the same sign.

3   Prearranging Integer Programs

For LPs, we studied under which conditions prearrangement can be performed efficiently. We envision applications of this work for example in very time critical, real-time systems like sensor networks where changing volumes of data need to be managed very quickly which does not allow to re-solve larger LPs. However, one may argue that, since LPs can be solved in polynomial time, the number of applications is limited in which reducing the complexity to sublinear time is really necessary. Therefore, in this section we study integer linear programs that are NP-hard to solve in general.

Definition 5 (Uncertain Integer Programs). *The* uncertain integer program (UIP) S_{IP} *consists of the set of all IPs of the form*

$$\max p \cdot x$$

$$s.t. \begin{cases} Ax \leq b \\ x \geq 0 \\ x \quad integer, \end{cases}$$

such that $\underline{A} \leq A \leq \overline{A}, \underline{b} \leq b \leq \overline{b}$ *and* $\underline{p} \leq p \leq \overline{p}$.

A prearrangement *of* S_{IP} *with parameter* ϵ *is a set* X *of integer vectors such that for every IP* $I = (A, b, p)$ *in* S_{IP}, *there exists an* $x \in X$ *which is feasible for* I *and satisfies* $(1 + \epsilon)p \cdot x \geq OPT(I)$.

Range-positiveness and IPs with bounded uncertainties (UIPs) can be defined analogue to Definition 3 and Definition 4. First, we show that the conditions that need to hold to allow for an efficient prearrangement for UIPs are strictly stronger than for ULPs:

Theorem 4. *There exist UIPs which cannot be prearranged efficiently, but whose linear relaxations can.*

Proof: For all $M \in \mathbb{N}$, $M > 1$, consider the following UIP:

$$\max x_1 + 2\epsilon x_2$$

$$s.t. \begin{cases} x_3 \leq b_1 \\ Mx_2 - x_3 \leq b_2 \\ x_1, x_2 \leq 1 \\ x_1, x_2, x_3 \geq 0 \\ x_1, x_2, x_3 \quad integer, \end{cases}$$

where $b_1, b_2 \in [1, \ldots, M - 1]$.

Since only two components can change, the UIPs above have 2-0-bounded uncertainties, of course. Note also, that the constraint matrix A and the objective p (with respect to which the UIP is range-positive) are not allowed to change at all and that the UIP is range-positive wrt b. Consequently, according to Theorem 3, its corresponding linear relaxation ULP can be prearranged efficiently.

Now, for all b_1, b_2 such that $b_1 + b_2 < M$, the optimal objective value is 1. As soon as $b_1 + b_2 = M$ the objective suddenly raises to $1 + 2\epsilon$. The optimal solutions for all the $M - 1$ scenarios where $b_1 + b_2 = M$ are all different from each other and no one solution that is feasible for 2 of them achieves an objective value of more than 1. Consequently, a feasible prearrangement for this class of UIPs must contain at least M solutions, which is exponential in the size of the input.   □

The intractability in the above example is solely caused by infeasibility of solutions that could otherwise represent several scenarios. We achieve our first positive result for UIPs for cases where feasibility is not an issue.

Theorem 5. *Assume we are given* $k \in \mathbb{N}$, $\epsilon > 0$ *and* $S_{BIP} = (\underline{A}, \overline{A}, \underline{b}, \overline{b}, \underline{p}, \overline{p})$ *with* k-ϵ-*bounded uncertainties that is range-positive wrt* p *and for which* $\underline{A} = \overline{A}$ *and* $\underline{b} = \overline{b}$. *Then* S_{BIP} *can be prearranged efficiently in the sense of Theorem 1.*

Proof: We use the same algorithm as in Theorem 1 with the modification that we set $\delta := \epsilon$. Since $\underline{A} = \overline{A}$ and $\underline{b} = \overline{b}$, we have that $I_A = \emptyset$ and $I_b = \emptyset$. Therefore, for each element $(k_i)_{i \in I_p}$, we consider the IP defined by

$$p_i = \begin{cases} \underline{p}_i & \text{if } i \notin I_p \\ \underline{p}_i(1+\delta)^{k_i} & \text{if } i \in I_p \end{cases},$$

$b_i = \underline{b}_i$ and $a_{ij} = \underline{a}_{ij}$. The lookup of x' works in the same way as for ULPs. Since constraint matrix and right hand side are constant, feasibility of the returned solution is not an issue in this case. Regarding the quality of the returned solution, let us denote the optimal solution for P by x. Then, the same inequality as in the last case of Theorem 3 proves the desired accuracy.

Finally, we analyze the efficiency of this algorithm. The set X has size at most $\binom{n}{k} \log^k(C)/\epsilon$. Constructing X amounts to solving $|X|$ LPs from S_{BIP}. The computation of the representing tuple can then be performed in time $O(k \log \log C)$. Then, the total lookup time is in $O(\log(1/\epsilon) + \log(|S_{BIP}|) + k)$ where $|S_{BIP}|$ denotes the input size of the UIP. Consequently, the lookup can be performed in time sublinear in the size of the input after a polynomial number of IPs have been solved in the prearrangement phase. □

Again, note that the size of the prearrangement decreases to $O(\log^k(C)/\epsilon)$ when we know which k parameters in the objective can change more than marginally. We have seen before that even changing only two components in the constraint structure of the problem can yield to intractable prearrangement problems. The interesting question arises whether we can afford any change within A or b at all. The following result answers this question affirmatively.

Theorem 6. *Let S_{BIP} be a UIP with all integer coefficients in the matrix and the right hand side. Assume also that exactly one matrix or right hand side coefficient c can vary: $\underline{c} \leq c \leq \overline{c}$, and such that the optimal solution for the most restricted problem in S_{BIP} has a strictly positive value $\underline{z} > 0$, while the optimal solution to the least restricted problem has value $\overline{z} < \infty$. Then we can compute efficiently a prearrangement X of size $|X| \in O(mn(\log(\overline{z}/\underline{z})/\log(1+\epsilon)))$.*

Proof: The prearrangement algorithm that we propose is quite simple: We try to assemble a sequence of scenarios P_i where P_0 solves the most restricted problem (i.e. when c is a coefficient in b we consider the problem with $c = \underline{c}$, otherwise the one where $c = \overline{c}$). Then, for $i > 0$ we want P_i to be the problem where $\underline{c} \leq c \leq \overline{c}$ is minimal (maximal) and $P_i \geq (1+\epsilon)P_{i-1}$, where c is a coefficient in b (A). We iterate this process until no subsequent P_i can be found anymore. Clearly, the number of problems that we assemble that way is bounded by $mn(\log(\overline{z}/\underline{z})/\log(1+\epsilon))$. Also, this set of solutions is a feasible prearrangement in that an ϵ-approximation for every possible scenario exists in our set. The question remains how we can find the subsequent P_i's efficiently. This is where we exploit the fact that only one coefficient can change: We simply perform a binary search for the critical value of c where the next level in the objective is reached. The total number of problems that need to be solved is then in $O(mn(\log(\overline{z}/\underline{z})\log(\overline{c}-\underline{c})/\log(1+\epsilon)))$, which is polynomial in the size of the input. □

Interestingly, the search for a critical coefficient in b or A where a jump in the objective occurs can be combined with the result that we achieved Theorem 5: for each problem considered in the algorithm for UIPs with changing objectives we can perform a binary search on the one changing component in A or b and therefore prearrange UIPs with bounded uncertainties where at least all but one unknown coefficient are located in the range-positive objective function.

The last result that we report is probably the most important of all, even though it is really a simple implication of our results on ULPs. When looking back at the example where the change of two coefficients yielded an intractable integer prearrangement problem, we find that the gap in the objective value between integer solution and the best solution of the linear relaxation of the problem is quite large. It is straightforward that we can approximate the accuracy defined by that LP-IP gap by using the algorithm from Theorem 1. Formally, we achieve:

Theorem 7. *Assume we are given $k \in \mathbb{N}$, $\epsilon > 0$, and a UIP S_{BIP} with k-$\epsilon/4$-bounded uncertainties where each part (A, b, or p) where potential changes can occur is range-positive. If $g > 0$ is a lower bound on the minimal ratio between an IP in S_{BIP} and its linear continuous relaxation, then we can efficiently compute a prearrangement that achieves an approximation guarantee of $(1 + \epsilon)/g$.*

4   Practical Considerations

In this final section, we would like to make a few comments on the practicability of the algorithms presented here. While at the end of the discussion on ULPs we found that, if solving the problem is too easy, in many practical applications it may not be necessary at all to plan ahead. After reading our last section that tackles UIPs, one may now argue that solving a polynomial number of NP-hard problems in the prearrangement phase may be asked too much. Just like humans usually only plan ahead in this manner when a lot is at stake, we will probably only want to go through this hassle when having a near optimal solution in some uncertain environment is critical. Or, for example, when we design a system that needs to work well in changing environments and where the time that is spent to "prepare" the system (for example before the software is launched) is much less important than a good performance later.

While the work presented here is theoretical in nature, we would like to note that what we really did was to analyze and adapt a strategy that is used in practice not only by human beings, but also by existing computer systems, for instance for parametric query optimization in distributed data base systems as cited in [10]. Therefore, prearrangement must be viewed as a method that is used in practice — but to the best of our knowledge it has not been analyzed theoretically before. Our analysis revealed some sufficient conditions under which efficient prearrangement is possible at all. Moreover, we give a strategy how the scenarios for which one should plan ahead ought to be selected.

We discussed already how knowledge about the concrete non-marginal, uncertain parameters can be leveraged to decrease the size of a prearrangement significantly. Note that, in practice, we can further reduce the computational effort needed by compromising on the solution quality. By reversing the arguments in this paper, if a limit on the maximum size of the prearrangement is given, our method also allows to pick scenarios so that the worst-case error made is minimized. For instance, in the introductory example of the delivery company, we may find that there is time to compute 120 problems on the four computers in their office. Then, our method can be used to decide for which scenarios we should prepare a schedule. Moreover, when stochastic information is available which and how coefficients in the problem can differ, then by sampling over the parameters $((k_i)_{i \in I_p}, (\ell_i)_{i \in I_b}, (m_{ij})_{(i,j) \in I_A})$ we can precompute a set of solutions that maximizes the likelihood that an ϵ-approximate solution will be available.

For UIPs, another possible scenario is that an approximation algorithm for the problem at hand is available. Then we can precompute approximate solutions rather than solving each IP to optimality. Our prearrangement algorithm will then approximate with the performance guarantee of the approximation algorithm that is being used plus the additional error caused by the LP-IP gap. Even more, if the analysis of the approximation algorithm itself is based on the LP-IP gap then we even achieve that same approximation guarantee through prearrangement, even if only approximate solutions have been precomputed!

Furthermore, our method is naturally parallelizable and it also works when the LP-IP gap is not known. That way, if we have a complex problem to solve for which we believe that the LP provides a good upper bound, then we have a method at hand that prepares us for the future in a very reasonable manner.

As a final remark, it is a matter of future work to determine whether linear programming sensitivity analysis could allow us to significantly reduce the work that needs to be conducted for prearrangement. In practice one would expect to reduce the computational effort considerably by realizing that many parts of an LP can change significantly before affecting the optimal solution at all.[1] It is questionable, however, whether this important practical improvement could actually result in better worst case guarantees like the ones that were the target of this study.

5 Conclusions

We investigated one of the most common ways how humans cope with uncertainty, namely by the provision of backup plans. While this approach is frequently being used in time-critical computer systems, to the best of our knowledge no theoretical analysis has been provided before. Our assumption is that there exists an extensive time period before the day of operation where the problem is still associated with uncertainty. Then, when the real problem that needs to be accommodated is revealed to us, there is very little time available before a solution needs to realized.

Within this scenario, we studied uncertain linear and integer linear programs. We showed that bounding the number of uncertain coefficients is essential to allow for an efficient prearrangement, whereby we do not necessarily need to know exactly which coefficients will change more than marginally while this knowledge can be used to reduce the computational efforts. For uncertain linear programs, we specified a method for choosing a polynomial number of scenarios whose pre-solution guarantees that an ϵ-approximation of the real problem will be available. For this method, it is essential

---

[1] We owe this observation to an anonymous referee.

that every part of the LP (constraint matrix, right hand side, or objective) that is subject to potential changes is range-positive. For uncertain integer programs, we found analogous results as for LPs when only the range-positive objective function is subject to changes or when the desired approximation guarantee is not larger than the LP-IP gap. Especially when it is known in advance which uncertain parameters can change more than just marginally, our method yields to prearrangements of very manageable size.

There are some open questions: It can be shown that our method breaks down when value ranges include 0 or when both negative and positive ranges are mixed within the same part of the LP. Is there another way of choosing scenarios that can accommodate these cases? Are there other conditions under which an efficient prearrangement of uncertain integer programs is possible?

References

1. I. Althoefer, F. Berger, S. Schwarz. Generating True Alternatives with a Penalty Method. http://www.minet.uni-jena.de/MathNet/reports/shadows/-02-04report.html, 2002.
2. A. Ben-Tal, A. Nemirovski. Robust Optimization - Methodology and Applications. *Mathematical Programming Series B*, 92:453–480, 2002.
3. J. Birge, F. Louveaux. Introduction to Stochastic Programming. *Springer*, 1997.
4. A. Borodin, R. El-Yaniv. Online Computation and Competitive Analysis. *Cambridge University Press*, 1998.
5. J.W. Chinneck and K. Ramadan. Linear programming with interval coefficients. *Journal of the Operational Research Society*, 51(2):209–220, 2000.
6. K. Elbassioni, I. Katriel. Multiconsistency and Robustness with Global Constraints. *CP-AI-OR*, 2005.
7. D. Eppstein. Bibliography on k shortest paths and other "k best solutions" problems. http://www.ics.uci.edu/~eppstein/bibs/kpath.bib, 2001.
8. C. Guestrin, D. Koller, C. Gearhart, N. Kanodia. Generalizing Plans to New Environments in Relational MDPs. *IJCAI*, 2003.
9. E. Hebrard, B. Hnich, T. Walsh. Super Solutions in Constraint Programming. *CP-AI-OR*, LNCS 3011:157–172, 2004.
10. A. Hulgeri, S. Sudarshan. Parametric Query Optimization for Linear and Piecewise Linear Cost Functions. *VLDB*, 167–178, 2002.
11. M. Lagoudakis, R. Parr. Least-Squares Policy Iteration. *Journal of Machine Learning Research*, 4:1107–1149, 2003.
12. I. Sameith. On the Generation of Alternative Solutions for Discrete Optimization Problems with Uncertain Data. http://www.minet.uni-jena.de/MathNet/reports/-shadows/04-01report.html, 2004.
13. B. Verweij, S. Ahmed, A. Kleywegt, G. Nemhauser, A. Shapiro. The Sample Average Approximation Method Applied to Stochastic Routing Problems: A Computational Study. *Computational Optimization and Applications*, 24(2-3):289–333, 2003.
14. T. Vossen, M. Ball, A. Lotem, D.S. Nau. On the Use of Integer Programming Models in AI Planning. *IJCAI*, 1999.
15. N. Yorke-Smith. Reliable Constraint Reasoning with Uncertain Data. PhD thesis, IC-Parc, Imperial College London, June 2004.
16. N. Yorke-Smith and C. Gervet. Tight and Tractable Reformulations for Uncertain CSPs. Proceedings of CP'04 Workshop on Modelling and Reformulating Constraint Satisfaction Problems, Toronto, Canada, September 2004.

Progressive Solutions: A Simple but Efficient Dominance Rule for Practical RCPSP

András Kovács[1,2] and József Váncza[1]

[1] Computer and Automation Research Institute,
Hungarian Academy of Sciences
[2] Cork Constraint Computation Centre,
University College Cork, Ireland
{akovacs, vancza}@sztaki.hu

Abstract. This paper addresses the solution of practical resource-constrained project scheduling problems (RCPSP). We point out that such problems often contain many, in a sense similar projects, and this characteristic can be exploited well to improve the performance of current constraint-based solvers on these problems. For that purpose, we define the straightforward but generic notion of progressive solution, in which the order of corresponding tasks of similar projects is deduced a priori. We prove that the search space can be reduced to progressive solutions. Computational experiments on two different sets of industrial problem instances are also presented.

1 Introduction

The practical value of constraint-based scheduling hinges both on the representation power of the models and the efficiency of the solution techniques. Solution performance, in turn, depends on whether the solver can recognize and take advantage of the structural properties of the problem at hand.

Generic models, though different (like flow shop, job shop, resource-constrained project scheduling, etc.), hide some eventual structural properties of specific real-life problem instances. Making such properties explicit by adding extra features to the model is an option, but it comes together also with specialized solution techniques. Just to the contrary, in this paper we suggest to detect and exploit some hidden structural properties within the boundaries of a generic model. We introduce the simple notion of *progressive pairs* to characterize similar patterns of activities of a scheduling problem. Similarity will be defined both in terms of temporal relations and resource requirements. Typically, progressive pairs are inherent in practical discrete manufacturing problems where products or components of similar/same type are produced in parallel, by using the same technology and a common pool of resources.

We take the classical model of resource-constrained project scheduling problem (RCPSP) [3], and demonstrate our approach on the objective of minimizing

J.C. Beck and B.M. Smith (Eds.): CPAIOR 2006, LNCS 3990, pp. 139–151, 2006.

the makespan.[1] When detecting and exploiting progressive pairs, we rely on no extra domain-specific information.

The results presented here are based on our previous works that suggested the application of *consistency preserving* transformations to exploit some structural properties of constraint programs [7]. Earlier experiments with a combination of symmetry breaking techniques and so-called freely completable solutions convinced us that the performance of generic constraint-based methods can considerably be improved on *practical* problem instances. Now we take a more general approach to rule out dominated solutions by constraints added *before* the search process.

In the sequel we give an overview of relevant works related to symmetry breaking and the application of dominance rules. Following the definition of the RCPSP model, Sect. 4 presents the idea of progressive solutions together with the basic definitions, theorems and proofs. Sect. 5 describes how we detect this structural property among the projects in an RCPSP instance, while Sect. 6 summarizes the results of our experiments on two industrial data sets. Finally, conclusions are drawn.

2   Related Work

Recently, considerable efforts have been made to explore various classes of consistency preserving transformations in constraint programming. These transformations reduce the search space while ensuring that at least one (optimal) solution remains, if the original problem was solvable. Hence, they essentially extend the traditional toolbox of constraint programmers that mostly consists of equivalence preserving transformations. Such transformations – like constraint propagation or shaving – guarantee that the original and the transformed problems have exactly the same set of solutions.

The most intensively studied branch of consistency preserving transformations is doubtlessly *symmetry breaking*. *Symmetry* is a bijective function f defined on the bindings of the variables such that for each variable binding α, $f(\alpha)$ is a solution iff α is a solution, too. Breaking this symmetry means excluding all but one of the symmetric equivalents. The foremost of all symmetry breaking techniques is the addition of symmetry breaking constraints to the model *before* *search*. More sophisticated methods, such as the Symmetry Breaking During Search (also called Symmetry Excluding Search) and the Symmetry Breaking via Dominance Detection prune symmetric branches of the search tree *during* *search*. All of these general frameworks require an explicit declaration of the symmetries in the form of symmetry functions or a dominance checker. See [11] for a recent overview of symmetry breaking techniques.

---

[1] Note that while this objective is often criticized by practitioners, it really helps to squeeze a given amount of work into a pre-defined time frame. In a hierarchical production planning and scheduling setting, where the primary goal of scheduling is to generate an executable solution that complies with a segment of the production plan, makespan is a useful criterion [13].

A wider class of consistency preserving transformations is constituted by the *dominance rules*. They define properties of a problem that must be satisfied by at least one of its (optimal) solutions. By now, little work has been done to apply dominance rules in constraint programming. The recent paper [12] calls the attention to the application of dominance rules and defines novel dominance rules for three different problems.

At the same time, dominance rules are widely used in operations research and project scheduling. For instance, dominance rules of different strength and computational complexity are described for RCPSP with the criteria of minimal makespan in [3]. Dominance rules, as well as methods for the insertion of redundant precedence constraints are proposed for the problem of minimizing the number of late jobs on a single machine in [2].

3   Notations

Below, we define progressive solutions for resource-constrained project scheduling problems with the criterion of minimizing makespan, and give an outlook on possible extensions at the end of the paper. Hence, let T denote a set of non-preemptive *tasks*. Each task $t \in T$ has a fixed *duration* d_t, and requires ρ_t^r units of each renewable cumulative *resource* $r \in R$ during the whole length of its execution. The number of available units of resource r at a time, i.e., the *capacity* of r is denoted by q_r. Tasks can be connected by end-to-start *precedence constraints* $(t_1 \to t_2)$ that state that task t_1 must end before the start of task t_2. We assume that there is no directed circle in the graph of precedences.

Then, the objective is to find non-negative start times $start_t$ for the task $t \in T$, such that all precedence and resource constraints are satisfied, and the *makespan*, i.e., the maximum of the end times $end_t = start_t + d_t$ is minimal.

Although they are not allowed in the original problem definition, we will use the notion of *start-to-start precedences* as well, denoted by $(t_1 \dashrightarrow t_2)$, meaning that $start_{t_1} \leq start_{t_2}$. For brevity, we call the maximal sets of tasks connected by precedence constraints *projects*.

4   Progressive Solutions of Scheduling Problems

Factories often produce several pieces of the same product, or products belonging to the same product family during their short-term scheduling horizon. As a consequence, their detailed scheduling problems may include many, in a sense similar projects. This chapter is devoted to show that in such cases, a valid ordering of tasks belonging to similar projects can be deduced by off-line inference. These investigations will allow us to insert precedence constraints in the constraint-based model of the scheduling problem *before search*, and hence, to reduce the search space.

4.1   The Underlying Idea

As a simple example, suppose that two identical projects, P and Q are to be executed within the scheduling horizon (besides arbitrary other projects). By

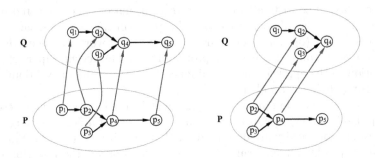

Fig. 1. a) and b) Examples of similar projects in the scheduling problem

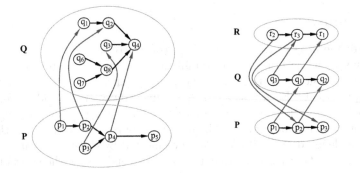

Fig. 2. a) and b) Two more examples of similar projects in the scheduling problem

identical, we mean that for each index i, tasks p_i and q_i in Fig. 1.a. (and in all subsequent figures) have equal durations and resource requirements. Now, it is easy to see that if there exists a solution to this scheduling problem, then there exists one in which each task of P precedes its corresponding task in Q.

Now, assume that some tasks belonging to project P have already been executed before the start of the current scheduling horizon, hence, there is no match of q_1 in P. Similarly, some tasks of Q suffice to be done after the end of the horizon, resulting in no corresponding task for p_5 in Q. Again, tasks of P can precede tasks of Q, see Fig. 1.b. The third example in Fig. 2.a. depicts a case where P and Q are different members of the same product family. P requires an additional finishing operation (p_5) that has no match in Q, while there is an extra component built in Q (q_6, q_7, q_8). Finally, Fig. 2.b. has a theoretical significance, since it shows an example where the inferred precedence constraints form directed circles between projects, but not between tasks.

Below, we formally define our notion of similarity between projects and present how all this makes possible the reduction of the solution space.

4.2   Progressive Solutions

Definition 1. *Two sets of tasks P and Q are defined isomorphic, and will be denoted by $P \equiv Q$, iff there exists a bijection $\beta : P \leftrightarrow Q$ such that for each pair of tasks $p \in P$ and $q \in Q$*

$$
\begin{aligned}
\beta(p,q) &\Rightarrow \forall r \in R : \rho_p^r = \rho_q^r \ \wedge \ d_p = d_q, \ and \\
\beta(p_1,q_1) \wedge \beta(p_2,q_2) &\Rightarrow (p_1 \to p_2) \Leftrightarrow (q_1 \to q_2).
\end{aligned}
$$

Definition 2. *Given two projects P and Q, we call them a progressive pair iff there exists a $P^* \subseteq P$ and a $Q^* \subseteq Q$ such that $P^* \equiv Q^*$, and there are no incoming precedences to P^* and no outgoing precedences from Q^*. This relation will be denoted by $P \rightrightarrows Q$ (see Fig. 3).*

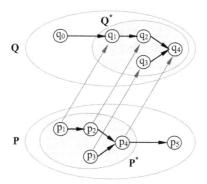

Fig. 3. The progressive pair $P \rightrightarrows Q$

Furthermore, to avoid the ambiguous situations where $P \rightrightarrows Q$ and $P \leftleftarrows Q$ hold simultaneously for two isomorphic projects $P \equiv Q$, we label the projects by unique identifiers $L(.)$. Now, we say that two isomorphic projects constitute a progressive pair $P \rightrightarrows Q$ only if $L(P) < L(Q)$.

Definition 3. *A solution of a scheduling problem is called progressive, iff for each progressive pair $P \rightrightarrows Q$, the execution of P precedes Q, in the formal sense that for each pair of tasks $p \in P^*$ and $q \in Q^*$ such that $\beta(p,q)$, $p \dashrightarrow q$ holds. We will refer to this type of start-to-start precedence constraints as progressive constraints.*

Note that if at least one of the resources required by p and q is unary, then $(p \dashrightarrow q) \Leftrightarrow (p \to q)$.

Theorem 1. *If an RCPSP problem has a solution, then it also has a progressive solution with minimal makespan.*

We start the proof by the following simple lemma.

Lemma 1. *Given an RCPSP problem with no directed circles of precedence constraints, the insertion of progressive precedence constraints does not create a directed circle of (end-to-start and start-to-start) precedences between the tasks.*

Proof: Let us label the tasks $t \in T$ by $l(t) = |Pred(t)| - |Succ(t)|$, where $Pred(t)$ and $Succ(t)$ are the sets of predecessors and successors (direct and indirect) of t in the original problem, respectively. Notice that $l(t_1) < l(t_2)$ holds for all precedences $(t_1 \to t_2)$ in the original problem, and $l(t_1) \le l(t_2)$ for all the inserted progressive constraints $(t_1 \dashrightarrow t_2)$.

Now, let us assume that there is a directed circle of precedences C. According to the above, C consists of progressive constraints only, with $l(t_1) = l(t_2)$. Then, by the definition of the progressive pairs, $Pred(t_1) \equiv Pred(t_2)$ and $Succ(t_1) \equiv Succ(t_2)$. This also implies that all the projects traversed by C are isomorphic. It is a contradiction, because by definition $L(P) < L(Q)$ must hold for each subsequent pair of projects P and Q traversed by C.   □

Now, we prove Theorem 1 by an algorithm that departs from an arbitrary optimal solution, and through iteratively swapping pairs of tasks, generates a progressive solution with the same makespan. In each step of the algorithm, a progressive pair $P \rightrightarrows Q$ is selected, such that some of the progressive constraints between P and Q are violated in the actual schedule S. Then, the algorithm computes a modified schedule S' by swapping all the pairs of tasks in P and Q which violate the progressive constraints as follows.

$$\forall p \in P, q \in Q : \beta(p,q) \wedge start_p^S > start_q^S \;\; \Rightarrow \;\; start_p^{S'} = start_q^S, \text{ and} \\ start_q^{S'} = start_p^S.$$

For all other tasks $t \in T$, $start_t^{S'} = start_t^S$.

Lemma 2. *S' is feasible, and its makespan equals the makespan of S.*

Proof: All resource capacity constraints are satisfied in S', because only pairs of tasks with equal durations and resource requirements were swapped. In order to show that precedence constraints $p_1 \to p_2$, where $p_1, p_2 \in P^*$, cannot be violated in S' either, we introduce q_1 and q_2 to denote the two tasks in Q for which $\beta(p_1, q_1)$ and $\beta(p_2, q_2)$ hold. Then,

- if neither the pair (p_1, q_1), nor (p_2, q_2) were swapped, then the start times of p_1 and p_2 are unchanged in S' w.r.t. S, and S is feasible;
- If the pair (p_1, q_1) was swapped, but (p_2, q_2) not, then
  $end_{p_1}^{S'} = end_{q_1}^S < end_{p_1}^S \le start_{p_2}^S = start_{p_2}^{S'}$;
- If the pair (p_2, q_2) was swapped, but (p_1, q_1) not, then
  $end_{p_1}^{S'} = end_{p_1}^S \le end_{q_1}^S \le start_{q_2}^S = start_{p_2}^{S'}$;
- If both (p_1, q_1) and (p_2, q_2) were swapped, then
  $end_{p_1}^{S'} = end_{q_1}^S \le start_{q_2}^S = start_{p_2}^{S'}$.

Precedence constraints pointing from P^* to $P \setminus P^*$ and those within $P \setminus P^*$ are also satisfied, because only tasks of P^* were moved earlier, and tasks of Q^*

later in the schedule. The proof is analogous for precedence constraints in Q, and trivial for the precedence constraints between tasks of $T \setminus (P \cup Q)$, because those tasks were not moved.   □

The above step is iterated until there are no more progressive constraints violated.

Proof of Theorem 1: The algorithm halts when it has found a progressive schedule. According to Lemma 2, this schedule is feasible, and has an optimal makespan. Furthermore, this is reached in finitely many steps, because the algorithm performs a brick sort over the tasks, according to the partial ordering defined by the progressive constraints.   □

5   Computing the Progressive Pairs

Computing the progressive pairs in essence requires a pairwise comparison of the projects, and checking whether they have appropriate isomorphic subsets of tasks. The computational efficiency of these algorithms is of special importance, because no polynomial-time algorithm is known for deciding whether two general graphs are isomorphic [5]. Furthermore, we did not find a generic way to decrease the number of necessary isomorphism tests below $O(n^2)$.

Despite all this, our experiments suggest that in practical cases, PPs can be computed fast enough. On the one hand, efficient graph isomorphism algorithms are known for many classes of graphs, including trees [1, pp. 84–86], planar graphs, and graphs of bounded valence [4, 5, 8]. The algorithms also return a matching. In fact, the graph structures we encountered in our recent industrial applications belonged to the class of in-trees, in the most general case. On the other hand, often some kind of meta-knowledge about the projects (drawing numbers of parts and assemblies, product family codes, etc.) can be exploited, too.

In our pilot system, we assumed that the precedence graphs of projects form in-trees. For each pair of projects P and Q, and for each task $p \in P$, we took the sub-tree P^p of the precedence tree of P rooted at the node corresponding to p. We checked if P^p is isomorphic with a sub-tree of Q also containing the root of Q. Clearly, a positive answer of the isomorphism test means that a progressive pair $P \rightrightarrows Q$ has been found with $P^* \equiv P^p$ (see procedures `ComputeProgressivePairs` and `CheckIfProgressive` in Fig. 4.). We implemented a simple depth-first search to perform the isomorphism tests (procedure `TryMatching`). Note that `TryMatching` considers only the precedence constraints present in the original problem formulation, but not the progressive constraints inserted beforehand.

Although this algorithm has an exponential worst-case complexity, it proved efficient enough for even the largest practical instances we tackled (see the next section for details). This was possible because the vast majority of the isomorphism tests could return false immediately, due to the different durations or resource requirements of tasks in the roots of the examined sub-trees.

```
1 PROCEDURE ComputeProgressivePairs()
2   FORALL project P
3     FORALL project Q : Q ≠ P
4       CheckIfProgressive(P, Q)

5 PROCEDURE CheckIfProgressive(P, Q)
6   q := last task of Q
7   FORALL task p ∈ P, ordered by the increasing distance of p
-           from the root of P
8     IF NOT ((p is the last task in P) AND (L(P) > L(Q))) THEN
9       IF p ≡ q THEN
10        M := TryMatching(Pᵖ, Q, {< p, q >})
11        IF M ≠ ∅ THEN
12          Add progressive pair P ⇉ Q with matching M
13          RETURN

14 PROCEDURE TryMatching(P, Q, M)
15   p, p₀ := Select a pair of task from P such that (p → p₀),
-            p has no match in M, but p₀ has a match in M
16   IF there is no such p, p₀
17     RETURN M
18   q₀ := Match of p₀ in M
19   FORALL q : (q → q₀) AND q has no match in M
20     IF p ≡ q THEN
21       M' := TryMatching(P, Q, M ∪ {< p, q >})
22       IF M' ≠ ∅
23         RETURN M'
24   RETURN ∅
```

Fig. 4. An algorithm for computing the progressive pairs when projects form in-trees

6   Experiments

We performed computational experiments on two different sets of industrial problems with two purposes. First, to estimate to what extent the tasks in a scheduling problem can be ordered by off-line inference, and second, to measure how much the inferred precedence constraints can speed up the solution process.

Our first set of data derives from an industrial partner that manufactures mechanical parts of high complexity for the energy industry. Their products can be ordered into four product families. Members of the same family share a similar structure, but differ in various parameters. The overall number of different end products is ca. 40, but this number may grow in the future. A project, aimed at the fabrication of one end product, consists of up to a few hundred tasks. Since the bill of materials of the products are tree-structured, the precedence relations within a project also form an in-tree. Tasks require one unit of a machine resource (unary or cumulative) and one unit of a human resource (cumulative) for their execution. There are altogether ca. 100 different resources in the plant. For more details on this scheduling problem, the readers are referred to [6].

The other set of problem instanced originates from the ILOG MascLib library [10]. MascLib contains industrial and generated benchmarks classified according to the complexity of the scheduling model. For our experiments, we used the *No-Calendar General Shop* (NCGS) problem class. Although the authors of the library suggest the usage of more realistic criteria – combinations of non-performance, earliness/tardiness, setup and mode costs – we simplified these instances to standard job-shop problems and minimized makespan. Therefore, we disregarded the due times of the tasks and the option of not performing them, while preserving their durations, resource requirements, and the precedence relations. The library contains 26 NCGS instances, but 13 of them differ only in the tardiness and non-performance costs from the others, and 7 others do not contain progressive pairs at all – we believe these were the generated instances. Hence, we performed experiments on the 6 remaining instances.[2]

As the first step of the experiments, we detected the progressive pairs in the problem instances. The results are presented in Table 1. The first group of rows stands for the instances from the industrial partner, while the second group for the NCGS instances. Columns *Tasks*, *Projects*, and *Resources* give information about the size of the problem instance, while column *PPs* indicate the number of progressive pairs of projects found. *EtS* and *StS* displays the number of inferred end-to-start and start-to-start progressive precedence constraints. The last two columns contain the *order strength (OS)* [9] without (OS^-) and with (OS^+) the inferred constraints. *OS* was calculated as the number of precedence constraints within one resource, divided by the number of task pairs competing for a resource, i.e.,

$$OS = \frac{\sum_{r \in R} |Prec_r|}{\sum_{r \in R} \frac{|T_r|(|T_r|-1)}{2}},$$

where $T_r = |t \in T : \rho_t^r \geq 1|$ and $Prec_r = \{(t_1 \rightarrow t_2) \text{ or } (t_1 \dashrightarrow t_2) : t_1, t_2 \in T_r\}$.[3] Hence, OS is 0 when there are no ordering constraints within the resources, and 1 if the tasks are completely ordered. The results show that in all the instances, the inferred progressive constraints could considerably reduce the search space. In the case of the NCGS instances, all the inserted precedence constraints were of the end-to-start type, since these problems contained unary machines only. The time needed to find the progressive pairs did not exceed 1 second even for the largest problem instances.

In order to measure the effect of the inferred ordering decisions on algorithm performance, we fed these instances into ILOG Scheduler 5.1. We used ILOG's

---

[2] We also experimented with a third set of data that came from the automotive industry. These were job-shop problems with ca. 50 unary resources and hundreds of projects, each containing at most 6 sequentially ordered tasks. Progressive pairs could be found in these instances as well, but the resource loads were so unbalanced that it was easy to find optimal solutions even without the progressive constraints.

[3] Including all the edges in the transitive closure of the precedence graph.

Table 1. Order strength without and with progressive constraints

	Tasks	Projects	Resources	PPs	EtS	StS	OS^-	OS^+
p1	3511	97	95	308	17605	864	0.011	0.090
p2	2767	80	95	242	10674	641	0.014	0.093
p3	1470	70	95	132	2729	273	0.030	0.105
p4	1753	80	95	148	2833	225	0.025	0.077
p5	2472	89	95	196	3389	339	0.017	0.050
p6	2570	91	95	181	3653	323	0.019	0.058
p7	1133	70	95	134	1495	212	0.068	0.227
p8	769	68	95	122	293	248	0.052	0.084
p9	1620	85	95	160	2723	299	0.027	0.094
p10	1677	71	95	156	491	348	0.024	0.033
p11	1471	69	95	129	337	277	0.026	0.034
p12	585	71	95	143	232	221	0.032	0.069
p13	1786	83	95	187	2918	353	0.024	0.082
p14	1240	72	95	220	1570	230	0.067	0.201
p15	947	45	95	92	1223	97	0.088	0.269
NCGS_21	60	16	5	39	147	-	0.000	0.377
NCGS_31	75	19	5	42	162	-	0.000	0.277
NCGS_54	260	45	10	476	1783	-	0.007	0.399
NCGS_55	260	45	10	588	1921	-	0.007	0.427
NCGS_75	1250	41	30	40	1600	-	0.042	0.128
NCGS_81	2500	72	30	302	12210	-	0.022	0.184

default *branch-and-bound* search with the *setting times* branching strategy and the *edge-finding* algorithm for the propagation of resource constraints. For all instances, the solution process was stopped when the optimality of a solution was proven, or after 600 seconds passed without improvement. In the latter case, we computed a lower bound by pure constraint propagation. The tests were run on a 1.6 GHz Pentium IV computer under Windows 2000 operating system.

The results achieved without and with the presence of the progressive constraints are shown in Table 2, where each row stands for one problem instance. *UB* and *LB* stand for the best found upper and lower bounds, respectively. *Error* was calculated as $(UB - LB)/LB$, while '-' denotes optimality. Values displayed in columns *Nodes* and *Time* were measured only until the best solution was found. The solver often generated significantly more search nodes until the timeout, but displaying those figures in the table would not be informative.

The figures show that progressive constraints facilitated both finding better solutions and proving tighter lower bounds. While improved lower bounds and better pruning is an evident outcome of a tighter formulation, the presence of the progressive constraints also had a positive effect on the branching heuristic. This is clearly shown by better first solutions for 10 of the 21 problem instances. All in all, the addition of the progressive constraints decreased the gap between the solutions found and the lower bounds by 60% on average, and made

Table 2. Effect on algorithm performance

	Tasks	Without PP					With PP				
		UB	LB	Error (%)	Nodes	Time (sec)	UB	LB	Error (%)	Nodes	Time (sec)
p1	3511	372	341	9.09	3511	651	345	342	0.88	3511	646
p2	2767	289	247	17.00	2767	29	251	248	1.21	2767	29
p3	1470	276	276	-	1470	11	276	276	-	1470	15
p4	1753	252	252	-	19719	1891	252	252	-	7347	553
p5	2472	290	276	5.07	2472	21	276	276	-	2472	36
p6	2570	269	230	16.96	2570	25	254	230	10.43	2570	23
p7	1133	254	254	-	5686	148	254	254	-	1133	8
p8	769	264	264	-	2309	31	264	264	-	770	2
p9	1620	343	334	2.69	1620	9	343	334	2.69	3240	328
p10	1677	304	284	7.04	1677	9	295	284	3.87	6924	406
p11	1471	320	307	4.23	3450	286	310	307	0.98	4469	273
p12	585	358	349	2.58	585	1	356	349	2.01	585	1
p13	1786	370	366	1.09	4073	483	373	366	1.91	5409	475
p14	1240	383	373	2.68	5504	147	376	373	0.80	4885	541
p15	947	233	222	4.95	5232	107	232	222	4.50	3267	77
NCGS_21	60	2872	2799	2.61	9304	0	2872	2854	0.63	13469	0
NCGS_31	75	3412	3339	2.19	9552	0	3348	3348	-	10235	1
NCGS_54	260	1105	1105	-	260	0	1105	1105	-	260	0
NCGS_55	260	975	975	-	260	0	975	975	-	260	0
NCGS_75	1250	1164	1028	13.23	1200404	1310	1122	1044	7.47	17649	82
NCGS_81	2500	2220	1902	16.72	5956417	20479	2014	1902	5.89	5001	234

possible finding optimal solutions for two previously unsolvable instances (p5 and NCGS_31).

At the same time, the presence of progressive constraints had a negative impact on the solution process in the case of two instances: p9, where equivalent solutions were found, but with less search without the progressive constraints, and p13, where better solution could be constructed without them.[4] This effect is caused by the adverse interaction of the inserted constraints with the search strategy, an unfavorable phenomenon well known from the literature of symmetry breaking [11].

7   Conclusions

In this work we focused on the solution efficiency of constraint-based scheduling on practical RCPSP instances. For that purpose, we suggested a method for

---

[4] There is an apparent disproportion between search nodes and time for p9 and p10: the number of nodes doubles while time multiplies by ca. 40. This is due to the fast processing of nodes until a first solution is found (where the number of nodes equals the number of tasks), and heavier computation later, with a valid upper bound.

detecting progressive pairs, and transforming the original problem into a tighter constrained formulation by the application of a novel dominance rule. It was proven that the proposed transformation preserves the consistency of the original problem.

Our hypothesis was that practical scheduling problems do have components with inherent temporal and resource-related similarities. The experiments confirmed that the simple but generic notion of progressive solutions is appropriate to capture this structural property. Further on, applying progressive constraints made the scheduling problems almost in each case easier to solve.

The proposed method naturally extends to richer scheduling models, including earliest start and latest finish times, setup times, or various other criteria, such as the minimization of tardiness costs. Finally, one has still to investigate if the harmful interaction of the progressive constraints and the search heuristic can be eliminated, likewise it is done by advanced techniques of symmetry breaking.

Acknowledgement

This research has been supported by the grants NKFP 2/010/2004 and OTKA T046509.

References

1. A.V. Aho, J.E. Hopcroft, and J.D. Ullman. The Design and Analysis of Computer Algorithms. Addison-Wesley, 1974.
2. Ph. Baptiste, L. Peridy, and E. Pinson. A Branch and Bound to Minimize the Number of Late Jobs on a Single Machine with Release Time Constraints. European Journal of Operational Research, 144(1), pp. 1–11, 2003.
3. E.L. Demeulemeester and W.S. Herroelen. Project Scheduling: A Research Handbook. Kluwer Academic Publishers, 2002.
4. J.E. Hopcroft and R.E. Tarjan. A V^2 Algorithm for Determining Isomorphism of Planar Graphs. Information Processing Letters 1, pp. 32–34, 1971.
5. B. Jenner, J. Köbler, P. McKenzie, and J. Torán. Completeness Results for Graph Isomorphism. Journal of Computer and System Sciences 66(3), pp. 549–566, 2003.
6. A. Kovács. Novel Models and Algorithms for Integrated Production Planning and Scheduling. PhD Thesis, Budapest University of Technology and Economics, 2005. http://www.sztaki.hu/~akovacs/thesis/
7. A. Kovács and J. Váncza. Completable Partial Solutions in Constraint Programming and Constraint-based Scheduling. In Proc. of the 10th International Conference on Principles and Practice of Constraint Programming, Springer LNCS 3258, pp. 332–346, 2004.
8. E. Luks. Isomorphism of Bounded Valence Can Be Tested in Polynomial Time. Journal of Computer and System Sciences 25, pp. 42–46, 1982.
9. A.A. Mastor. An Experimental and Comparative Evaluation of Production Line Balancing Techniques. Management Science 16, pp. 728–746, 1970.
10. W. Nuijten, T. Bousonville, F. Focacci, D. Godard, and C. Le Pape. Towards an Industrial Manufacturing Scheduling Problem and Test Bed. In Proc. of the 9th Int. Conf. on Project Management and Scheduling, pp. 162–165, 2004.

11. K.E. Petrie and B.M. Smith. Comparison of Symmetry Breaking Methods in Constraint Programming. In Proc. of the 5th International Workshop on Symmetry and Constraint Satisfaction Problems, 2005.
12. S.D. Prestwich and J.C. Beck. Exploiting Dominance in Three Symmetric Problems. In Proc. of the 4th International Workshop on Symmetry and Constraint Satisfaction Problems, pp. 63–70, 2004.
13. J. Váncza, T. Kis, and A. Kovács. Aggregation – The Key to Integrating Production Planning and Scheduling. CIRP Annals – Manufacturing Technology 53(1), pp. 377–380, 2004.

AND/OR Branch-and-Bound Search for Pure 0/1 Integer Linear Programming Problems

Radu Marinescu and Rina Dechter

School of Information and Computer Science
University of California, Irvine, CA 92697-3425
{radum, dechter}@ics.uci.edu

Abstract. *AND/OR search spaces* have recently been introduced as a unifying paradigm for advanced algorithmic schemes for graphical models. The main virtue of this representation is its sensitivity to the structure of the model, which can translate into exponential time savings for search algorithms. In this paper we extend the recently introduced AND/OR Branch-and-Bound algorithm (AOBB) [1] for solving pure 0/1 Integer Linear Programs [2]. Since the variable selection can have a dramatic impact on search performance, we introduce a new dynamic AND/OR Branch-and-Bound algorithm able to accommodate variable ordering heuristics. The effectiveness of the dynamic AND/OR approach is demonstrated on a variety of benchmarks for pure 0/1 integer programming, including instances from the MIPLIB library, real-world combinatorial auctions and random uncapacitated warehouse location problems.

1 Introduction

A *constraint optimization problem* is the minimization/maximization of an objective function subject to a set of constraints on the possible values of a set of independent decision variables. An important class of constraint optimization problems are the Integer Linear Programming problems (ILP) [2] where the objective is to optimize a linear function of integer-valued variables, subject to a set of linear equality or inequality constraints defined on subsets of variables. The classical approach to solving ILPs is the *branch-and-bound* method [3] which maintains the best solution found so far, while discarding partial solutions which cannot improve on the best.

The AND/OR search space for graphical models [4] is a newly introduced framework for search that is sensitive to the independencies in the model, often resulting in exponentially reduced complexities. It is based on a pseudo-tree that captures independencies in the graphical model, resulting in a search tree exponential in the depth of the pseudo-tree, rather than in the number of variables.

The AND/OR Branch-and-Bound algorithm (AOBB) is a new search method that explores the AND/OR search tree for solving optimization tasks in graphical models [1]. In this paper we present an extension of the algorithm for solving optimization problems from the class of pure 0/1 Integer Linear Programs [2]. A pure 0/1 integer linear program is a linear program where all the decision variables are restricted to be either 0 or 1 at the optimal solution.

J.C. Beck and B.M. Smith (Eds.): CPAIOR 2006, LNCS 3990, pp. 152–166, 2006.

Since variable selection can have a dramatic impact on search performance [2], we introduce a *dynamic* AND/OR Branch-and-Bound search algorithm that combines the AND/OR decomposition principle with variable ordering heuristics. There are two orthogonal approaches to incorporating dynamic orderings into AOBB. The first one improves AOBB by applying an independent semantic variable ordering heuristic whenever the partial order dictated by the decomposition principle allows. The second, orthogonal approach gives priority to the semantic variable ordering heuristic and applies problem decomposition as a secondary principle. We demonstrate empirically the efficiency of the dynamic AND/OR Branch-and-Bound approach on several benchmarks for pure 0/1 integer linear programming problems, including test instances from the MIPLIB library, combinatorial auctions simulating radio spectrum allocation and random uncapacitated warehouse location problems.

The paper is organized as follows. In Section 2 we present background on constraint optimization problems and integer linear programming. Section 3 presents the AND/OR search space as well as an efficient heuristic for constructing low depth balanced pseudo-trees. In Section 4 we introduce the AND/OR Branch-and-Bound algorithm, specialized for solving pure 0/1 integer linear programs. In Section 5 we introduce the dynamic AND/OR Branch-and-Bound algorithm. Section 6 shows our empirical evaluation and Section 7 concludes.

2   Background

2.1   Constraint Optimization Problems

A finite *Constraint Optimization Problem* (COP) is a four-tuple $\langle \mathcal{X}, \mathcal{D}, \mathcal{C}, z \rangle$, where $\mathcal{X} = \{X_1, ..., X_n\}$ is a set of variables, $\mathcal{D} = \{D_1, ..., D_n\}$ is a set of finite domains, $\mathcal{C} = \{C_1, ..., C_m\}$ is a set of constraints on the variables and z is a global cost function (i.e. objective function) to be optimized. The scope of a constraint C_i, denoted $scope(C_i) \subseteq \mathcal{X}$, is the set of arguments of C_i. Constraints can be expressed *extensionally*, through relations, or *intentionally*, by a mathematical formula (equality or inequality). An optimal solution to a COP is a complete value assignment to all the variables such that every constraint is satisfied and the objective function is minimized or maximized.

With every COP instance we can associate a *constraint graph G* which has a node for each variable and connects any two nodes whose variables appear in the scope of the same constraint. The *induced graph* of G relative to an ordering d of its variables, denoted $G^*(d)$, is obtained by processing the nodes in reverse order of d. For each node all its earlier neighbors are connected, including neighbors connected by previously added edges. Given a graph and an ordering of its nodes, the *width* of a node is the number of edges connecting it to nodes lower in the ordering. The *induced width* of a graph, denoted $w^*(d)$, is the maximum width of nodes in the induced graph.

2.2   Integer Linear Programming

A *Linear Program* (LP) consists of a set of continuous variables and a set of linear constraints (equalities or inequalities). The goal is to optimize a global linear cost function subject to the constraints. One of the standard forms of a linear program is:

$$min\{c^\top x \mid Ax \leq b, x \geq 0\} \tag{1}$$

where $c \in \mathbb{R}^n$, $b \in \mathbb{R}^m$, $A \in \mathbb{R}^{m \times n}$ and $x \in \mathbb{R}^n$. Here c represents the cost vector and x is the vector of decision variables. The vector b and the matrix A define the m linear constraints. Linear programs are usually solved by Dantzig's SIMPLEX method [5].

A *Mixed Integer Linear Programming* (MILP) problem is a linear program where some of the decision variables are constrained to have only integer values at the optimal solution. An important special case is a decision variable x_i that is integer with $0 \leq x_i \leq 1$. This forces x_i to be either 0 or 1 at the solution. Variables like x_i are called *0/1* or *binary integer variables*. Subsequently, a MILP problem with binary integer variables is also called a *0/1 Mixed Integer Linear Programming* problem. A *pure 0/1 Integer Linear Programming* problem is a MILP where all the decision variables are binary. Pure 0/1 ILPs can formulate many practical problems such as capital budgeting [6], cargo loading [7], processor allocation in distributed systems [8] or combinatorial auctions [9, 10].

Clearly, any pure 0/1 integer linear program can be viewed as a finite COP instance $\langle \mathcal{X}, \mathcal{D}, \mathcal{C}, z \rangle$ with linear constraints and a linear objective function. In the remaining of the paper we will consider a *minimization* problem defined by $z = \sum_{i=1}^{n} c_i X_i$ subject to m linear constraints $\mathcal{C} = \{C_1, ..., C_m\}$, over n binary decision variables $\mathcal{X} = \{X_1, ..., X_n\}$.

2.3 Branch-and-Bound Search for Constraint Optimization

Branch-and-Bound (BB) is a general *search* method for solving constraint optimization problems [3]. It traverses the search tree defined by the problem, where internal nodes represent partial assignments and leaf nodes denote complete ones, which may or may not be optimal. During the traversal, which is usually *depth first*, BB maintains un *upper bound ub*, the cost of the best solution found so far. At each internal node the algorithm computes a *lower bound lb* on the optimal extension of the current partial assignment. When $lb \geq ub$, the current best cost cannot be improved and the algorithm *backtracks* pruning the subtree below the current node. Otherwise, the algorithm moves forward and tries to instantiate the next variable in the ordering.

In the context of pure 0/1 integer linear programs, the lower bound of a subproblem is obtained by solving its linear relaxation (i.e. relaxing the integrality restrictions). In this case the branching process can fail at a particular node for one of the following reasons: (i) the LP solution can be integer; or (ii) the LP problem can be infeasible; or (iii) the lower bound exceeds the upper bound (for more details see [2, 3]).

3   AND/OR Search Spaces

The classical way to do search is to instantiate variables one at a time, following a static/dynamic variable ordering. In the simplest case, this process defines a search tree (called here OR search tree), whose nodes represent states in the space of partial assignments. The traditional search space does not capture independencies that appear in the structure of the underlying graphical model. Introducing AND states into the search

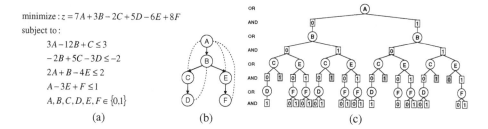

$$\text{minimize}: z = 7A + 3B - 2C + 5D - 6E + 8F$$

subject to :

$$3A - 12B + C \le 3$$
$$-2B + 5C - 3D \le -2$$
$$2A + B - 4E \le 2$$
$$A - 3E + F \le 1$$
$$A, B, C, D, E, F \in \{0,1\}$$

(a)   (b)   (c)

Fig. 1. The AND/OR search space

space can capture the structure, decomposing the problem into independent subproblems by conditioning on values [11, 4]. The AND/OR search space is defined using a backbone *pseudo-tree*.

Definition 1 (pseudo-tree). *Given an undirected graph $G = (V, E)$, a directed rooted tree $T = (V, E')$ defined on all its nodes is called* pseudo-tree *if any arc of G which is not included in E' is a back-arc, namely it connects a node to an ancestor in T.*

3.1 AND/OR Search Trees

Given a COP instance $\langle \mathcal{X}, \mathcal{D}, \mathcal{C}, z \rangle$, its constraint graph G and a pseudo-tree T of G, the associated AND/OR search tree S_T has alternating levels of OR nodes and AND nodes. The OR nodes are labeled by X_i and correspond to the variables. The AND nodes are labeled by $\langle X_i, x_i \rangle$ and correspond to value assignments in the domains of the variables. The structure of the AND/OR tree is based on the underlying pseudo-tree T of G. The root of the AND/OR search tree is an OR node, labeled with the root of T.

The children of an OR node X_i are AND nodes labeled with assignments $\langle X_i, x_i \rangle$, consistent along the path from the root, $path(x_i) = (\langle X_1, x_1 \rangle, ..., \langle X_{i-1}, x_{i-1} \rangle)$. The children of an AND node $\langle X_i, x_i \rangle$ are OR nodes labeled with the children of variable X_i in T. In other words, the OR states represent alternative ways of solving the problem, whereas the AND states represent problem decomposition into independent subproblems, all of which need be solved. When the pseudo-tree is a chain, the AND/OR search tree coincides with the regular OR search tree.

A *solution subtree* Sol_{S_T} of S_T is an AND/OR subtree such that: (i) it contains the root of S_T; (ii) if a nonterminal AND node $n \in S_T$ is in Sol_{S_T} then all of its children are in Sol_{S_T}; (iii) if a nonterminal OR node $n \in S_T$ is in Sol_T then exactly one of its children is in Sol_{S_T}.

Example 1. For illustration consider the pure 0/1 integer program with 6 decision variables A, B, C, D, E, F and 4 linear constraints $C_1(A, B, C)$, $C_2(B, C, D)$, $C_3(A, B, E)$, $C_4(A, E, F)$ from Figure 1(a). The objective function to be minimized is $z = 7A+B-2C+5D-6E+8F$. The pseudo-tree arrangement of the constraint graph, together with the back-arcs (dotted lines) are given in Figure 1(b). Figure 1(c) shows the corresponding AND/OR search tree (for AND nodes we only denote the value, namely $\langle A, 0 \rangle$ is written as $\boxed{0}$ child of A). The shaded nodes represent dead-ends (i.e. inconsistent values).

The AND/OR search tree can be traversed by a depth-first search algorithm that is guaranteed to have a time complexity exponential in the depth of the pseudo-tree and can operate in linear space. The arcs from X_i to $\langle X_i, x_i \rangle$ are annotated by appropriate *labels* of the objective function. The nodes in S_T can be associated with *values*, defined over the subtrees they root.

Definition 2 (label). *Given a COP instance with objective function $z = \sum_{i=1}^{n} c_i X_i$ and a corresponding AND/OR search tree S_T, the label $l(X_i, x_i)$ of the arc from the OR node X_i to the AND node $\langle X_i, x_i \rangle$ is defined as $l(X_i, x_i) = c_i \cdot x_i$.*

Definition 3 (value). *The value $v(n)$ of a node $n \in S_T$ is defined recursively as follows: (i) if $n = \langle X_i, x_i \rangle$ is a terminal AND node then $v(n) = l(X_i, x_i)$; (ii) if $n = \langle X_i, x_i \rangle$ is an internal AND node then $v(n) = l(X_i, x_i) + \sum_{n' \in succ(n)} v(n')$; (iii) if $n = X_i$ is an internal OR node then $v(n) = min_{n' \in succ(n)} v(n')$, where $succ(n)$ are the children of n in S_T.*

Clearly, the value of each node can be computed recursively, from leaves to root.

Proposition 1. *Given an AND/OR search tree S_T of a COP instance $\mathcal{P} = (\mathcal{X}, \mathcal{D}, \mathcal{C}, z)$, the value $v(n)$ of a node $n \in S_T$ is the minimal cost solution to the subproblem rooted at n, subject to the current variable instantiation along the path from root to n. If n is the root of S_T, then $v(n)$ is the minimal cost solution to $\mathcal{P}$.*

Therefore, we can traverse the AND/OR search tree in a depth-first manner to compute the value of the root. This approach would require linear space, storing only the current partial solution subtree. The algorithm expands alternating levels of OR and AND nodes, periodically evaluating the values of the nodes along the current path. It terminates when the root node is evaluated with the optimal cost.

Theorem 1 (complexity). *The complexity of an algorithm that traverses an AND/OR search tree in a depth-first manner is linear space and time is $O(n \cdot exp(h))$, where h is the depth of the pseudo-tree associated with the constraint graph. When the constraint graph has induced width w, the algorithm can be bounded by $O(n \cdot exp(w \cdot log(n)))$.*

3.2 Pseudo-Trees Based on Recursive Hypergraph Decomposition

The performance of the AND/OR tree search algorithms is influenced by the quality of the pseudo-tree. Finding the minimal depth pseudo-tree is a hard problem [11, 12]. In this section we describe a heuristic for generating a low depth balanced pseudo-tree, based on the recursive decomposition of a hypergraph.

Definition 4 (hypergraph). *Given a COP instance $\langle \mathcal{X}, \mathcal{D}, \mathcal{C}, z \rangle$, its hypergraph $\mathcal{H} = (V, E)$ has a vertex $v_i \in V$ for each constraint in $\mathcal{C}$ and each variable in $\mathcal{X}$ is an edge $e_j \in E$ connecting all the constraints in which it appears.*

Definition 5 (hypergraph separators). *Given a hypergraph $\mathcal{H} = (V, E)$, a hypergraph separator decomposition is a triple $(\mathcal{H}, \mathcal{S}, \mathcal{R})$ where: (i) $\mathcal{S} \subset E$, and the removal of $\mathcal{S}$ separates $\mathcal{H}$ into k disconnected components (subgraphs) $\mathcal{H}_1, ..., \mathcal{H}_k$; (ii) $\mathcal{R}$ is a relation over the size of the disjoint subgraphs (i.e. balance factor).*

It is well known that the problem of generating optimal hypergraph partitions is hard. However heuristic approaches were developed over the years. A good approach is packaged in hMeTiS[1]. We will use this software as a basis for our pseudo-tree generation. This idea and software were also used by [13] to generate low width decomposition trees. Generating a pseudo-tree using hMeTiS is fairly straightforward. The vertices of the hypergraph are partitioned into two balanced (roughly equal-sized) parts, denoted by $\mathcal{H}_{left}$ and $\mathcal{H}_{right}$ respectively, while minimizing the number of hyperedges across. A small number of crossing edges translates into a small number of variables shared between the two sets of constraints. $\mathcal{H}_{left}$ and $\mathcal{H}_{right}$ are then each recursively partitioned in the same fashion, until they contain a single vertex. The result of this process is a tree of hypergraph separators which is also a pseudo-tree of the original model since each separator corresponds to a subset of variables chained together.

4 AND/OR Branch-and-Bound Search

AND/OR Branch-and-Bound (AOBB) was recently proposed by [1] as a depth-first Branch-and-Bound that explores an AND/OR search tree for solving optimization tasks in graphical models. In this section we review briefly the static version of the algorithm.

4.1 Lower Bounds on Partial Trees

At any stage during search, a node n along the current path roots a current *partial solution subtree*, denoted by $G_{sol}(n)$, to the corresponding subproblem. By the nature of the search process, $G_{sol}(n)$ must be connected, must contain its root n and will have a *frontier* containing all those nodes that were generated but not yet expanded. The leaves of $G_{sol}(n)$ are called *tip* nodes. Furthermore, we assume that there exists a *static* heuristic evaluation function $h(n)$ underestimating $v(n)$ that can be computed efficiently when node n is first generated.

Given the current partially explored AND/OR search tree S_T, the *active path* $\mathcal{AP}(t)$ is the path of assignments from the root of S_T to the current tip node t. The *inside context* $in(\mathcal{AP})$ of $\mathcal{AP}(t)$ contains all nodes that were fully evaluated and are children of nodes on $\mathcal{AP}(t)$. The *outside context* $out(\mathcal{AP})$ of $\mathcal{AP}(t)$, contains all the frontier nodes that are children of the nodes on $\mathcal{AP}(t)$. The *active partial subtree* $\mathcal{APT}(n)$ rooted at a node $n \in \mathcal{AP}(t)$ is the subtree of $G_{sol}(n)$ containing the nodes on $\mathcal{AP}(t)$ between n and t together with their OR children. We can define now a *dynamic heuristic evaluation function* of a node n relative to $\mathcal{APT}(n)$, as follows.

Definition 6 (dynamic heuristic evaluation function). *Given an active partial tree* $\mathcal{APT}(n)$, *the* dynamic heuristic evaluation function *of* n, $f_h(n)$, *is defined recursively as follows: (i) if* $\mathcal{APT}(n)$ *consists only of a single node* n, *and if* $n \in in(\mathcal{AP})$ *then* $f_h(n) = v(n)$ *else* $f_h(n) = h(n)$; *(ii) if* $n = \langle X_i, x_i \rangle$ *is an AND node, having OR children* $m_1, ..., m_k$ *then* $f_h(n) = max(h(n), l(X_i, x_i) + \sum_{i=1}^{k} f_h(m_i))$; *(iii) if* $n = X_i$ *is an OR node, having an AND child* m, *then* $f_h(n) = max(h(n), f_h(m))$.

---

[1] http://www-users.cs.umn.edu/ karypis/metis/hmetis

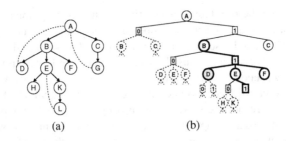

Fig. 2. A partially explored AND/OR search tree

We can show that:

Theorem 2. *(1)* $f_h(n)$ *is a* lower bound *on the minimal cost solution to the subproblem rooted at* n, *namely* $f_h(n) \leq v(n)$; *(2)* $f_h(n) \geq h(n)$, *namely the dynamic heuristic function is tighter than the static one.*

Example 2. For illustration consider the pseudo-tree in Figure 2(a) and the partially explored AND/OR search tree in Figure 2(b). The active path has tip node $\langle E, 1 \rangle$ and represents the partial assignment $A = 1, B = 1, E = 1$. The shaded nodes at the left of the active path belong to the inside context (their corresponding subtrees have already been explored). The outside context includes the nodes $\{C, F\}$, which are also in the search frontier. For the active partial subtree rooted at B (highlighted), the lower bound $f_h(B)$ on $v(B)$ is computed recursively as follows: $f_h(B) = max(h(B), f_h(\langle B, 1 \rangle))$, where $f_h(\langle B, 1 \rangle) = max(h(\langle B, 1 \rangle), l(B, 1) + v(D) + f_h(E) + h(F))$. Similarly, $f_h(E) = max(h(E), f_h(\langle E, 1 \rangle)) = max(h(E), h(\langle E, 1 \rangle))$, since $f_h(\langle E, 1 \rangle) = h(\langle E, 1 \rangle)$.

4.2 Static AND/OR Branch-and-Bound

AOBB can calculate a *lower bound* on $v(n)$ for any node n on the active path, by using $f_h(n)$. It also maintains an *upper bound* on $v(n)$ which is the current minimal cost solution subtree rooted at n. If $f_h(n) \geq ub(n)$ then the search can be safely terminated below the tip node of the active path.

Figure 3 shows AOBB. The algorithm assumes the *global* linear objective function $z = \sum_{i=1}^{n} c_i X_i$. The following notation is used: $(\mathcal{X}, \mathcal{D}, \mathcal{C})$ is the problem with which the procedure is called, T is a pseudo-tree arrangement of the underlying constraint graph, st is current partial solution subtree being explored (initially $st =$ NULL), in (resp. out) represents the inside (resp. outside) context of the active path. These contexts are constantly updated during search. Variables are selected statically according to the pseudo-tree T (indicated by the input parameter $vo =$ SVO).

If the set $\mathcal{X}$ is empty, then the result is trivially computed (line 1). Else, AOBB selects a variable X_i (i.e. expands the OR node X_i) and iterates over its values (lines 3-5) to compute the OR value $v(X_i)$. Each value a defines the current subproblem $P = (X_i = a, \mathcal{X}, \mathcal{D}, \mathcal{C})$ that is decomposed into a set of q independent subproblems $P_k = (\mathcal{X}_k, \mathcal{D}_k, \mathcal{C}_k, z_k)$, with $k = 1..q, q > 0$, one per child X_k of X_i in the pseudo-tree T. Each subproblem P_k is defined by the subset of variables $\mathcal{X}_k$ corresponding

function: AOBB $(vo, st, T, \mathcal{X}, \mathcal{D}, \mathcal{C}, z)$
1 **if** $\mathcal{X} = \emptyset$ **then return** 0;
2 **else**
3       $X_i \leftarrow$ SelectVar $(vo, T, \mathcal{X})$;
4       $v(X_i) \leftarrow \infty$;
5   **foreach** $a \in D_i$ **do**
6            $st' \leftarrow st \cup (X_i, a)$;
7   **foreach** $k = 1..q$ **do**
8                 $h(X_k) \leftarrow$ LB $(\mathcal{X}_k, \mathcal{D}_k, \mathcal{C}_k)$;
9   UpdateContext (out, X_k, $h(X_k)$);
10   **end**
11           $h(X_i, a) \leftarrow c_i a + \sum_{k=1}^{q} h(X_k)$;
12   **if** $\neg$FindCut $(X_i, a, h(X_i, a))$ **then**
13                $v(X_i, a) \leftarrow 0$;
14   **foreach** $k = 1..q$ **do**
15                     $val \leftarrow$ AOBB $(vo, st', T, \mathcal{X}_k, \mathcal{D}_k, \mathcal{C}_k, z_k)$;
16                     $v(X_i, a) \leftarrow v(X_i, a) + val$;
17   **end**
18                $v(X_i, a) +$ label (i, a);
19   UpdateContext (in, $v(X_i, a)$);
20                $v(X_i) \leftarrow min(v(X_i), v(X_i, a))$;
21   **end**
22   **end**
23   **return** $v(X_i)$;
24 **end**

Fig. 3. AND/OR Branch-and-Bound

to the descendants of X_k in T including X_k, the subset of constraints and constraint projections $\mathcal{C}_k$ involving the variables in $\mathcal{X}_k$, subject to the current instantiation along the active path, and a *local* objective function denoted by $z_k = \sum_{X_j \in \mathcal{X}_k} c_j X_j$, which corresponds to the projection of z on the variables in $\mathcal{X}_k$. For each P_k, the algorithm computes a static heuristic function $h(X_k)$ which underestimates z_k (line 8). The outside context of the active path is updated in line 9. The static heuristic function of P is $h(X_i, a)$ (line 11), computed as the sum of independent lower bounds including the projection on X_i and value a of the objective function (i.e. the label of the corresponding AND node $\langle X_i, a \rangle$).

Upon instantiating X_i with value a (i.e. expanding the AND node $\langle X_i, a \rangle$) (line 12), AOBB successively updates the *dynamic heuristic evaluation function* $f_h(m)$ for every ancestor node m along the active path (procedure FindCut implements Definition 6). If $f_h(m) \geq ub(m)$, for some ancestor node m, then the algorithm backtracks and tries the next value in the domain of X_i. As the algorithm recursively solves independent subproblems (line 14) the AND value $v(X_i, a)$ accumulates the results (line 16). The inside context of the active path is also updated with the actual solution of the current subproblem (line 19). Once all subproblems are solved, the OR value $v(X_i)$ is also updated (line 20). After trying all feasible values of variable X_i, the cost of the optimal solution to the problem rooted by X_i remains in $v(X_i)$, which is returned (line 23).

In the context of pure 0/1 integer linear programming problems, the static heuristic function $h(X_k)$ for each subproblem P_k is obtained by solving its linear relaxation (i.e. relaxing the integrality restrictions). If the respective linear program is infeasible, then $h(X_k)$ is set to ∞. For illustration, consider again the pure 0/1 integer program from Figure 1(a). Let $A = 0$ and $B = 0$ be the current partial assignment of the active path in the AND/OR search tree from Figure 1(c). The subproblem P_C rooted at node C in the search tree corresponds to minimizing the local objective function $z_C = -2C + 5D$, subject to the constraints and constraint projections involving variables C and D only (i.e. $C \leq 3$ and $5C - 3D \leq -2$, respectively). Moreover, if the subproblem P_k rooted at node $k = X_k$ in the search tree has an integer solution (i.e. the solution to the linear relaxation of P_k has no fractional variables), then there is no need to search the subtree below k. In this case, $v(k) = h(X_k)$ and this is the value returned for P_k.

5   Dynamic AND/OR Branch-and-Bound

It is well known that the variable ordering can dramatically influence search performance [2, 14]. In this section we go beyond static orderings and introduce a new *dynamic* AND/OR Branch-and-Bound algorithm that incorporates variable ordering heuristics used in the classic OR search space.

We distinguish two classes of variable ordering heuristics. *Graph*-based heuristics (e.g. pseudo-tree arrangements) that try to maximize problem decomposition and *semantic*-based heuristics (e.g. min pseudo cost, min reduced cost) that aim at shrinking the search space. These two forces are orthogonal, namely we can use one as the primary goal and break ties based on the other. Moreover, we can use each class statically or dynamically. We present next two ways of combining efficiently these two classes of heuristics.

5.1   Partial Variable Ordering (PVO)

The first approach, called *AND/OR Branch-and-Bound with Partial Variable Ordering* (AOBB+PVO) combines the static graph-based decomposition given by a pseudo-tree

---

function: SelectVar $(vo, st, T, \mathcal{X})$
1 **switch** vo **do**
2   **case** SVO
3   **if** $st = NULL$ **then** $next \leftarrow$ GetPseudoTreeRoot (T);
4   **else** $next \leftarrow$ GetPseudoTreeChild (st, T)
5   **case** PVO
6        $candidates \leftarrow$ GetPseudoTreeVarGroup (st, T);
7        $next \leftarrow$ SelectBestCandidate $(candidates)$;
8   **case** DVO
9        $next \leftarrow$ SelectBestCandidate $(\mathcal{X})$;
10 **end**
11 **return** $next$;

Fig. 4. Variable selection procedure

with a dynamic semantic ordering heuristic. Let us illustrate the idea with an example. Consider the pseudo-tree from Figure 1(a) inducing the following variable group ordering: {A,B}, {C,D}, {E,F}; which dictates that variables {A,B} should be considered before {C,D} and {E,F}. Variables in each group can be dynamically ordered based on a second, independent semantic heuristic. Notice that after variables {A,B} are instantiated, the problem decomposes into two independent components that can be solved separately.

AOBB+PVO is similar to its precursor AOBB described in Figure 3 in the sense that it is also guided by a pre-computed pseudo-tree. The partial variable ordering strategy, indicated by the input parameter $vo = PVO$, is implemented by the SelectVar procedure from Figure 4. The algorithm selects the next variable X_i as the best scoring uninstantiated variable from the current variable group of the the pseudo-tree T.

5.2 Dynamic Variable Ordering (DVO)

The second, orthogonal approach to PVO called *AND/OR Branch-and-Bound with Dynamic Variable Ordering* (AOBB+DVO), gives priority to the dynamic semantic ordering heuristic and applies static problem decomposition as a secondary principle during search.

AOBB+DVO is also based on the algorithm from Figure 3. It instantiates variables dynamically using a semantic ordering heuristic while constantly updating the problem graph structure. Specifically, after variable X_i is selected by procedure SelectVar, AOBB+DVO tentatively removes X_i from the graph and, if disconnected components are detected their corresponding subproblems are then solved separately and the results combined in an AND/OR manner (lines 14-20). It is easy to see that in this case a variable may have the best semantic heuristic to tighten the search space, yet, it may not yield a good decomposition for the remaining of the problem, in which case the algorithm would explore primarily an OR space.

6 Experiments

In this section we evaluate empirically the performance of the AND/OR Branch-and-Bound algorithms on several benchmarks for pure 0/1 integer linear programming including problem instances from the MIPLIB library[2], combinatorial auctions and uncapacitated warehouse location problems. All our experiments were done on a 2.4GHz Pentium IV with 2GB of RAM, running Windows XP. Our C++ implementation of the AND/OR algorithms was based on the open source lp_solve library[3].

We consider three classes of depth-first AND/OR Branch-and-Bound (AOBB) algorithms described in the previous sections and denoted by AOBB+SVO (i.e. static AND/OR Branch-and-Bound), AOBB+PVO and AOBB+DVO, respectively. For comparison, we include results obtained with the classic OR depth-first Branch-and-Bound (BB) available in the lp_solve library. All competing algorithms used a semantic variable ordering heuristic based on *reduced costs* (i.e. dual values) [2]. Specifically, the

---

[2] Available at http://miplib.zib.de/miplib2003.php
[3] lp_solve 5.5.0.6 is available at http://groups.yahoo.com/group/lp_solve/

Table 1. Results for MIPLIB problem instances

| miplib | n | (w*,h) | BB | | AOBB | | | | | |
| | | | | | SVO | | PVO | | DVO | |
			time	nodes	time	nodes	time	nodes	time	nodes
p0033	33	(18, 20)	6.53	18,081	0.59	1,893	**0.39**	1,099	3.39	9,251
p0201	201	(120, 142)	37.41	15,575	57.88	25,284	**22.90**	8,988	42.46	14,463
lseu	89	(53, 69)	153.90	368,573	39.74	87,537	**38.94**	86,073	152.55	336,953

next fractional variable to instantiate has the smallest reduced cost. Ties are broken lexicographically.

We report the average effort, as CPU time (in seconds) and number of nodes visited (which is equivalent to the number of time the SIMPLEX routine was called to solve the linear relaxation of the current subproblem), required for proving optimality of the solution. We also record the number of variables (n), the depth of the pseudo-trees (h) and the induced width of the graphs (w*) obtained for the test instances. As AOBB+SVO and AOBB+PVO algorithms use a non-deterministic algorithm for generating the pseudo-tree, the running time may vary significantly from one run to the next. We therefore ran these algorithms 5 times on each benchmark and provide an average of those runs. The best performance points are highlighted in all test cases.

6.1   MIPLIB Library

MIPLIB is a library of mixed integer linear programming instances that is commonly used for benchmarking integer programming algorithms. For our purpose we selected 3 pure 0/1 integer linear instances of increasing difficulty. Table 1 reports a summary of the experiment. We see immediately that, overall, AOBB+PVO is the best performing algorithm, both in terms of CPU time and number of nodes visited. AOBB+DVO does indeed explore a smaller search space than BB in all test cases, but due to its computational overhead these savings do not reflect in the running time.

6.2   Combinatorial Auctions

In **combinatorial auctions** (CA), an auctioneer has a set of goods, $M = \{1, 2, ..., m\}$ to sell and the buyers submit a set of bids, $\mathcal{B} = \{B_1, B_2, ..., B_n\}$. A bid is a tuple $B_j = \langle S_j, p_j \rangle$, where $S_j \subseteq M$ is a set of goods and $p_j \geq 0$ is a price. The winner determination problem is to label the bids as winning or loosing so as to maximize the sum of the accepted bid prices under the constraint that each good is allocated to at most one bid. We used the following pure 0/1 integer formulation of the problem:

$$max \sum_{j=1}^{n} p_j x_j \qquad (2)$$

$$\text{s.t.} \sum_{j|i \in S_j} x_j \leq 1 \; i \in \{1..m\}$$
$$x_j \in \{0, 1\} \qquad j \in \{1..n\}$$

Table 2. Results for combinatorial auction problem instances

auction	(w*,h)	BB			AOBB							
					SVO			PVO			DVO	
		time	nodes	wins	time	nodes	wins	time	nodes	wins	time	nodes
reg-upv-b200g50	(145, 162)	**1.71**	602	4	3.28	938	0	2.98	888	6	1.97	602
reg-upv-b250g75	(166, 190)	16.27	3,472	0	7.32	1,209	3	**6.30**	1,110	7	18.36	3,472
reg-upv-b300g100	(173, 204)	63.29	7,997	2	52.75	4,855	4	**45.61**	4,801	4	69.18	7,997
reg-npv-b200g50	(140, 161)	1.27	443	1	1.78	514	1	**1.15**	302	8	1.45	443
reg-npv-b250g75	(160, 187)	**5.53**	1,150	2	5.97	1,085	4	5.96	1,144	4	6.24	1,150
reg-npv-b300g100	(172, 206)	58.61	7,342	1	21.54	1,904	4	**16.35**	1,748	5	63.74	7,342

Table 2 shows results for experiments with combinatorial auctions drawn from the `regions` distribution of the CATS 2.0 test suite [10]. The suffixes `npv` and `upv` indicate that the bid prices were drawn from either a normal or uniform distribution. These problem instances simulate the auction of radio spectrum in which a government sells the right to use specific segments of spectrum in different geographical areas (for more details see [10]). We looked at moderate size auctions by varying the number of bids between 200 and 300, and the number of goods between 50 and 100. The number of bids is also the number of variables in the ILP model. For each value combination of bids and goods we drawn randomly 10 auctions from the respective distribution. For each algorithm we also report the number of wins out of the 10 runs. These instances are highly connected with induced widths over 150. For this problem class AOBB+PVO outperforms its competitors, exploring the smallest search space. If we look for example at the 300 bid problem instances from the `reg-npv` distribution, AOBB+PVO is about 4 times faster than BB, exploring a search space 4 times smaller. Notice that AOBB+DVO explores the same number of nodes as BB, showing that in this case the dynamic semantic variable ordering heuristic does not generate decomposable subproblems.

6.3   Uncapacitated Warehouse Location Problem

In the **uncapacitated warehouse location problem** (UWLP) a company considers opening m warehouses at some candidate locations in order to supply its n existing stores. The objective is to determine which warehouse to open, and which of these warehouses should supply the various stores, such that the sum of the maintenance and supply costs is minimized. Each store must be supplied by exactly one warehouse. The typical 0/1 integer formulation of the problem is as follows:

$$min \sum_{j=1}^{n} \sum_{i=1}^{m} c_{ij} x_{ij} + \sum_{i=1}^{m} f_i y_i \qquad (3)$$

s.t. $\sum_{i=1}^{m} x_{ij} = 1 \; \forall j \in \{1..n\}$
$\quad x_{ij} \leq y_i \qquad \forall j \in \{1..n\}, \forall i \in \{1..m\}$
$\quad x_{ij} \in \{0,1\} \quad \forall j \in \{1..n\}, \forall i \in \{1..m\}$
$\quad y_i \in \{0,1\} \qquad \forall i \in \{1..m\}$

Table 3. Results for 10 uncapacitated warehouse location problem instances

uwlp	(w*,h)	BB		SVO		AOBB PVO		DVO	
		time	nodes	time	nodes	time	nodes	time	nodes
uwlp50-200-a	(50, 123)	6.27	27	15.72	70	**6.28**	12	7.23	27
uwlp50-200-b	(50, 123)	11.34	53	17.22	60	**5.78**	12	11.75	53
uwlp50-200-c	(50, 123)	73.66	469	15.78	58	**5.83**	10	77.94	469
uwlp50-200-d	(50, 123)	836.52	4,309	27.94	116	**11.97**	26	904.15	4,309
uwlp50-200-e	(50, 123)	2501.75	11,973	32.69	80	**16.98**	28	2990.19	12,733
uwlp50-200-f	(50, 123)	43.36	237	18.70	64	**8.03**	20	45.99	237
uwlp50-200-g	(50, 123)	1328.40	6,905	27.89	84	**8.53**	20	1515.48	7,265
uwlp50-200-h	(50, 123)	76.88	331	25.20	84	**13.70**	30	88.38	331
uwlp50-200-i	(50, 123)	224.33	1,003	46.06	194	**17.17**	50	367.14	1,533
uwlp50-200-j	(50, 123)	7737.65	31,003	28.03	64	**9.13**	10	9276.98	33,415

where f_i is the cost of opening a warehouse at location i and c_{ij} is the cost of supplying store j from the warehouse at location i.

Table 3 displays the results obtained on 10 randomly generated UWLP problem instances[4] with 50 warehouses and 200 stores. The warehouse opening and store supply costs were chosen uniformly randomly between 0 and 1000. These are large problems with 10,050 variables and 10,500 constraints. The semantic variable ordering heuristic that worked best in this case selects the next fractional variable whose value is closest to 0.5 (ties are broken lexicographically). We can see that AOBB+PVO dominates in all test cases, outperforming the classic BB with several orders of magnitude in terms of both running time and size of the search space explored. In uwlp50-200-e for example, one of the hardest instances, AOBB+PVO causes a speed-up of 147 over the classic OR Branch-and-Bound algorithm, exploring a search tree 428 times smaller. This is due to the problem's structure partially captured by a shallow pseudo-tree with depth 123. AOBB+DVO has a similar performance as BB on all test instances (it is slower than BB due to its computational overhead), indicating that these problems do not break into disconnected components when the semantic variable ordering heuristic has higher priority than problem decomposition.

7   Conclusion

In this paper we extended the AND/OR Branch-and-Bound search algorithm for solving pure 0/1 integer linear programming problems. The contribution of the paper is two-fold. First, we restricted the algorithm to a static variable ordering induced by a pseudo-tree of the constraint graph. Since the order in which variables are selected for instantiation can influence dramatically the search performance, we then proposed a dynamic version of the AND/OR Branch-and-Bound that incorporates variable ordering heuristics. We looked at two orthogonal approaches to incorporating dynamic orderings

---

[4] Problem generator from http://www.mpi-sb.mpg.de/units/ag1/projects/benchmarks/UflLib/

into AOBB. On one hand, AOBB+PVO augments a static pseudo-tree based problem decomposition with a dynamic semantic variable ordering heuristics. On the other hand, AOBB+DVO gives priority to the dynamic semantic variable ordering heuristic while constantly updating the graph structure and solving separately, in an AND/OR manner, disconnected components that may be discovered during search. Our empirical evaluation demonstrated on a variety of benchmark problems for pure 0/1 integer linear programming that AOBB+PVO is a promising candidate solver, outperforming the classic BB with several orders of magnitude in terms of both running time and size of the search space explored.

Our dynamic AND/OR approach leaves room for future improvements, which are likely to make it more effective in practice. For instance, it can be modified to explore the search tree in a *best-first* manner, rather than depth-first. This is desirable in the sense that no optimal tree search algorithm can guarantee expanding fewer nodes [15]. We also mention that the *Branch-and-Cut*, a more modern algorithm that generates *cutting planes* to tighten the LP relaxation of the current subproblem, can be adapted to traverse an AND/OR search tree. Finally, the AND/OR algorithms can be easily adapted for solving *mixed* 0/1 ILPs, where only a subset of the decision variables is restricted to integer values. In that case, the AND/OR search space is based on a *partial* pseudo-tree which spans only the integer variables.

Related Work. AOBB is related to the Branch-and-Bound method proposed by [16] for acyclic AND/OR graphs and game trees, as well as the pseudo-tree search algorithm proposed in [17] for boosting Russian Doll search. The optimization method developed in [18] for semi-ring CSPs can also be interpreted as an AND/OR graph search algorithm. Problem decomposition based on hypergraph separators was also explored by [19] and [20] for solving large real-world SAT problem instances.

Acknowledgments

We would like to thank the anonymous reviewers for commenting on an earlier version of the paper. This work has been partially supported by the NSF grant IIS-0412854.

References

1. R. Marinescu and R. Dechter. And/or branch-and-bound for graphical models. *In International Joint Conference on Artificial Intelligence (IJCAI'05)*, pages 224–229, 2005.
2. G. Nemhauser and L. Wolsey. *Integer and combinatorial optimization.* Wiley, 1988.
3. E. Lawler and D. Wood. Branch-and-bound methods: A survey. *Operations Research*, 14(4):699–719, 1966.
4. R. Dechter and R. Mateescu. And/or search spaces for graphical models. *UCI-ICS Techical Report*, 2006.
5. G.B. Dantzig. Maximization of a linear function of variables subject to linear inequalities. *Activity Analysis of Production and Allocation*, 1951.
6. M. Vasquez and J. Hao. A hybrid approach for the 0/1 multidimensional knapsack approach. *In International Joint Conference on Artificial Intelligence (IJCAI'01)*, pages 328–333, 2001.

7. W. Shih. A branch-and-bound method for the multiconstraint 0/1 knapsack problem. *Journal of the Operational Research Society*, 30:369–378, 1979.
8. B. Gavish and H. Pirkul. Allocation of data bases and processors in a distributed computing system. *Management of Distributed Data Processing*, 31:215–231, 1982.
9. T. Sandholm. An algorithm for optimal winner determination in combinatorial auctions. *In International Joint Conference on Artificial Intelligence (IJCAI'99)*, pages 542–547, 1999.
10. K. Leyton-Brown, M. Pearson, and Y. Shoham. Towards a universal test suite for combinatorial auction algorithms. *In ACM Electronic Commerce*, pages 66–76, 2000.
11. E. Freuder and M. Quinn. Taking advantage of stable sets of variables in constraint satisfaction problems. *In International Joint Conference on Artificial Intelligence (IJCAI'85)*, pages 1076–1078, 1985.
12. R. Bayardo and D. Miranker. On the space-time trade-off in solving constraint satisfaction problems. *In International Joint Conference on Artificial Intelligence (IJCAI'95)*, pages 558–562, 1995.
13. A. Darwiche. Recursive conditioning. *Artificial Intelligence*, 126(1-2):5–41, 2001.
14. Rina Dechter. *Constraint Processing*. MIT Press, 2003.
15. J. Pearl. *Heuristics: Intelligent search strategies for computer problem solving*. Addison-Welsey, 1984.
16. L. Kanal and V. Kumar. *Search in artificial intelligence*. Springer-Verlag., 1988.
17. J. Larrosa, P. Meseguer, and M. Sanchez. Pseudo-tree search with soft constraints. *In European Conference on Artificial Intelligence (ECAI'02)*, pages 131–135, 2002.
18. P. Jegou and C. Terrioux. Decomposition and good recording for solving max-csps. *In European Conference on Artificial Intelligence (ECAI'04)*, pages 196–200, 2004.
19. J. Huang and A. Darwiche. A structure-based variable ordering heuristic. *In International Joint Conference on Artificial Intelligence (IJCAI'03)*, pages 1167–1172, 2003.
20. W. Li and P. van Beek. Guiding real-world sat solving with dynamic hypergraph separator decomposition. *In International Conference on Tools with Artificial Intelligence (ICTAI'04)*, pages 542–548, 2004.

The Timetable Constrained Distance Minimization Problem

Rasmus V. Rasmussen[1] and Michael A. Trick[2]

[1] Department of Operations Research, University of Aarhus, Ny Munkegade,
Building 1530, 8000 Aarhus C, Denmark
[2] Tepper School of Business, Carnegie Mellon University, Pittsburgh PA 15213, USA

Abstract. The Timetable Constrained Distance Minimization Problem
is a sports scheduling problem applicable for tournaments where the total
travel distance must be minimized. In this paper we define the problem
and present an integer programming and a constraint programming formulation for the problem. Furthermore, we describe a hybrid integer pro-
gramming/constraint programming approach and a branch and bound
algorithm for solving the Timetable Constrained Distance Minimization
Problem. Finally, the computational performances of the four solution
methods are tested and compared.

Keywords: Timetabling, Integer Programming, Constraint Programming, Sports scheduling.

1 Introduction

Sports scheduling has proven to be a research area containing a large number of highly applicable problems which are very hard to solve. Furthermore,
good solutions may lead to great savings in travel costs due to reduced travel
distances or may increase the revenue earned from TV stations since special requirements can be satisfied. From an operations research perspective this makes
sports scheduling constitute an ideal research area.

Two problems have dominated the field of sports scheduling. The first is
the problem of designing a schedule which minimizes the total travel distance.
This problem is applicable for many sports leagues in America, since teams
often travel from one opponent to the next without returning home. The second
problem is to design a schedule which minimizes the number of *breaks*. A break
is two consecutive home games or two consecutive away games. This problem is
applicable in most European sports leagues where teams normally return home
after each away game and prefer an alternating pattern of home and away games.

Both the break and the distance problem have been solved by decomposing
the problem into a number of subproblems. See for instance Croce and Oliveri [2],
Henz [6, 7], Henz, Müller and Thiel [8], Nemhauser and Trick [11], Rasmussen and
Trick [12], Russell and Leung [14] and Schaerf [15]. Typically, a subproblem determines when the teams meet (meeting problem) and another subproblem decides
when the teams play home and away (home-away problem). The order of these

J.C. Beck and B.M. Smith (Eds.): CPAIOR 2006, LNCS 3990, pp. 167–181, 2006.
© Springer-Verlag Berlin Heidelberg 2006

subproblems vary but Trick [16] suggests that the subproblem which is most constrained or contains the most important constraints should be solved first.

When a TV network is willing to pay extra in case a number of games is scheduled on specific dates, we might want to solve the meeting problem before we solve the home-away problem. This decomposition has been widely used when breaks are minimized and in this case the home-away problem is known as the *Break Minimization Problem*. This problem has attracted a considerable amount of research and solution methods are presented by Elf, Júnger and Rinaldi [5], Miyashiro and Matsui [10], Régin [13] and Trick [16]. On the other hand we have not found results for the corresponding problem when travel distances are minimized instead of breaks.

In this paper, we define the *Timetable Constrained Distance Minimization Problem* (TCDMP) to be the problem of assigning home/away values to a fixed timetable so as to minimize the total distance traveled by all the teams. The following two applications emphasize the importance of the problem.

1. When sports leagues want to increase the revenue earned from TV networks, they must be able to schedule high-quality games at certain dates. This means that the meeting problem must be solved before the home-away problem. However, teams want a schedule which minimizes travel costs and therefore, the TCDMP must be solved after the meeting problem is solved.
2. A famous problem within the sports scheduling community is the Traveling Tournament Problem (TTP) [3] which is very simple to state and yet has proven very hard to solve. The TCDMP is a subproblem of the TTP and effective solution methods for the TCDMP could be useful in some of the heuristic solution methods for the TTP.

After having defined the TCDMP, we present an integer programming (IP) and a constraint programming (CP) formulation and we outline two solution methods which both outperform the IP and the CP model. The first solution method combines IP and CP and utilizes the strengths of both techniques by using CP for solving feasibility problems and IP for solving an optimization problem. The second solution method is a branch and price algorithm. See Barnhart, Johnson, Nemhauser, Savelsbergh and Vance [1] for a detailed description. This method has also been applied to the TTP by Easton, Nemhauser and Trick [4] and is to date the best exact solution method for the TTP.

In the following section, we give a short introduction to sports scheduling, define TCDMP and present the IP and the CP formulations for the problem. In Sect. 3 we outline the hybrid IP/CP approach and Sect. 4 explains the branch and price algorithm. Computational results are presented in Sect. 5 and we give some final remarks in Sect. 6.

2   Problem Formulation

In this section we will give a formal definition of the TCDMP and present an IP and a CP formulation of the problem. However, before we define the problem, let us give a brief introduction to some of the sports scheduling terminology.

We consider *double round robin tournaments* with an even number of teams. A double round robin tournament consists of two games between all pairs of teams leading to a total of $n(n-1)$ games in a tournament with n teams. We assume that these games are distributed evenly over $2(n-1)$ *time slots* such that all teams play exactly one game in each slot.

All teams have a home venue and play one game against all other teams at this venue. These games are called *home games* while games played at another venue are called *away games*. If a team plays two consecutive home games or away games, we say that the team has a *break* in the last of the two slots.

In Fig. 2.1, we give an example of a schedule for a double round robin tournament. The columns correspond to slots while the rows correspond to teams and the entrance (i,s) gives the opponent of team i in slot j. In case i plays away against j, this is shown by an @ and we have highlighted breaks using boldface. A schedule for a round robin tournament can be separated into two components, a *timetable* and a *pattern set*. The timetable gives the opponent of each team in each slot without considering venues. The pattern set consists of a *pattern* for each team and these patterns are vectors with an entrance for each slot, saying whether the corresponding team plays home or away. The combination of a timetable and a pattern set gives a schedule for the tournament. Figure 2.2 shows the timetable and the pattern set for the schedule shown in Fig. 2.1. In the pattern set, 1 represents a home game while 0 represents an away game. Using this terminology we are now able to define the TCDMP.

Definition 1 (TCDMP). *Given a timetable for a double round robin tournament with n teams, a distance matrix specifying the distances between the venues and an upper bound UB on the number of consecutive home and consecutive away games, find a feasible pattern set which minimizes the total distance traveled by all teams.*

Slot	1	2	3	4	5	6	7	8	9	10
Team 1	@6	3	@5	2	@4	6	@3	5	@2	4
Team 2	@5	6	4	@1	3	**5**	@6	@4	1	@3
Team 3	4	@1	6	**5**	@2	@4	1	@6	@5	2
Team 4	@3	5	@2	**@6**	1	3	@5	2	**6**	@1
Team 5	2	@4	1	@3	6	@2	4	@1	3	@6
Team 6	1	@2	**@3**	4	@5	@1	2	**3**	@4	5

Fig. 2.1. Schedule for a double round robin tournament with 6 teams

Slot	1	2	3	4	5	6	7	8	9	10
Team 1	6	3	5	2	4	6	3	5	2	4
Team 2	5	6	4	1	3	5	6	4	1	3
Team 3	4	1	6	5	2	4	1	6	5	2
Team 4	3	5	2	6	1	3	5	2	6	1
Team 5	2	4	1	3	6	2	4	1	3	6
Team 6	1	2	3	4	5	1	2	3	4	5

Slot	1	2	3	4	5	6	7	8	9	10
Team 1	0	1	0	1	0	1	0	1	0	1
Team 2	0	1	1	0	1	**1**	0	0	1	0
Team 3	1	0	1	**1**	0	0	1	0	0	1
Team 4	0	1	0	**0**	1	1	0	1	**1**	0
Team 5	1	0	1	0	1	0	1	0	1	0
Team 6	1	0	**0**	1	0	**0**	1	1	0	1

(a)	(b)

Fig. 2.2. (a) Timetable, (b) Pattern set

We have modelled the problem using both IP and CP and the models are presented in the following two subsections. In the rest of the paper, we let n denote the number of teams, while T denotes the set of teams. The set of slots is denoted S, and we let $S^0 = S \cup \{0\}$. The distance matrix is represented by D, and entrance $D_{i_1 i_2}$ contains the distance between the venue of team i_1 and the venue of team i_2. TT denotes the timetable and entrance TT_{is} gives the opponent of team i in slot s. Notice that $D_{TT_{is} TT_{is+1}}$ is the travel distance of team i between slots s and $s+1$ if team i plays away in both slots.

2.1   Integer Programming Formulation

To formulate the problem as an IP model, we use a binary variable h_{is} for each $i \in T$ and each $s \in S$. h_{is} equals 1 if team i plays home in slot s and it equals 0 if it plays away. To calculate the total travel distance, we use an integer variable, d_{is} for each $i \in T$ and each $s \in S^0$, which is equal to the distance team i travels between slot s and slot $s+1$. We use the dummy slots 0 and $2n-1$ to make sure that all teams start and end home. This gives the following IP model.

$$\min \sum_{i \in T} \sum_{s \in S^0} d_{is} \tag{2.1}$$

$$\text{s.t.} \quad d_{is} \geq (1 - h_{is} - h_{is+1}) D_{TT_{is} TT_{is+1}} \quad \forall i \in T, \ \forall s \in S^0 \tag{2.2}$$

$$d_{is} \geq (h_{is} - h_{is+1}) D_{i TT_{is+1}} \quad \forall i \in T, \ \forall s \in S^0 \tag{2.3}$$

$$d_{is} \geq (-h_{is} + h_{is+1}) D_{TT_{is} i} \quad \forall i \in T, \ \forall s \in S^0 \tag{2.4}$$

$$h_{i0} = 1 \quad \forall i \in T \tag{2.5}$$

$$h_{i2n-1} = 1 \quad \forall i \in T \tag{2.6}$$

$$\sum_{s \in S} h_{is} = n - 1 \quad \forall i \in T \tag{2.7}$$

$$h_{i_1 s} + h_{i_2 s} = 1 \quad \forall i_1, i_2 \in T, \ i_1 < i_2, \ \forall s \in S, \ TT_{i_1 s} = i_2 \tag{2.8}$$

$$h_{is_1} + h_{is_2} = 1 \quad \forall i \in T, \ \forall s_1, s_2 \in S, \ s_1 < s_2, \ TT_{is_1} = TT_{is_2} \tag{2.9}$$

$$\sum_{s=\hat{s}}^{\hat{s}+UB} h_{is} \leq UB \quad \forall i \in T, \ \forall \hat{s} \in \{1, \ldots, 2(n-1) - UB\} \tag{2.10}$$

$$\sum_{s=\hat{s}}^{\hat{s}+UB} h_{is} \geq 1 \quad \forall i \in T, \ \forall \hat{s} \in \{1, \ldots, 2(n-1) - UB\} \tag{2.11}$$

$$h_{is} \in \{0, 1\} \quad \forall i \in T, \ \forall s \in \{0, \ldots, 2n-1\} \tag{2.12}$$

$$d_{is} \in \mathbb{Z}_+ \quad \forall i \in T, \ \forall s \in S^0 \tag{2.13}$$

Constraints (2.2) - (2.4) give lower bounds on the distance team i travels between slots s and $s+1$ when i plays away in the two slots, when it plays home and away and when it plays away and home, respectively. Constraints (2.5) and (2.6) ensure that all teams start and end at home and constraints (2.7) make sure that all teams have exactly $n-1$ home games. Constraints (2.8) require that, when teams i_1 and i_2 meet in slot s, one of the teams must play home and the other must play away, while constraints (2.9) make sure that team i plays one home game and one away game in two slots with the same opponent. Finally, the constraints (2.10) and (2.11) give upper bounds on the number of consecutive home games and the number of consecutive away games.

2.2 Constraint Programming Formulation

When formulating the problem as a CP model, we use variables similar to the variables used in the IP model, but the CP model allows us to reformulate the constraints. In particular, we are able to formulate the constraints (2.7), (2.10) and (2.11) as a single constraint called *sequence* and we can use logical expressions to determine the travel distance. This gives the following CP model.

$$\min \sum_{i \in T} \sum_{s \in S^0} d_{is} \tag{2.14}$$

$$\text{s.t.} \quad (h_{is} = 1) \wedge (h_{is+1} = 1) \Rightarrow (d_{is} = 0) \qquad \forall i \in T, \ \forall s \in S^0 \tag{2.15}$$

$$(h_{is} = 0) \wedge (h_{is+1} = 1) \Rightarrow (d_{is} = D_{TT_{is}i}) \qquad \forall i \in T, \ \forall s \in S^0 \tag{2.16}$$

$$(h_{is} = 1) \wedge (h_{is+1} = 0) \Rightarrow (d_{is} = D_{iTT_{is+1}}) \qquad \forall i \in T, \ \forall s \in S^0 \tag{2.17}$$

$$(h_{is} = 0) \wedge (h_{is+1} = 0) \Rightarrow (d_{is} = D_{TT_{is}TT_{is+1}}) \qquad \forall i \in T, \ \forall s \in S^0 \tag{2.18}$$

$$sequence(1, UB, UB - 1, [h_{i1}, \ldots, h_{i2n-2}], [1], [n - 1]) \qquad \forall i \in T \tag{2.19}$$

$$h_{is_1} \neq h_{is_2} \qquad \forall i \in T, \ \forall s_1, s_2 \in S, \ s_1 < s_2, \ TT_{is_1} = TT_{is_2} \tag{2.20}$$

$$h_{is} \neq h_{TT_{is}s} \qquad \forall i \in T, \ \forall s \in S \tag{2.21}$$

$$h_{i0} = 1 \qquad \forall i \in T \tag{2.22}$$

$$h_{i2n-1} = 1 \qquad \forall i \in T \tag{2.23}$$

$$h_{is} \in \{0, 1\} \qquad \forall i \in T, \ \forall s \in \{0, \ldots, 2n - 1\} \tag{2.24}$$

$$d_{is} \in \mathbb{Z}_+ \qquad \forall i \in T, \ \forall s \in S^0 \tag{2.25}$$

The constraints (2.15) - (2.18) determine the travel distance of team i between slots s and $s + 1$, depending on whether team i plays home or away in the two slots. The *sequence constraints* (2.19) are global constraints with the syntax $sequence(min, max, width, varVec, valueVec, cardVec)$ where min, max and $width$ are numbers, $varVec$ is a vector of variables and $valueVec$ and $cardVec$ are vectors with the same index set. The constraints are satisfied if each entrance in $cardVec$ equals the number of variables in $varVec$, which are equal to the corresponding entrance in $valueVec$ and for each value in $valueVec$ at least min and at most max variables are equal to this value in any subsequence of length $width$. In our case the constraint says that team i must play exactly $n - 1$ home games and in $UB + 1$ consecutive slots it cannot play less than one home game or more than UB home games. Constraints (2.20) state that team i must play one home game and one away game in two slots where it meets the same opponent and constraints (2.21) require that the opponent of team i plays home (away) if team i plays away (home). Constraints (2.22) and (2.23) make sure that all teams start and end home.

In the following sections, we present a hybrid IP/CP approach and a branch and price algorithm which are both able to outperform the IP and CP models shown in this section. A Benders decomposition approach similar to the method presented in [12] has also been implemented but the reduction in time used to solve the master problem cannot offset the additional iterations which are required compared to the hybrid IP/CP approach.

3   Hybrid IP/CP Approach

The first of the specialized solution methods is a hybrid IP/CP approach which decomposes the problem into two phases. Phase 1 generates all feasible patterns for each team in the tournament and Phase 2 finds the optimal pattern set by assigning each team to one of the patterns found in Phase 1.

CP is in general very effective at solving feasibility problems and it has the advantage of being able to find all solutions to a specific problem instead of a single solution. These characteristics makes a CP model ideal for finding feasible patterns in Phase 1. On the other hand, IP is typically stronger than CP when it comes to optimization problems, since the linear relaxation can be used to prune suboptimal solutions. Therefore IP is used in Phase 2 to choose the optimal patterns from the patterns generated in Phase 1. The details of the two phases are explained below.

3.1   Phase 1

In order to generate all feasible patterns, we use a CP model for each team and find all feasible solutions to each of the models. Each pattern must contain exactly $n - 1$ home games and satisfy the upper bound on the number of consecutive home games and consecutive away games. Furthermore, a pattern for a specific team i must satisfy that, for all pairs of slots s_1 and s_2 where $TT_{is_1} = TT_{is_2}$, the pattern has both a home game and an away game.

To formulate a CP model for finding all feasible patterns of team i, we use a binary variable h_s for each $s \in S$. As in the earlier sections, $h_s = 1$ implies a home game in slot s and $h_s = 0$ implies an away game. The CP model for team i looks as follows.

solve:

$$sequence(1, UB, UB + 1, [h_1, \ldots, h_{2(n-1)}], [1], [n - 1]) \tag{3.1}$$
$$h_{s_1} \neq h_{s_2} \qquad \forall s_1, s_2 \in S, \ \ s_1 < s_2, \ \ TT_{is_1} = TT_{is_2} \tag{3.2}$$
$$h_s \in \{0, 1\} \qquad \forall s \in S \tag{3.3}$$

Constraint (3.1) corresponds to constraint (2.19) in the CP formulation and makes sure that the number of home games is correct and the upper bound on consecutive home games and consecutive away games is satisfied. Constraints (3.2) correspond to the constraints (2.20) and make sure that the patterns are feasible with respect to the timetable.

For each team $i \in T$, we let P_i denote the set of feasible patterns and for each $j \in P_i$, we let h_{js} represent the entrance h_s of pattern j. We also calculate the distance team i must travel, if it uses pattern j, and denote it d_{ij}. The distance can be calculated since the timetable gives us the opponent of each slot and the pattern tells if team i plays home or away.

3.2 Phase 2

In Phase 2, we must find an optimal allocation of each team i to a pattern $j \in P_i$, such that the total travel distance is minimized and the pattern set is feasible with respect to the timetable.

To formulate an IP model for this problem, we use a binary variable x_{ij} for each $i \in T$ and each $j \in P_i$. The variable is 1 if team i is assigned to pattern j and 0 otherwise. We also use the home-away parameter h_{js} for each pattern j and each slot s and the distance parameter d_{ij} for each team i and each pattern $j \in P_i$.

$$\min \sum_{i \in T} \sum_{j \in P_i} d_{ij} x_{ij} \tag{3.4}$$

$$\text{s.t.} \sum_{j \in P_i} x_{ij} = 1 \quad \forall i \in T \tag{3.5}$$

$$\sum_{i \in \{i_1, i_2\}} \sum_{j \in P_i} h_{js} x_{ij} = 1 \quad \forall i_1, i_2 \in T, \; i_1 < i_2, \; \forall s \in S, \; TT_{i_1 s} = i_2 \tag{3.6}$$

$$x_{ij} \in \{0, 1\} \quad \forall i \in T, \; \forall j \in P_i \tag{3.7}$$

Constraints (3.5) are the assignment constraints saying that all teams must be assigned to a feasible pattern and constraints (3.6) make sure that when 2 teams meet, one of the teams play home and the other plays away. These two constraints are enough to ensure a feasible pattern set with respect to the timetable since we know from Phase 1 that all the teams play one home game and one away game in two slots where they meet the same opponent.

4 Branch and Price

In addition to the hybrid IP/CP approach, we present a branch and price algorithm to solve the TCDMP. This method has successfully been applied to the TTP and is to date the best exact solution method for the TTP.

The branch and price algorithm assigns teams to patterns by solving a linear programming (LP) problem which is restricted to only contain a subset of the feasible patterns instead of all the feasible patterns. This problem is known as the master problem.

The solution of the master problem may be fractional since we use an LP problem and it might not be optimal since we consider only a subset of the patterns. The optimality issue is handled by solving a pricing problem which finds patterns to the master problem with negative reduced costs. These patterns are added to the master problem and it is re-solved. If no patterns with negative reduced costs exist and the solution is fractional, the algorithm uses branch and bound to obtain an integer solution.

Before describing the details of the algorithm, let us present an outline with references to the relevant sections. UB^M denotes an upper bound on the master problem and N is the node set of the branch and bound tree.

Initialization. Find a feasible pattern set (Sect. 4.1).

 Initialize UB^M to the solution value of the initial feasible solution.

 Initialize N to a single node.

Step 1. If $N \neq \emptyset$, choose $\hat{\eta}$ from N and let $N = N \setminus \hat{\eta}$ (Sect. 4.2).

 Otherwise, stop.

Step 2. Solve the master problem and go to Step 3 (Sect. 4.3).

Step 3. Solve the pricing problem for each team (Sect. 4.4). If no patterns with negative reduced costs exist, go to Step 4.

 Otherwise, add patterns to the master problem and go to Step 2.

Step 4. If the solution value is greater than UB^M, go to Step 1.

 Otherwise, if the solution is fractional, go to Step 5.

 Otherwise, update UB^M to the solution value and go to Step 1.

Step 5. branch, add the new nodes to N and go to Step 1 (Sect. 4.5).

4.1   Initial Feasible Pattern Set

In order to find an initial feasible pattern set, we use a CP model. The model corresponds to the CP model presented in Sect. 2.2 but, in this model, we ignore the travel distance, since we are only looking for a feasible solution. This gives the following CP model where the variable h_{is}, for each $i \in T$ and each $s \in S$, is 1 if team i plays home in slots s and 0 if it plays away.

solve:

$$sequence(1, UB, UB + 1, [h_{i1}, \cdots, h_{i2(n-1)}], [1], [n-1]) \qquad \forall i \in T \qquad (4.1)$$

$$h_{is_1} \neq h_{is_2} \qquad \forall i \in T, \ \forall s_1, s_2 \in S, \ s_1 < s_2, \ TT_{is_1} = TT_{is_2} \qquad (4.2)$$

$$h_{is} \neq h_{TT_{is}s} \qquad \forall i \in T, \ \forall s \in S \qquad (4.3)$$

$$h_{is} \in \{0, 1\} \qquad \forall i \in T, \ \forall s \in S \qquad (4.4)$$

The constraints (4.1) - (4.3) are similar to the constraints (2.19) - (2.21) from the CP model in Sect. 2.2. The rest of the constraints from the CP model in Sect. 2.2 can be ignored, since they are all related to the travel distance.

If the model is infeasible, it means that no feasible pattern set exists and we are done. Otherwise, we store the patterns used by each team and calculate the travel distances. As in Sect. 3 we let P_i denote patterns which are feasible for team i but in this context P_i does not necessarily contain all the feasible patterns for team i. When an initial feasible solution has been found, P_i is initialized to contain the pattern team i uses and the travel distance is calculated.

In the rest of the section, we use the notation that, for each team $i \in T$, the parameter h_{js} for each pattern $j \in P_i$ and each slot $s \in S$ is 1 if pattern j has a home game in slot s and 0 if it has an away game. We also let d_{ij} denote the travel distance of team i if it uses pattern j from P_i.

4.2   Node Selection Strategy

We have implemented two node selection strategies which are used to choose nodes from the branch and bound tree. The first strategy is the well-known *depth first strategy*. This strategy chooses one of the child nodes of the current strategy if any exists and otherwise it backtracks. The strategy corresponds to a *last in, first out* (LIFO) strategy since it always chooses the last node which has been added.

The second strategy is a *best lower bound strategy* which chooses the node with the lowest lower bound. In our case, we use the objective value of the parent node as lower bound and therefore the strategy chooses the node with the smallest parent value. Ties are broken arbitrarily.

4.3   Master Problem

The master problem of the branch and price algorithm is almost identical to the linear relaxation of the problem solved in Phase 2 of the hybrid IP/CP approach. The only differences are that the problem is restricted since P_i does not contain all the feasible patterns of team i, and that a number of branching constraints are added. The number of branching constraints corresponds to the level of the current node in the branch and bound tree.

The branching strategy and the branching constraints will be further discussed in Sect. 4.5 but we need some notation to formulate the master problem. We let B^η denote the set of branching constraints present in node η and we let i_b, s_b and v_b denote the team, the slot and the value of branching constraint $b \in B^\eta$. If v_b equals 1, it means that team i_b must play home in slot s_b and it must play away if v_b equals 0.

Since we use the linear relaxation, the variable x_{ij} gives the fraction of team i which is assigned to pattern j from P_i. Now, we can state the master problem corresponding to node $\eta \in N$ as follows.

$$\min \sum_{i \in T} \sum_{j \in P_i} d_{ij} x_{ij} \tag{4.5}$$

$$\text{s.t.} \sum_{j \in P_i} x_{ij} = 1 \quad \forall i \in T \tag{4.6}$$

$$\sum_{i \in \{i_1, i_2\}} \sum_{j \in P_i} h_{js} x_{ij} = 1 \quad \forall i_1, i_2 \in T, \ i_1 < i_2, \ \forall s \in S, \ TT_{i_1 s} = i_2 \tag{4.7}$$

$$\sum_{j \in P_{i_b}} h_{js_b} x_{i_b j} = v_b \quad \forall b \in B^\eta \tag{4.8}$$

$$x_{ij} \in R_+ \quad \forall i \in T, \ \forall j \in P_i \tag{4.9}$$

The constraints (4.6) - (4.7) are similar to constraints (3.5) - (3.6) from Sect. 3.2 and constraints (4.8) are the branching constraints.

In the following, we refer to the optimal solution of the master problem as $\bar{x}$ and we let $\bar{P}_i = \{j \in P_i : \bar{x}_{ij} > 0\}$ denote the set of patterns to which a fraction of team i is assigned.

In case the master problem is infeasible, we need to check if this is because of missing patterns or because the branching constraints make the problem infeasible. To do this, we solve a CP model similar to the model for finding an initial feasible solution presented in Sect. 4.1 - but with the branching constraints added. If this model has a feasible solution, we add the patterns used by each of the teams and re-solve the master problem. Otherwise, we return to Step 1 of the algorithm and choose a new node in the search tree.

4.4   Pricing Problem

When the master problem has been solved we use a pricing problem for finding patterns with negative reduced costs. For a general optimization problem

$$\min cx$$
$$\text{s.t.}\quad Ax = b$$
$$x \geq 0$$

where c and x are n-vectors, A is an $m \times n$ matrix and b is an m-vector, the reduced cost of a variable x_i is $c_i - \bar{u}A_i$ when $\bar{u}$ is an optimal dual solution.

We let $\bar{u}_i^1$ for all $i \in T$, $\bar{u}_{i_1 i_2 s}^2$ for all $i_1, i_2 \in T$, $i_1 < i_2$ and $s \in S$ where $TT_{i_1 s} = i_2$ and $\bar{u}_b^3$ for all $b \in B^\eta$ denote optimal dual variables corresponding to constraints (4.6), (4.7) and (4.8) from the master problem, respectively. Furthermore, we let $\bar{u}_{i_2 i_1 s}^2 = \bar{u}_{i_1 i_2 s}^2$ for all $i_1, i_2 \in T$ with $i_1 < i_2$ and $s \in S$ where $TT_{i_1 s} = i_2$. Then the reduced cost of a pattern $\hat{j}$ used by team $\hat{i}$ in node η can be written as follows.

$$d_{\hat{i}\hat{j}} - \bar{u}_{\hat{i}}^1 - \sum_{\substack{i \in T \\ TT_{\hat{i}s} = i}} \sum_{\substack{s \in S}} \bar{u}_{\hat{i}is}^2 h_{\hat{j}s} - \sum_{b \in B^\eta} \bar{u}_b^3 h_{js_b}$$

In order to find patterns with negative reduced costs, we use a pricing problem for each team. The pricing problem finds the pattern with the smallest reduced cost and the pattern is added to the master problem if the reduced cost is negative.

To solve the pricing problem, we use an IP model since we want to minimize the reduced cost. Alternatively, a CP model could be used to generated patterns with negative reduced costs but we have obtained the best results when the reduced cost is minimized.

To formulate the IP model for a team $\hat{i}$, we use a binary variable h_s for each $s \in \{0, \ldots, 2n-1\}$ and an integer variable d_s for each $s \in S^0$. The variable h_s is 1 if the pattern has a home game in slot s and 0 if it has an away game, while d_s is the travel distance of team $\hat{i}$ between slot s and $s + 1$. The IP model for team $\hat{i}$ is presented below.

$$\min \sum_{s \in S^0} d_s - \bar{u}_{\hat{i}}^1 - \sum_{i \in T} \sum_{\substack{s \in S \\ TT_{\hat{i}s} = i}} \bar{u}_{\hat{i}is}^2 h_s - \sum_{b \in B^\eta} \bar{u}_b^3 h_{s_b} \tag{4.10}$$

$$\text{s.t. } d_s \geq (1 - h_s - h_{s+1}) D_{TT_{\hat{i}s} TT_{\hat{i}s+1}} \qquad \forall s \in S^0 \tag{4.11}$$

$$d_s \geq (h_s - h_{s+1}) D_{\hat{i} TT_{\hat{i}s+1}} \qquad \forall s \in S^0 \tag{4.12}$$

$$d_s \geq (-h_s + h_{s+1}) D_{TT_{\hat{i}s} \hat{i}} \qquad \forall s \in S^0 \tag{4.13}$$

$$h_0 = 1 \tag{4.14}$$

$$h_{2n-1} = 1 \tag{4.15}$$

$$\sum_{s \in S} h_s = n - 1 \tag{4.16}$$

$$h_{s_1} + h_{s_2} = 1 \qquad \forall s_1, s_2 \in S, \ s_1 < s_2, \ TT_{\hat{i}s_1} = TT_{\hat{i}s_2} \tag{4.17}$$

$$\sum_{s=\hat{s}}^{\hat{s}+UB} h_s \leq UB \qquad \forall \hat{s} \in \{1, \dots, 2(n-1) - UB\} \tag{4.18}$$

$$\sum_{s=\hat{s}}^{\hat{s}+UB} h_{is} \geq 1 \qquad \forall \hat{s} \in \{1, \dots, 2(n-1) - UB\} \tag{4.19}$$

$$h_s \in \{0, 1\} \qquad \forall s \in \{0, \dots, 2n - 1\} \tag{4.20}$$

$$d_s \in \mathbb{Z}_+ \qquad \forall s \in S^0 \tag{4.21}$$

The objective function calculates the reduced cost of the pattern given by the h_s variables. Constraints (4.11) - (4.13) are used to calculate the distance, constraints (4.14) and (4.15) make sure that $\hat{i}$ starts and ends home and constraints (4.16) - (4.19) state the general constraints for a pattern. We refer to the IP model presented in Sect. 2.1 for further explanation of the constraints.

The pricing problem is solved for each team $\hat{i} \in T$ and in case the solution value is less than zero, we add the pattern given by the h_s variables to $P_{\hat{i}}$. When the pricing problem has been solved for all teams, we re-solve the master problem if patterns have been added and otherwise, we go to Step 4 of the algorithm.

4.5 Branching Strategy

In case the optimal solution of the master problem is fractional and no patterns with negative reduced costs exist, the algorithm branches to obtain an integer solution. Instead of branching on one of the fractional x values from the master problem, we use higher order branching. We choose a team $\hat{i}$ and a slot $\hat{s}$ and create two new nodes by letting team $\hat{i}$ play home in slot $\hat{s}$ in one of the nodes and away in the other.

In order to find $\hat{i}$ and $\hat{s}$, we have implemented two strategies. The first strategy starts by finding the team $\hat{i}$ and pattern $\hat{j}_1$ which result in the highest fractional $\bar{x}_{\hat{i}\hat{j}_1}$ value such that

$$\bar{x}_{\hat{i}\hat{j}_1} = \max\{\bar{x}_{ij} | i \in T, j \in \bar{P}_i : 0 < \bar{x}_{ij} < 1\}.$$

Then it finds the pattern $\hat{j}_2$ which results in the second highest fractional value for team $\hat{i}$ such that

$$\bar{x}_{\hat{i}\hat{j}_2} = \max\{\bar{x}_{\hat{i}j}|j \in \bar{P}_{\hat{i}} : 0 < \bar{x}_{\hat{i}j} < 1 \wedge j \neq \hat{j}_1\}.$$

Finally, it finds a slot $\hat{s}$ where the two patterns $\hat{j}_1$ and $\hat{j}_2$ have a difference such that $h_{\hat{j}_1\hat{s}} \neq h_{\hat{j}_2\hat{s}}$.

The second strategy starts by finding the team $\hat{i}$ and pattern $\hat{j}_1$ such that the variable $\bar{x}_{\hat{i}\hat{j}_1}$ is as close to 0.5 as possible. It then finds a second pattern $\hat{j}_2$ different from $\hat{j}_1$ such that $\bar{x}_{\hat{i}\hat{j}_2}$ is as close to 0.5 and finally it finds a slot $\hat{s}$ such that $h_{\hat{j}_1\hat{s}} \neq h_{\hat{j}_2\hat{s}}$.

When we have found the branching team $\hat{i}$ and the branching slot $\hat{s}$ we can formulate the two branching cuts

$$\sum_{j \in P_{i_b}} h_{js_b}x_{i_bj} = 0 \qquad \sum_{j \in P_{i_b}} h_{js_b}x_{i_bj} = 1$$

where $i_b = \hat{i}$ and $s_b = \hat{s}$.

Now, we are ready to add two nodes η_1 and η_2 to the search tree. Assuming that the current node is node η, we let B^{η_1} and B^{η_2} be equal to B^{η} and add the first of the two branching constraints to B^{η_1} and the second to B^{η_2}.

5 Computational Results

In order to explore the computational complexity of the TCDMP and to compare the proposed solution methods, we have tested all 4 methods on 60 instances ranging from 6 to 16 teams.

At the homepage http://mat.gsia.cmu.edu/TOURN/, Michael Trick's benchmark problems for the TTP can be found. These problems have been studied intensively and a number of solutions are presented at the web page. By letting $UB = 3$, using the presented distance matrices and permuting the slots of the presented TTP solutions, we have generated instances of the TCDMP.

For each even number of teams from 6 to 16 we have generated 10 instances by permuting slots of the following solutions. For 6 teams we have used the solution of Easton May 7, 1999; for 8 teams, the solution of Easton January 27, 2000; for 10 teams, the solution of Langford, June 13, 2005 and for 12, 14 and 16 teams, the solution of Zhang Xingwen August 28, 2002. All the tests have been performed on an Intel Xeon 2.67 GHz processor with 6 GB RAM. The IP and CP models have been solved by using OPL Studio [9] with the callable libraries CPLEX and Solver. The hybrid IP/CP approach and the branch and price algorithm have been implemented in OPL script.

The computational results are presented in Table 5.1. For each number of teams and each solution method, the table shows the number of instances solved and the minimum, average and maximum time used on the solved instances. We have used a time limit of 1800 seconds and instances which have not been solved

Table 5.1. Computational Results

n	Solution method	Number solved	Time (s)		
			Min.	Avg.	Max.
6	IP	10	0.42	22.51	217.18
6	CP	10	8.03	17.00	24.94
6	IP/CP	10	0.06	0.07	0.08
6	BP-df-1	10	0.83	1.59	3.34
6	BP-bv-1	10	0.82	1.27	2.29
6	BP-df-2	10	0.84	1.74	3.43
6	BP-bv-2	10	0.83	1.25	2.19
8	IP	10	60.72	397.35	954.56
8	CP	0	-	-	-
8	IP/CP	10	0.41	0.44	0.47
8	BP-df-1	10	2.87	11.74	21.65
8	BP-bv-1	10	2.88	6.99	11.90
8	BP-df-2	10	2.83	10.05	22.94
8	BP-bv-2	10	2.82	6.15	8.19
10	IP	1	1273.74	1273.74	1273.74
10	CP	0	-	-	-
10	IP/CP	10	1.79	2.02	2.34
10	BP-df-1	10	7.23	27.01	157.56
10	BP-bv-1	10	7.05	16.55	46.50
10	BP-df-2	10	6.98	24.66	64.49
10	BP-bv-2	10	7.03	15.51	44.34
12	IP	0	-	-	-
12	CP	0	-	-	-
12	IP/CP	10	10.16	12.19	18.19
12	BP-df-1	10	24.29	281.60	1386.09
12	BP-bv-1	10	23.77	130.31	434.97
12	BP-df-2	10	23.80	243.23	1038.98
12	BP-bv-2	10	23.56	151.74	650.79
14	IP	0	-	-	-
14	CP	0	-	-	-
14	IP/CP	10	35.52	38.07	42.83
14	BP-df-1	10	49.62	95.00	248.32
14	BP-bv-1	10	48.61	91.71	240.47
14	BP-df-2	10	48.69	165.40	750.18
14	BP-bv-2	10	48.74	72.78	164.82
16	IP	0	-	-	-
16	CP	0	-	-	-
16	IP/CP	10	153.80	197.16	260.47
16	BP-df-1	9	122.47	866.13	1770.94
16	BP-bv-1	9	119.37	827.90	1645.91
16	BP-df-2	8	121.42	662.81	1377.21
16	BP-bv-2	9	119.38	631.30	1382.13

within this time are not included in the average. Since we have 2 node selection strategies and two branching strategies for the branch and price algorithm, we present four versions of this solution method: BP-df-1, BP-df-2, BP-bv-1 and BP-bv-2. The terms df and bv refer to depth first and best value node selection, respectively, while 1 and 2 refer to the first and the second branching strategy.

Table 5.1 shows that, even though CP is better than IP at solving instances with 6 teams, it is not able to solve any of the instances with 8 teams. IP is doing a little better and is able to solve all the instances with less than 10 teams and a single instance with 10 teams. The fact that IP is able to handle larger instances than CP is no surprise, since IP models often excel compared to CP models when optimization problems are considered.

Both the hybrid IP/CP approach and the branch and price algorithm clearly outperform the IP and CP models. We see that the hybrid IP/CP approach shows the best results and, in addition to being the fastest on average, it is also the most stable of the solution methods. The only drawback of this method is a rather large memory consumption since all patterns are generated initially. For instances with 16 teams, it generates up to 85000 patterns and uses approximately 200 MB of memory.

The computation times of the branch and price algorithm are highly dependent on the size of the branching tree and we see that there is an order of magnitude in difference between the minimum and maximum time. This means that the algorithm is only competitive with the hybrid IP/CP approach when an integer solution is found relatively fast in the branching tree.

In addition to the tests presented here, we have tested the solution methods on instances with 18 teams but none of the methods were able to solve these instances within the given time limit. In this case the hybrid IP/CP approach generated up to 246000 patterns. Still, the tests have shown that the hybrid IP/CP approach is capable of solving the problem for practical applications like the National League Baseball which consists of 14 teams.

6   Conclusion

In this paper we have defined the TCDMP which applies when double round robin tournaments with a fixed timetable are scheduled and the total travel distance must be minimized.

We have modelled the problem both as an IP model and as a CP model and presented two specialized solution methods. The solution methods have been tested and the computational results show that the problem can be solved effectively for instances with up to 16 teams by using a hybrid IP/CP approach.

In addition to being an interesting problem of its own, the TCDMP also opens for new research directions. The performances of the hybrid IP/CP approach opens for the opportunity to include the TCDMP in heuristics for the Traveling Tournament Problem. By solving the TCDMP, the heuristics could be able to solve a large neighborhood effectively.

References

1. C. Barnhart, E.L. Johnson, G.L. Nemhauser, M.W.P. Savelsbergh, P.H. Vance, Branch-and-price: Column generation for solving huge integer programs, Operations Research 46 (3)(1998)316-329.
2. F.D. Croce, D. Oliveri, Scheduling the Italian football league: An ILP-based approach, Computers & Operations Research 33 (7)(2006)1963-1974.
3. K. Easton, G. Nemhauser, M. Trick, The traveling tournament problem: Description and benchmarks, in: Proceedings CP'01, Lecture Notes in Computer Science 2239 (2001)580-584.
4. K. Easton, G. Nemhauser, M. Trick, Solving the traveling tournament problem: A combined integer programming and constraint programming approach, in: E. Burke, P. De Causmaecker (Eds.), (PATAT 2002), Lecture Notes in Computer Science 2740, Springer, 2003, 100-109.
5. M. Elf, M. Jünger, G. Rinaldi, Minimizing breaks by maximizing cuts, Operations Research Letters, 31 (2003)343-349
6. M. Henz, Constraint-based round robin tournament planning, in: D. De Schreye (Eds.), Proceedings of the International Conference on Logic Programming, Las Cruces, New Mexico, MIT Press (1999)545-557.
7. M. Henz, Scheduling a major college basketball conference - revisited, Operations Research 49 (2001)163-168.
8. M. Henz, T. Müller, S. Thiel, Global constraints for round robin tournament scheduling, European Journal of Operational Research 153 (1)(2004)92-101.
9. ILOG, ILOG OPL Studio 3.7, Language Manual, 2003.
10. R. Miyashiro, T. Matsui, Semidefinite programming based approaches to the break minimization problem, Computers & Operations Research 33 (7)(2006)1975-1982.
11. G.L. Nemhauser, M.A. Trick, Scheduling a Major College Basketball Conference, Operations Research 46 (1)(1998)1-8.
12. R.V. Rasmussen, M.A. Trick, A Benders approach for the constrained minimum break problem, European Journal of Operational Research (to appear).
13. J.C. Régin, Minimization of the number of breaks in sports scheduling problems using constraint programming, Constraint programming and large scale discrete optimization, DIMACS Series in Discrete Mathematics and Theoretical Computer Science 57 (2001)115-130.
14. R.A. Russell and J.M.Y. Leung, Devising a Cost Effective Schedule for a Baseball League, Operations Research 42 (4)(1994)614-625.
15. A. Schaerf, Scheduling Sport Tournaments using Constraint Logic Programming, Constraints 4 (1999)43-65.
16. M.A. Trick, A Schedule-Then-Break Approach to Sports Timetabling, in: Burke, E.K., Erben, W. (Eds.), (Practice and Theory of Automated Timetabling III) Lecture Notes in Computer Science 2079, Springer-Verlag, Berlin Heidelberg New York (2001)242-253.

Conflict-Directed A* Search for Soft Constraints

Martin Sachenbacher[1] and Brian C. Williams[2]

[1] LMU München, Oettingenstraße 67, 80538 München, Germany
sachenba@pms.ifi.lmu.de
[2] MIT CSAIL, 32 Vassar Street, Cambridge, MA 02139, USA
williams@mit.edu

Abstract. As many real-world problems involve user preferences, costs, or probabilities, constraint satisfaction has been extended to optimization by generalizing hard constraints to soft constraints. However, as techniques such as local consistency or conflict learning do not easily generalize to optimization, solving soft constraints appears more difficult than solving hard constraints. In this paper, we present an approach to solving soft constraints that exploits this disparity by re-formulating soft constraints into an optimization part (with unary objective functions), and a satisfiability part. This re-formulation is exploited by a search algorithm that enumerates subspaces with equal valuation, that is, plateaus in the search space, rather than individual elements of the space. Within the plateaus, familiar techniques for satisfiability can be exploited. Experimental results indicate that this hybrid approach is in some cases more efficient than other known methods for solving soft constraints.

1 Introduction

Many real-world problems are naturally framed as optimization problems where the task is to find assignments to variables that optimize user preference, cost, or probability. Therefore, constraint satisfaction problems (CSPs) have been extended from satisfaction to optimization by the notion of soft constraints. One general framework for soft constraints are valued constraint satisfaction problems (VCSPs) [22, 1], which augment CSPs with a valuation structure and generalize many earlier notions such as fuzzy CSPs, probabilistic CSPs, or partial constraint satisfaction.

For the case of solving CSPs, techniques such as local consistency filtering [16] and conflict (nogood) learning [5] have proven to be very effective. Substantial progress has been made in extending these techniques to the more general case of soft constraints [2, 7]; however, the optimization case still appears far more difficult than the satisfaction case.

In practical applications, the constraints often exhibit structure or regularities that can be exploited in order to make optimization feasible. For instance, approaches based on tree decomposition [8, 12] exploit favorable properties of the constraint graph (limited width) to break down the problem into lower-dimensional subproblems.

J.C. Beck and B.M. Smith (Eds.): CPAIOR 2006, LNCS 3990, pp. 182–196, 2006.

In this paper, we present an approach to exploit a form of structure that is specific to optimization problems. It has been underlying algorithmic approaches in the area of model-based reasoning and diagnosis [26, 9] for quite some time. Model-based reasoning aims at describing the behavior of physical systems in terms of formal models, where some variables capture preferences (such as the failure probability of a component, or the cost of repairing it), and constraints capture consistency (such as the physically possible behavior of a component). As the number of variables defining preferences is typically small compared to the overall number of variables, there often exist large sets of assignments that have equal valuation, that is, large "plateaus" in the search space. Williams [27] presents an algorithmic approach called *conflict-directed A** that exploits this fact by coupling together optimization and satisfaction techniques. The algorithm enumerates plateaus (parts of the search space with the same valuation) in best-first order, and subsequently checks if there exists a consistent solution within the plateau; information about infeasible assignments is re-used between the plateaus in the form of conflicts. This approach can be more efficient than enumerating individual elements of the search space, because depending on the problem, there can be fewer plateaus than total elements in the search space.

In this paper, we generalize upon these ideas, and extend their applicability from model-based reasoning applications to the general case of soft constraints. Our approach consists of factoring VCSPs into a set of (unary) soft constraints that carry all the information about valuations of assignments, and a set of hard constraints that do not carry valuations but just need to be satisfied. Like in [18, 19], this re-formulation is based on introducing additional variables capturing the cost of violating constraints; a special case of this re-formulation is taking the dual of the problem [14]. Following the terminology of [27], we call the resulting hybrid representation *optimal CSP*; it makes explicit the optimization and satisfiability aspects of a problem. The idea is then to algorithmically exploit the separation into hard and soft problem parts by applying optimization techniques to the optimization part, and satisfiability techniques to the satisfiability part. Specifically, for a reasonably small soft constraint part, we can use A* search [10], which is optimal in the number of search nodes visited, but would typically be infeasible to apply on the original VCSP problem due to its memory requirements. For the hard constraint part, we can draw on well-established, efficient techniques for satisfiability problems, such as local consistency and conflict learning; in this paper, we build on an existing SAT solver for this purpose.

The paper is organized as follows: We review the definitions of valued CSPs [22] and optimal CSPs [27] and present a method for transforming between them. The transformation yields a separation into hard constraints and unary soft constraints. We then present a variant of conflict-directed A* that exploits this re-formulation by searching over sets of assignment with equal valuation, rather than searching over individual assignments of the variables in the problem. We give experimental results demonstrating that this algorithm can outperform other methods for solving valued CSPs, and finally we indicate some directions for future work.

2   Valued CSPs

A classical *constraint satisfaction problem* (CSP) is a triple (X, D, C) with variables $X = \{x_1, \ldots, x_n\}$, finite domains $D = \{\text{dom}(x_1), \ldots, \text{dom}(x_n)\}$, and constraints $C = \{c_1, \ldots, c_m\}$. Each constraint $c_j \in C$ is a relation $c_j \subseteq \Pi_{x_i \in \text{var}(c_j)} \text{dom}(x_i)$ over variables $\text{var}(c_j) \subseteq X$. An assignment t to variables $\text{var}(c_j)$ *satisfies* the constraint if $t \in c_j$, and *violates* it otherwise.

Definition 1 (Valuation Structure [22]). *A valuation structure is a tuple* $(E, \leq, \oplus, \bot, \top)$ *where E is a set of valuations, totally ordered by $\leq$ with a minimum element $\bot \in E$ and a maximum element $\top \in E$, and $\oplus$ is an associative, commutative, and monotonic binary operation with identity element $\bot$ and absorbing element $\top$.*

The set of valuations E expresses different levels of constraint violation, such that $\bot$ means satisfaction and $\top$ means unacceptable violation. The operation $\oplus$ is used to combine (aggregate) several valuations. A constraint is *hard*, if all its valuations are either $\bot$ or $\top$.

Definition 2 (Valued Constraint Satisfaction Problem [22]). *A valued constraint satisfaction problem (VCSP) consists of a classical CSP (X, D, C) with valuation structure $(E, \leq, \oplus, \bot, \top)$, and a mapping ϕ from C to E which associates a valuation with each constraint.*

For example, the problem of diagnosing the polycell circuit in Fig. 1 [27] can be framed as a VCSP with variables $X = \{a, b, c, d, e, f, g, x, y, z\}$. Each variable models a boolean signal and has domain $\{0, 1\}$. The VCSP has five ternary constraints $f_{o1}, f_{o2}, f_{o3}, f_{a1}, f_{a2}$ corresponding to gates in the circuit, and four unary constraints f_c, f_d, f_f, f_g corresponding to observations. The constraints $f_{o1}, f_{o2}, f_{o3}, f_{a1}, f_{a2}$ express that the respective gates are performing their boolean functions. The constraints f_c, f_d, f_f, f_g express that variables $c, d,$ and g are observed to be 1, and variable f is observed to be 0. The valuation structure $(\mathbb{N}_0^+ \cup \infty, +, \leq, 0, \infty)$ captures the cost of violating a constraint, which we assume to be 1 for the constraints f_{o1}, f_{o2}, f_{o3}, to be 2 for the constraints f_{a1} and f_{a2}, and to be ∞ for the constraints modeling the observations.

Given a VCSP, the problem is to find an assignment t to X which mimimizes the combined valuation of all violated constraints, $\bigoplus_{\{c_j \in C \mid t[\text{var}(c_j)] \notin c_j\}} \phi(c)$. For the boolean polycell example, the minimum valuation of an assignment is 1, corresponding to a fault of a single OR gate.

3   Optimal CSPs

Since solving VCSPs is more complex than solving classical CSPs, an algorithmic approach that is based on splitting the VCSP into a set of classical (hard) constraints and a set of valued (soft) constraints can be useful.

In the following, we consider a specialization of this approach where the constraints are divided into hard constraints and unary soft constraints. In [27], this type of optimization problem is called *optimal CSP*.

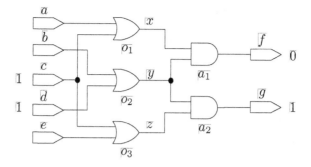

Fig. 1. The boolean polycell example consists of three OR gates and two AND gates. Variables c, d, f, and g are observed as indicated.

Definition 3 (Optimal CSP). *An optimal CSP (OCSP) consists of a classical CSP (X, D, C), with valuation structure $(E, \leq, \oplus, \perp, \top)$, and a set U of unary functions $u_j : \mathrm{dom}(y_j) \to E$ defined over a subset $Y \subseteq X$ of the variables. The variables in Y are called decision variables, and the variables in $X \setminus Y$ are called non-decision variables.*

An OCSP can be viewed as a special case of a VCSP where soft constraints (constraints with valuation $\phi(c_j) < \top$) must be unary. A solution to an OCSP is an assignment to Y with minimal total valuation (combined valuations of functions u_j), such that there exists an extension to all variables X that satisfies all the constraints in the CSP. Hence, whereas a solution to a VCSP is a single assignments to X, a solution to an OCSP is an assignment to the decision variables Y that can stand for a whole collection of assignments to X that have all the same valuation (plateau) and differ only with respect to the non-decision variables $X \setminus Y$. It is observed in [14] that a number of optimization problems can be directly expressed with hard and unary soft constraints, that is, as OCSPs; an example are combinatorial auctions [21].

4   Translation from Valued CSPs to Optimal CSPs

In general, a VCSP may have non-unary soft constraints and thus it does not necessarily have the form of an OCSP. However, it is possible to *transform* a VCSP into an OCSP with an equivalent optimal solution. This transformation is based on introducing additional variables (decision variables) that capture the cost of violating a constraint of the VCSP, analogous to the *hidden variable representation* described in [14] and the construction of *disjunctive constraints* for over-constrained problems described in [18, 19]. The translation demonstrates that OCSPs, though syntactically more restricted than VCSPs, actually have the same expressive power as VCSPs. OCSPs could therefore be viewed as a "normalization" of VCSPs that achieves our desired separation into a hard constraint part and a soft constraint part.

Introducing a decision variable for every constraint turns a VCSP with n variables and m constraints into an OCSP with $n + m$ variables. We can further reduce the size of the OCSP by observing that for any hard constraint c_j in the VCSP ($\phi(c_j) = \top$), choosing the value *false* for its corresponding decision variable y_j can never give rise to a solution of the OCSP because it will immediately lead to the valuation $\top$. Therefore, we do not need to introduce decision variables for hard constraints in the VCSP.

Definition 4 (Translation of VCSP to OCSP). *The translation of a VCSP* (X, D, C) *with valuation structure* $(E, \leq, \oplus, \perp, \top)$ *and mapping* ϕ *into an OCSP* (X', D', C') *with unary functions* U *over decision variables* $Y \subseteq X'$ *is defined as follows:*

- X' *consists of* X *and one decision variable* y_j *for each constraint* $c_j \in C$ *for which* $\phi(c_j) < \top$;
- D' *consists of* D *and the domain* {true, false} *for each decision variable* y_j;
- U *consists of one unary function* u_j *per decision variable* y_j. *The function maps the value* true *to* $\perp$ *and the value* false *to* $\phi(c_j)$;
- C' *consists of one constraint* c'_j *for each* $c_j \in C$. *If* $\phi(c_j) = \top$ *then* $c'_j = c_j$, *else* c'_j *is a relation over variables* $var(c'_j) = \text{var}(c_j) \cup y_j$. *An assignment* t *to* $var(c'_j) = \text{var}(c_j) \cup y_j$ *satisfies* c'_j *iff* $t[\text{var}(c_j)] \in c_j$ *and* $y_j = $ true *or* $t[\text{var}(c_j)] \notin c_j$ *and* $y_j = $ false.

For example, the translation of the VCSP for the boolean polycell circuit yields an OCSP with variables $\{a, b, c, d, e, f, g, x, y, z, y_1, y_2, \dots, y_5\}$. Variables $\{a, b, c, d, e, f, g, x, y, z\}$ are non-decision variables, and variables y_1 to y_5 are decision variables, obtained by extending the constraints $f_{o1}, f_{o2}, f_{o3}, f_{a1}, f_{a2}$ with an additional variable. There are five unary functions $u_1, u_2, \dots, u_5 \in U$, capturing the cost of violating the constraints $f_{o1}, f_{o2}, f_{o3}, f_{a1}, f_{a2}$. No decision variables need to be introduced for the hard constraints f_c, f_d, f_f, f_g corresponding to observations.

Theorem 1. *A VCSP and its translation to an OCSP have the same optimal solution.*

Note that for the special case of a VCSP that is actually a CSP (a VCSP where $\phi(c_j) = \top$ for all $c_j \in C$), the reduced translation is the CSP itself. Therefore, solving a CSP as an OCSP does not incur any overhead.

5   Solving OCSPs

The separation of valued CSPs into unary soft constraints and hard constraints can be algorithmically exploited by coupling together specialized algorithms for each part. In particular, for the hard constraint part, we can employ techniques that are highly optimized for satisfaction problems, and for the soft constraint part, we can employ techniques that work best for a relatively small optimization problem but would be infeasible for the original, bigger problem. This hybrid algorithmic approach can be more efficient than general solvers for soft constraints that do not make assumptions about how the valuations are distributed over the space of assignments.

5.1   Conflict-Directed A* Search

Williams and Ragno [27] describe such a hybrid approach for solving a subclass of OCSPs. The approach, called *conflict-directed A**, exploits the distinction between decision variables (which determine the valuation of an assignment) and non-decision variables (which determine only the consistency of an assignment) by treating them separately: it enumerates assignments to the decision variables (corresponding to plateaus) in best-first order. Once a complete assignment to the decision variables has been found, it is checked whether the CSP part can be satisfied by assigning the remaining, non-decision variables. If the CSP is satisfiable (corresponding to the plateau being non-empty), an optimal solution has been found. If the CSP is unsatisfiable (corresponding to the plateau being empty), one or more conflicts (inconsistent instantiations of variables, see [5]) are extracted and used to speed up and focus the further search. Depending on the problem structure, there can be fewer plateaus than individual elements of the search space, and therefore this two-step approach can be more efficient than enumerating the individual elements of the search space.

The enumeration of assignments is based on A* search [10], an instance of best-first search that uses a lower bound g for the partial assignment made so far, and an optimistic estimate h of the value that can be achieved when completing the assignment; at each point in the search, A* expands the assignment with the best combined value of g and h. A* search is *run-time optimal* [3] in that it visits a minimum number of search nodes (among all search methods having access to the same heuristics). Due to its memory requirements, which are worst-case exponential in the number of variables, A* search would hardly be feasible as a solution method for general VCSPs. However, as observed in [27], the memory requirements of A* search can be much more modest in the case of OCSPs, because only assignments to variables that have an associated cost (decision variables) need to be stored in the search queue; in addition, the conflicts further reduce the size of the queue.

In the following, we present a simplified variant of conflict-directed A* that is adapted to OCSPs obtained from VCSPs. As a heuristic estimate h for the cost of completing an assignment to the decision variables, we simply sum up the best possible valuation for each remaining (unassigned) decision variable; since the best possible valuation of a decision variable is $\perp$, it means h is equal to $\perp$. The pseudo-code of the algorithm is shown in Alg. 1. First, local consistency is established in the CSP part of the OCSP. If an inconsistency arises during local propagation, then the OCSP has no consistent solution (no assignment with valuation better than $\top$). Otherwise, the algorithm performs a best-first (A*) search over assignments to the decision variables Y of the OCSP, using a priority queue of (partial) assignments to Y that is ordered by their valuation. The A* search is based on two sub-procedures updateAssignment() and switchAssignment(), shown in Proc. 2 and Proc. 3, respectively. Procedure switchAssignment() establishes a (partial) assignment a to the decision variables from the queue, trying to reuse as much as possible the current search tree; it backtracks to the deepest point in the search tree up to which the current assignment to Y and a are the same. If an inconsistency occurs while trying to establish the assignment, then a conflict is extracted and added to the set of constraints, and the assignment is discarded. Next, updateAssignment() is used to assign decision variables that have only one value remaining, and extend the assignment (and in particular, its valuation) accordingly. Since this update might increase the valuation of the current assignment, it is now possible that it is no longer the best assignment; in this case, the assignment is pushed back into the queue. Otherwise (if the current assignment is still the best one), it is checked whether the assignment to the decision variables is complete. If the assignment is incomplete, the algorithm chooses a next decision variable y_i to assign and enqueues the two possible branches $y_i \leftarrow$ true and $y_i \leftarrow$ false. If the assignment to the decision variables is complete, then the algorithm uses procedure consistentAssignment() to check if the assignment is consistent with the CSP. To this end, consistentAssignment() tries to extend the assignment to $Y \subseteq X$ to an assignment to X by assigning the remaining (non-decision) variables $X \setminus Y$. In Proc. 4, this is done using depth-first search with conflict-directed backjumping. The current level of the search tree (which so far involves only decision variables) is frozen in variable decisionLevel, and whenever a conflict occurs that would require to backup higher than this level (backtrackLevel smaller than or equal to decisionLevel), the current assignment to the decision variables must be inconsistent and is discarded. Otherwise, the assignment is output as the next best solution.

For instance, for the boolean polycell example and the OCSP encoding in Def. 4, the algorithm has to assign five decision variables $y_1, y_2, \ldots, y_5$ corresponding to the constraints $f_{o1}, f_{o2}, f_{o3}, f_{a1}, f_{a2}$. Conflict-directed A* starts with an empty assignment to the decision variables. Propagation does not prune any values for the decision variables, so the algorithm assigns a decision variable. Assume the decision variables are assigned in the order $y_1, y_2, \ldots, y_5$. The algorithm thus creates two new assignments, $\langle y_1 \leftarrow$ true$\rangle$ with valuation 0 and $\langle y_1 \leftarrow$ false$\rangle$ with valuation 1, and puts them on the queue. The algorithm pops the assignment $\langle y_1 \leftarrow$ true$\rangle$ from the queue and establishes it using function switchAssignment(). Two new assignments, $\langle y_1 \leftarrow$ true, $y_2 \leftarrow$ true$\rangle$ with valuation 0 and $\langle y_1 \leftarrow$ true, $y_2 \leftarrow$ false$\rangle$ with valuation 1 are created and enqueued. When establishing the best assignment $\langle y_1 \leftarrow$ true, $y_2 \leftarrow$ true$\rangle$ using switchAssignment(), propagation forces y_4 to be false, and thus updateAssignment() refines the assignment to $\langle y_1 \leftarrow$ true, $y_2 \leftarrow$ true, $y_4 \leftarrow$ false$\rangle$ with valuation 2. Since a better assignment exists in the queue, this assignment is pushed back into the queue, and the next best assignment, say $\langle y_1 \leftarrow$ false$\rangle$ with valuation 1, is considered. Since this new assignment and the current assignment share no common prefix, switchAssignment() needs to backtrack up to y_1 in order to establish this assignment. After propagation, the updated assignment becomes $\langle y_1 \leftarrow \text{false}, y_4 \leftarrow \text{true} \rangle$ with valuation 1. The algorithm proceeds by assigning $y_2 \leftarrow \text{true}$ and $y_3 \leftarrow \text{true}$, at which point $y_5 \leftarrow \text{true}$ can be derived by propagation, and therefore a complete decision variable assignment $\langle y_1 \leftarrow \text{false}, y_2 \leftarrow \text{true}, y_3 \leftarrow \text{true}, y_4 \leftarrow \text{true}, y_5 \leftarrow \text{true} \rangle$ with valuation 1 is obtained. Procedure consistentAssignment() determines that this assignment is consistent (a satisfying assignment to the non-decision variables is e.g. $\langle a \leftarrow 1, b \leftarrow 1, c \leftarrow 1, d \leftarrow 1, e \leftarrow 0, f \leftarrow 0, g \leftarrow 1, x \leftarrow 0, y \leftarrow 1, z \leftarrow 1 \rangle$), and thus outputs value 1 as the optimal solution.

Theorem 2. *The conflict-directed A* algorithm in Alg. 1 computes the optimal solution of a given OCSP.*

Conflict-directed A* search can be further refined in a number of ways. [27, 15] describe extensions that reduce the size of the search queue by generating new entries only at a point where the current assignment to the decision variables becomes inconsistent, and an extension to the case of non-binary decision variables that generates only next best child assignments instead of all children at once. It is also easy to extend the algorithm such that it enumerates the solutions in best-first order, instead of computing only the optimal solution.

Algorithm 1. Conflict-directed A* for OCSPs

```
 1: if not (propagate() = conflict) then
 2:    queue ← ⟨∅, ⊥⟩
 3:    while queue ≠ ∅ do
 4:       ⟨a, value⟩ ← first(queue)
 5:       queue ← removeFirst(queue)
 6:       if switchAssignment(a) then
 7:          updateAssignment(⟨a, value⟩)
 8:          if assignment with better value exists in queue then
 9:             queue ← push(queue, ⟨a, value⟩)
10:          else
11:             if exists yᵢ ∈ Y, yᵢ = unknown then
12:                queue ← push(queue, ⟨a ∪ (yᵢ ← true), v⟩)
13:                queue ← push(queue, ⟨a ∪ (yᵢ ← false), v ⊕ φ(cᵢ)⟩)
14:             else
15:                if consistentAssignment() then
16:                   output value as best solution
17:                   exit
18:                end if
19:             end if
20:          end if
21:       end if
22:    end while
23: end if
24: output no solution
```

Procedure 2. UpdateAssignment($\langle a, \text{value} \rangle$)

1: **for all** $y_i \in Y$, $y_i \notin a$, $y_i \neq$ unknown **do**
2:    **if** $y_i =$ **true then**
3:       $\langle a, \text{value} \rangle \leftarrow \langle a \cup (y_i \leftarrow \textbf{true}), \text{value} \rangle$
4:    **else**
5:       $\langle a, \text{value} \rangle \leftarrow \langle a \cup (y_i \leftarrow \textbf{false}), \text{value} \oplus \phi(c_i) \rangle$
6:    **end if**
7: **end for**

Procedure 3. SwitchAssignment(a)

1: level $\leftarrow$ deepest level up to which a and current assignment are equal
2: backtrack(level)
3: **for** $(y_i \leftarrow \text{val}) \in a$ **do**
4:    **if** $y_i \neq$ val **then**
5:       **return false**
6:    **else if** $y_i =$ unknown **then**
7:       $y_i \leftarrow$ val
8:       level $\leftarrow$ level $+ 1$
9:       **if** propagate() $=$ conflict **then**
10:          CSP $\leftarrow$ CSP $\cup$ conflict
11:          **return false**
12:       **end if**
13:    **end if**
14: **end for**
15: **return true**

Procedure 4. ConsistentAssignment()

1: decisionLevel $\leftarrow$ level
2: **while** exists $x_i \in X \setminus Y$, $x_i =$ unknown **do**
3:    **choose** val $\in \text{dom}(x_i)$
4:    $x_i \leftarrow$ val
5:    level $\leftarrow$ level $+ 1$
6:    $\text{dom}(x_i) \leftarrow \text{dom}(x_i) -$ val
7:    **if** propagate() $=$ conflict **then**
8:       backtrackLevel $\leftarrow$ analyze(conflict)
9:       **if** backtrackLevel $\leq$ decisionLevel **then**
10:          **return false**
11:       **else**
12:          CSP $\leftarrow$ CSP $\cup$ conflict
13:          backtrack(backtrackLevel)
14:          level $\leftarrow$ backtrackLevel
15:       **end if**
16:    **end if**
17: **end while**
18: **return true**

6   Implementation

We have implemented the transformation of VCSPs into OCSPs and the conflict-directed A* search algorithm in C++. Conflict-directed A* search was implemented on top of zChaff [17], one of the most efficient complete solvers for boolean satisfiability (SAT) problems. The main reasons why we choose zChaff is that it offers (1) a highly optimized data-structure for local consistency (unit propagation), called *two-literal watching scheme*; (2) a method for extracting small conflicts from inconsistencies, based on so-called *unique implications points* (UIPs), which correspond to dominators in the implication graph; and (3) an efficient variable and value ordering heuristic called *variable state independent decaying sum* (VSIDS), which biases the search towards variables that occur in recently learned clauses, i.e., conflicts. (In addition, zChaff uses other techniques such as random restarts, which we do not exploit in our prototype).

Our prototypic implementation of conflict-directed A* adopts zChaff's local propagation scheme, its conflict extraction method, and its variable/value ordering heuristic for the non-decision variables. The decision variables are currently assigned in no specific order. Using a SAT solver as the underlying satisfiability engine means that the CSP part of the OCSP has to be first encoded as a SAT problem, by mapping variables to boolean variables, and mapping constraints to clauses in conjunctive normal form (CNF). For this purpose, we choose a logarithmic SAT encoding of the CSP [11], although other encodings are equally possible (see [25, 6] for two alternative encodings).

7   Experimental Results

We evaluated our prototype on various examples of valued CSPs, and compared its performance against other algorithms for solving soft constraints.

The algorithms we compared against are branch-and-bound with maintaining existential directional arc consistency (BB-MEDAC) [7], and cluster tree elimination (CTE) [4]. BB-MEDAC is a recently proposed search algorithm that combines depth-first branch-and-bound with a form of arc consistency generalized to soft constraints. In our experiments we used the implementation that is part of the TOOLBAR package [24]. CTE is an inference algorithm for both hard constraints and soft constraints that is based on decomposing the constraint graph into a tree structure, and solving it using dynamic programming. In our experiments, the tree was computed using a greedy min-fill heuristic.

All the examples shown below, apart from the random problems, are taken from the TOOLBAR repository. All experiments were performed under Windows XP using a 2.8 GHz Pentium 4 PC with 1 GB of Ram.

7.1   Academic Problems

First, we tried conflict-directed A* on three academic puzzles. Since these examples involve only hard constraints, the corresponding OCSPs do not contain any decision variables, and thus conflict-directed A* can solve these problems as efficiently as the underlying satisfiability engine (in our implementation, zChaff with the given SAT encoding). For all three algorithms, we used a time bound of 1 minute. Table 1 summarizes the results. Although these examples are relatively small, note that CTE fails to solve all but one of them within the given time bound.

Table 1. Results for academic puzzles (containing only hard constraints)

	CDA*	BB-MEDAC	CTE
zebra (25 variables, 19 constraints)	0.188 sec	0.016 sec	0.047 sec
send (11 variables, 32 constraints)	0.312 sec	0.031 sec	> 1 min
donald (15 variables, 51 constraints)	2.828 sec	0.156 sec	> 1 min

7.2   Random Problems

Next, we compared the algorithms on random Max-CSP problems. Max-CSPs are instances of VCSPs where each constraint has cost 1; thus, the task is to minimize the number of violated constraints. To generate the examples, we used a random binary constraint model with four parameters N, K, C, and T, where N is the number of variables, K the domain size, C the number of constraints, and T the tightness of each constraint (number of tuples having cost 1). Again, we used a time bound of 1 minute. Table 2 summarizes the results for six classes of random Max-CSP, averaged over 10 instances each.

Table 2. Results for random Max-CSPs (10 instances each)

(N, K, C, T)	CDA*	BB-MEDAC	CTE
(40, 4, 60, 4)	0.0346 sec	0.0092 sec	1.461 sec
(40, 4, 60, 8)	2.184 sec	0.022 sec	4.136 sec
(40, 4, 60, 12)	> 1 min	0.0468 sec	7.325 sec
(25, 4, 100, 4)	0.818 sec	0.0156 sec	> 1 min
(25, 4, 100, 8)	> 1 min	0.169 sec	> 1 min
(25, 4, 100, 12)	> 1 min	0.131 sec	> 1 min

For all these examples, BB-MEDAC converges very fast towards the optimal solution. Unfortunately, conflict-directed A* does not perform well for the denser and tighter instances. Further analysis of these cases reveals that the algorithm actually quickly finds small conflicts that could potentially guide the A* search towards the optimal solution, but then tries many assignments to the decision variables that are useless as they are not relevant to (i.e., do not resolve) those conflicts. Thus, we expect that using a similar variable ordering heuristic for the decision variables as for the non-decision variables (focusing on variables involved in conflicts) could substantially improve the performance of conflict-directed A* for these cases.

7.3   Real-World Problems

Finally, we evaluated the performance of our algorithm on four real-world circuit examples. These are obtained by turning SAT instances from the DIMACS challenge into Max-CSPs by making each clause a constraint with cost 1. For these examples, we used a time bound of 10 minutes. Table 3 summarizes the results.

Table 3. Results for DIMACS circuit examples

	CDA*	BB-MEDAC	CTE
ssa0432-003 (435 variables, 1027 constraints)	14.547 sec	> 10 min	1.219 sec
ssa7552-038 (1501 variables, 3575 constraints)	28.312 sec	> 10 min	142.969 sec
ssa2670-141 (986 variables, 2315 constraints)	101.765 sec	> 10 min	6.21 sec
ssa2670-130 (1359 variables, 3321 constraints)	233.89 sec	> 10 min	53.203 sec

CTE performs best for most of these examples; however, the run-times for CTE in Table 3 show only run-times of CTE itself and do not include the time for computing the tree decomposition, which takes longer than the run-time of CTE for some of the examples. Also, CTE requires significantly more memory than the other algorithms for most of the examples. BB-MEDAC, which performed best for the academic and random examples, cannot solve any of the DIMACS examples within the given time bound. In fact, even after 10 minutes of computation, its lower bound (best valuation found so far) is often far off the optimal solution. We suspect that this has to do with the fact that BB-MEDAC performs local propagation (existential directional arc consistency) for binary constraints only, and defers the propagation of non-binary constraints until they become binary. Thus, the propagation scheme is not effective for the DIMACS examples where almost all constraints are non-binary. In contrast, conflict-directed A* exploits efficient local propagation (zChaff's two literal scheme) for any hard constraints. In fact, for instance ssa7552-038, which has optimal cost 0, conflict-directed A* requires only one call to the SAT engine (zChaff) in order to solve it. The actual run-time of zChaff for this example is only a fraction of the run-time given in Table 3, indicating that the current implementation of conflict-directed A* wastes significant time constructing unnecessary search queue entries. We therefore expect that further improvements to the algorithm to reduce the size of the search queue by creating entries only as needed (as described in [27, 15]) will have a strong impact for these examples.

8   Discussion and Related Work

The idea of re-formulating problems into a part describing the cost of a solution and a part describing its feasibility is not new. Larrosa and Dechter [14] already found that dualization turns soft constraints into a set of hard constraints and unary soft constraints, without losing expressiveness. Petit et al. [18, 19] describe an approach to model optimization problems by specifying an additional variable (decision variable) for each constraint, capturing the cost of its violation; they show how this representation allows for additional expressiveness, for example, specifying "meta-constraints" between decision variables to control the distribution of violations. In contrast, our goal is to use re-formulation to make the structure in soft constraints more explicit; in particular, we believe separating optimization aspects from satisfaction aspects may provide a useful starting point for algorithmic development. Conflict-directed A* is an instance of such an approach, and it is inspired by research on model-based reasoning and diagnosis [26, 9], where problems can be naturally framed as a mixture of hard constraints and unary objective functions (that is, OCSPs).

From the perspective of viewing re-formulation as a process of "precompiling" preferences, the separation into unary soft constraints and hard constraints is only a special case; it is not actually required by the approach that the soft constraints are unary. Another useful view of the re-formulation into OCSPs is that of giving a "normal form" for soft constraints, which makes the degree to which the problem is an optimization problem vs. a satisfaction problem more explicit. It seems that research in soft constraints has so far focussed on expressive, unifying frameworks, but much less on such canonical representations. Optimal CSPs could provide a starting point in this direction.

A potential drawback of the re-formulation is that it may increase the size of the problem; since one decision variable is introduced for each soft constraint, the resulting OCSP may be much bigger than the original VCSP, especially if it has a high ratio of constraints to variables. However, even if the re-formulation incurs an increase in the problem size, the benefit of applying dedicated solvers to each part of the problem (as in conflict-directed A*) may still outweigh the increase in the search space. The identification of problem classes for which re-formulation is beneficial is a subject of further research.

As already indicated in Sec. 5.1, several improvements to conflict-directed A* are possible, in particular for switchAssignment(), the procedure that is most critical to the performance of the algorithm. The cost of switching between two A* search nodes (corresponding to two different assignments to the decision variables, i.e., two different CSPs) could be reduced by incremental techniques that allow for computing only the difference between two CSP instances. Truth maintenance systems (TMS) [13], which keep track of the dependencies in the implication graph, are frequently used in model-based reasoning and diagnosis for this purpose. However, the additional bookkeeping necessitated by the TMS creates a trade-off between between making the context switch more efficient and making the satisfiability check more efficient.

Another direction for future work is to combine conflict-directed A* search with structural (tree decomposition) methods. As can be seen from the experiments, the two approaches are fairly complementary to each other, and decomposing the problem into smaller subproblems can dramatically improve performance on examples with low tree width. The combination would involve an instance of conflict-directed A* running on every cluster in the tree, and a special set of decision variables that capture the cost of assignments to variables shared between clusters (separator variables). We are currently working on such a decomposed version of conflict-directed A*. Some earlier work on combining best-first search with tree decompositions can be found in [20], whereas [23] describes a method for (the simpler case of) combining depth-first search with tree decompositions.

In our implementation, we used a SAT solver (zChaff) to check consistency of the candidates (plateaus) enumerated by A* search, mainly for the reason that it provides an efficient implementation of local propagation and conflict extraction. Recently, the problem of extending SAT solvers to optimization counterparts where either the number of satisfied clauses must be maximized (max-SAT) or the clauses carry a weight to be maximized (weighted max-SAT) has received considerable attention [28]. Much of this work still focuses on extending the basic DPLL search algorithm that underlies most complete SAT solvers (especially the unit propagation and variable ordering heuristic) to this case, and does not yet exploit more advanced concepts like conflicts. Still, it would be interesting to compare such approaches to our method.

9 Conclusion

We presented an approach for transforming VCSPs into hard constraints and unary soft constraints (OCSPs), and exploiting this re-formulation by solving the optimization and satisfiability part separately using a combination of specialized algorithms. Because it can exploit structure in the search space by enumerating whole sets of assignments with equal valuations (plateaus) rather than just individual assignments, this hybrid approach can be more efficient than algorithms that work directly on the VCSP. We presented an instance of this approach, called conflict-directed A*, and its prototypic implementation on top of a SAT solver. The prototype can outperform other solvers for VCSPs on some problems of practical importance. Promising directions for future research include more sophisticated, incremental methods for the critical step of switching between plateaus, and incorporating structural decomposition methods.

References

1. Bistarelli, S., et al.: Semiring-based CSPs and Valued CSPs: Frameworks, Properties, and Comparison. Constraints 4 (3) (1999) 199–240
2. Cooper, M., and Schiex, T.: Arc consistency for soft constraints. Artificial Intelligence 154 (2004) 199-227
3. Dechter, R., Pearl, J.: Generalized Best-First Search Strategies and the Optimality of A*. Journal of the ACM 32 (3) (1985) 505–536
4. Dechter, R., Pearl, J.: Tree clustering for constraint networks. Artificial Intelligence 38 (1989) 353–366
5. Dechter, R.: Enhancement schemes for constraint processing: Backjumping, learning and cutset decomposition. Artificial Intelligence 41 (1990) 273-312.

6. Gent, I.P.: Arc consistency in SAT. Proc. ECAI-2002 (2002)
7. de Givry, S., Zytnicki, M., Heras, F., and Larrosa, J.: Existential arc consistency: Getting closer to full arc consistency in weighted CSPs. Proc. of IJCAI-2005 (2005)
8. Gottlob, G., Leone, N., Scarcello, F.: A comparison of structural CSP decomposition methods. Artificial Intelligence **124** (2) (2000) 243–282
9. W. Hamscher, W., Console, L., and de Kleer, J. (eds.): Readings in Model-Based Diagnosis, Morgan Kaufmann (1992)
10. Hart, P. E., Nilsson, N. J., and Raphael, B.: A formal basis for the heuristic determination of minimum cost paths. IEEE Trans. Sys. Sci. Cybern. **SSC–4** (2) (1968) 100-107.
11. Iwama, K, and Miyazaki, S.: SAT-variable complexity of hard combinatorial problems. IFIP World Computer Congress (1994) 253-258
12. Kask, K., et al.: Unifying Tree-Decomposition Schemes for Automated Reasoning. Technical Report, University of California, Irvine (2001)
13. de Kleer, J.: An Assumption based TMS, Artificial Intelligence **28** (1) (1986) 127–162
14. Larrosa, J., and Dechter, R.: On the Dual Representation of non-binary Semiring-based CSPs. Proceedings SOFT-2000 (2000)
15. Li, H., and Williams, B.C.: Generalized Conflict Learning for Hybrid Discrete/Linear Optimization, Proc. CP-2005 (2005)
16. Mackworth, A.: Constraint satisfaction. Encyclopedia of AI (second edition) **1** (1992) 285–293
17. Moskewicz, M., Madigan, C., Zhao, Y., Zhang, L., and Malik, S.: Chaff: Engineering an efficient SAT solver. In Proc. of the Design Automation Conference (DAC) (2001)
18. Petit, T., Régin, J.-C., and Bessière, C.: Meta-Constraints on Violations for Over-Constrained Problems, Proc. ICTAI-2000 (2000) 358–365.
19. Petit, T., Régin, J.-C., and Bessière, C.: Specific Filtering Algorithms for Over-Constrained Problems, Proc. CP-2001 (2001) 451–465.
20. Sachenbacher, M., and Williams, B.C.: On-demand Bound Computation for Best-First Constraint Optimization, Proc. CP-2004 (2004)
21. Sandholm, T.: An algorithm for optimal winner determination in combinatorial auctions. Proceedings IJCAI-1999 (1999)
22. Schiex, T., Fargier, H., Verfaillie, G.: Valued Constraint Satisfaction Problems: hard and easy problems. Proc. IJCAI-95 (1995) 631–637
23. Terrioux, C., Jégou, P.: Bounded Backtracking for the Valued Constraint Satisfaction Problems. Proc. CP-2003 (2003)
24. TOOLBAR http://carlit.toulouse.inra.fr/cgi-bin/awki.cgi/ToolBarIntro
25. Walsh, T.: SAT vs. CSP. Proc. CP-2000 (2000) 441-456
26. Weld, D.S., and de Kleer, J. (eds.): Readings in Qualitative Reasoning about Physical Systems, Morgan Kaufmann (1989)
27. Williams, B., Ragno, R.: Conflict-directed A* and its Role in Model-based Embedded Systems. Journal of Discrete Applied Mathematics, to appear.
28. Xing, Z., and Zhang, W.: MaxSolver: An efficient exact algorithm for (weighted) maximum satisfiability. Artificial Intelligence **164** (1–2) (2005) 47-80

Event-Driven Probabilistic Constraint Programming

S. Armagan Tarim[1], Brahim Hnich[2], and Steven D. Prestwich[1]

[1] Cork Constraint Computation Centre, University College Cork, Ireland
{a.tarim, s.prestwich}@4c.ucc.ie
[2] Izmir University of Economics, Faculty of Computer Science, Izmir, Turkey
hnich.brahim@gmail.com

Abstract. Real-life management decisions are usually made in uncertain environments, and decision support systems that ignore this uncertainty are unlikely to provide realistic guidance. We show that previous approaches fail to provide appropriate support for reasoning about reliability under uncertainty. We propose a new framework that addresses this issue by allowing logical dependencies between constraints. Reliability is then defined in terms of key constraints called "events", which are related to other constraints via these dependencies. We illustrate our approach on two problems, contrast it with existing frameworks, and discuss future developments.

1 Introduction

Real-life management decisions are usually made in uncertain environments. Random behavior such as the weather, lack of essential exact information such as the future demand, incorrect data due to errors in measurement, and vague or incomplete definitions, exemplify the theme of uncertainty in such environments.

It is generally impossible for any set of decisions to satisfy all the constraints under all circumstances. For instance, consider a probabilistic single-item distribution problem in which there are n independent suppliers with their given *probabilistic* supply capacities, and m different customers with known demands. It is realistic to assume that the deliveries are fixed in advance, by consideration of the probabilistic supply capacities. The need to fix the deliveries in advance has been at the heart of many problems such as the buying of raw materials on markets with fluctuating prices [3]. Thus the investigation of modeling approaches and solution algorithms is potentially important not only from a theoretical point of view, but also from the perspective of practical applications. It is quite unrealistic to ask for a plan that satisfies all demand and probabilistic supply constraints, irrespective of the unfolding of uncertainties.

To address this and related situations, we propose that one should determine in advance a distribution plan that satisfies customer demands as far as possible, under some probabilistic measure that accurately captures the user's notion of reliability. To address this important class of problems, we take a novel approach

J.C. Beck and B.M. Smith (Eds.): CPAIOR 2006, LNCS 3990, pp. 197–211, 2006.
© Springer-Verlag Berlin Heidelberg 2006

and develop a modeling framework that supports more reliable decisions in uncertain environments, yet reduces the cognitive burden on a decision-maker. Our *Event-Driven Probabilistic Constraint Programming (EDP-CP)* modeling framework allows users to designate certain probabilistic constraints (such as demand constraints) as *events* whose chance of satisfaction must be maximized, subject to hard probabilistic constraints (such as a lower bound on profit), and also logical dependencies among constraints (such as the dependency of demand constraints on the satisfaction of the probabilistic supply constraints). We shall show that the EDP-CP framework allows more realistic modelling of some problems than previous approaches.

The rest of this paper is organized as follows. In Section 2 we motivate the work. We define the new modelling framework in Section 3 and show how to compile any EDP-CP model into an equivalent constraint program in Section 4. In Section 5 we illustrate its flexibility and usefulness by studying two examples: probabilistic supply chain planning and production planning/capital budgeting. In Section 6 we survey related work. Finally, in Section 7 we summarise our work and discuss future directions.

2 Motivation

Our motivation for this work comes from an application in the supply chain management area: more precisely, addressing supply and demand uncertainties. The inherent difficulty in dealing with this class of probabilistic problems is mainly due to the fact that certain constraints — such as the ones imposing on complete satisfaction of customer demands — may be hinged on the satisfaction of others — such as supply constraints. The problem is particularly interesting when the latter constraints are exposed to uncertainty.

2.1 A Motivating Example

We provide a concrete example of the distribution problem to motivate the work. Figure 1 depicts a distribution system with three suppliers $S_{1,2,3}$ and

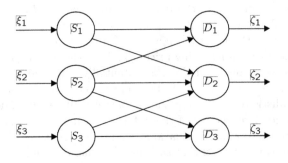

Fig. 1. Distribution Problem

three customers $D_{1,2,3}$. The scopes of the suppliers are $S_1 \rightarrow \{D_1, D_2\}$, $S_2 \rightarrow \{D_1, D_2, D_3\}$, $S_3 \rightarrow \{D_2, D_3\}$. The deterministic customer demands are $[8, 7, 4]$. The suppliers' probabilistic capacities are expressed as discrete probability density functions: $f_{S_1} = \{3(0.3), 7(0.5), 12(0.2)\}$, $f_{S_2} = \{6(0.4), 7(0.2), 10(0.4)\}$ and $f_{S_3} = \{3(0.3), 8(0.7)\}$, where values in parentheses represent probabilities. The objective is to obtain the most reliable distribution plan. We shall consider a series of models of increasing sophistication.

2.2   A Naive Model

Define decision variables $x_{s,c}$ where $s, c \in \{1, 2, 3\}$, denoting the planned supply from supplier s to customer c. Also define random variables ξ_i denoting the uncertain supply available to supplier i. A constant ζ_c denotes the deterministic demand of customer c. Any plan must satisfy the hard constraints

$$\sum_{s \in \mathbf{S}_c} x_{s,c} = \zeta_c$$

where $\mathbf{S}_c$ is the set of suppliers for customer c. There are also probabilistic constraints between decision and random variables:

$$\sum_{c \in \mathbf{C}_s} x_{s,c} \leq \xi_s$$

where $\mathbf{C}_s$ is the set of customers for supplier s. These probabilistic constraints are "soft": they may be violated in some scenarios. We therefore do not add them to the model (as with the deterministic constraints), but instead use them to define an objective function:

$$\max \ \sum_s E \left\{ \sum_{c \in \mathbf{C}_s} x_{s,c} \leq \xi_s \right\}$$

where $E\{C\}$ is the sum of the probabilities of the scenarios in which constraint C is satisfied. This model may be viewed as similar to a Probabilistic CSP [2], in which some contraints are hard, plus an optimization criterion that we wish to maximise the probability that other soft constraints are satisfied.

2.3   A Dependent-Chance Programming Model

A drawback of the above model is that the objective function does not measure plan reliability in a realistic way. For example, in any scenario in which supplier 2 cannot meet its demands (so that $x_{2,1} + x_{2,2} + x_{2,3} > \xi_2$) we cannot guarantee that *any* customers are supplied. This is therefore a worst-case plan for the given scenario, yet in the above model only one probabilistic constraint is violated under this scenario. A plan in which two or three probabilistic constraints are violated would be assigned a lower objective function value, but would be no less reliable.

Worse still, consider a similar problem in which supplier 1 supplies only customer 1, supplier 3 supplies only customer 3, and supplier 2 again supplies customers 1, 2 and 3. A plan in which suppliers 1 and 3 are unable to meet their demands under some scenario would be classed as less reliable than one in which supplier 2 is unable to meet its demand under the same scenario, because more probabilistic constraints are violated. However, the latter plan is less reliable: in the first plan customer 2 is satisfied, but in the second plan no customer is.

To improve the model we may define a more intelligent objective function: the reliability of a plan is now the sum of the reliabilities of three *events*, where an event is the satisfaction of a customer:

$$\max \quad \sum_c E\left\{\bigwedge_{s \in \mathbf{S}_c} \left(\sum_{c' \in \mathbf{C}_s} x_{s,c'} \leq \xi_s\right)\right\}$$

where $\wedge$ denotes logical conjunction: $E\{C \wedge C'\}$ is the sum of the probabilities of the scenarios in which both C and C' are satisfied. For example the reliability of satisfaction of customer 1 is the sum of the probabilities of the scenarios in which suppliers 1 and 2 both meet their demands. Under this objective function, our worst-case plan (in which supplier 2 cannot meet its demands) is assigned reliability 0 in the scenario, because the violated probabilistic constraint $x_{2,1} + x_{2,2} + x_{2,3} \leq \xi_2$ affects the reliability of each customer. Allowing logical connectives between constraints allows us to express the problem more accurately. This model is similar to a Dependent-Chance Programming [5] approach to a related problem.

2.4   An EDP-CP Model

However, the second model is still flawed. Consider a plan in which $x_{1,1} = 0$ so that customer 1 must recieve all supplies from supplier 2. The reliability of the satisfaction of customer 1 should now be independent of the ability of supplier 1 to meet its demand, but in the second model it is still dependent; this point was not considered in [5]. We should therefore refine the objective via further logical connectives between constraints:

$$\max \quad \sum_c E\left\{\bigwedge_{s \in \mathbf{S}_c} \left(x_{s,c} \neq 0 \rightarrow \sum_{c' \in \mathbf{C}_s} x_{s,c'} \leq \xi_s\right)\right\}$$

where $\rightarrow$ denotes logical implication: $E\{C \rightarrow C'\}$ is the sum of the probabilities of the scenarios in which either C is violated or C' is satisfied, or both. Because of this modification, under a scenario in which $x_{1,1} = 0$ there is no longer a penalty if

$$\sum_{c' \in \mathbf{C}_1} x_{s,c'} \leq \xi_1$$

is violated.

Table 1 presents some representative distribution plans and their corresponding probabilistic measure of realization. To gain more insight into this problem

Table 1. Representative distribution plans

Plan	Planned Delivery $S_i \rightarrow D_j$: (i,j)							Probabilistic
No	$(1,1)$	$(1,2)$	$(2,1)$	$(2,2)$	$(2,3)$	$(3,2)$	$(3,3)$	Measure
1	3	5	5	1	1	1	3	0.624
2	4	7	4	0	4	0	0	0.680
3	6	2	2	5	0	0	4	0.940
4	5	0	3	3	4	4	0	0.960
5	7	5	1	1	1	1	3	1.040
6	2	5	6	0	4	2	0	1.380
7	8	2	0	2	4	3	0	1.400
8	0	7	8	0	0	0	4	1.800
9	5	0	3	3	0	4	4	2.100
10	6	0	2	0	4	7	0	2.400

class we examine two different distribution plans, 1 and 5, given in Table 1. These two plans share common decisions, except at $S_1 \rightarrow D_1$ and $S_2 \rightarrow D_1$: Plan 1 assumes $[3, 5]$ whereas Plan 5 assumes $[7, 1]$, respectively. We now examine the reliabilities of these two plans. Plan 1 (5) requires a capacity value of 8 units (12 units) at S_1 to be feasible. Although the requisite capacities are different (8 and 12 units) in two plans, these values still have the same probability (0.2) of being realised. Therefore there is no difference between the two plans with regard to S_1. However, the same is not true for S_2. Although a comparison of $S_2 \rightarrow D_1$ for two plans shows that under all scenarios both shipments (1 unit and 5 units) are always feasible, the consideration of other shipments from S_2 gives a completely different picture. In this case, Plan 1 requires a capacity of 7 units ($[5,1,1]$) in total, whereas Plan 5 needs only 3 units ($[1,1,1]$). The corresponding probabilities are 0.6 and 1.0, respectively, thus Plan 5 is more reliable than Plan 1.

2.5   An Alternative EDP-CP Model

So far, the decision-maker's objective has been to maximize the plan reliability, defined in such a way that all violated plans are treated equally. In other words, plans in which not all customer demand constraints hold are considered equally unreliable, irrespective of the number of customers that are completely satisfied. An alternative objective is to satisfy as many customers as possible, that is to meet as many demand constraints as possible under probabilistic supply constraints. Clearly, this new objective may have a wider application.

In the first EDP-CP model any plan must satisfy the hard constraints on demands ζ_c, but a plan that reliably satisfies two customers might be more desirable than one that satisfies all three customers less reliably. We can model such a measure of plan reliability by removing the hard constraints and using them in the objective function instead:

$$\max \sum_c E\left\{ \bigwedge_{s \in \mathbf{S}_c} \left(x_{s,c} \neq 0 \rightarrow \left[\sum_{c' \in \mathbf{C}_s} x_{s,c'} \leq \xi_s \rightarrow \sum_{s' \in \mathbf{S}_c} x_{s',c} = \zeta_c \right] \right) \right\}$$

Under this new objective, the distribution plan [0,2,0,2,4,3,0] guarantees complete satisfaction of D_2 and D_3 with a reliability score of 2.0, whereas under the previous model it would have reliability score 0. The reliability scores for the plans given in Table 1 do not change. (Note that the first model can be modified to allow subsets of customers to be satisfied, simply by including the deterministic constraints in the objective function. But, like the first model, this would not accurately measure plan reliability.)

3   Event-Driven Probabilistic Constraint Programming

In this section we formalise the EDP-CP modeling framework.

3.1   Preliminaries

Recall that a constraint satisfaction problem (CSP) consists of a set of variables $\mathcal{X}$, each with a finite domain of values $D(x_i)$, and a set of constraints $\mathcal{C}$, each over a subset of $\mathcal{X}$ (denoted by $Scope(C)$) and specifying allowed combinations of values for given subsets of variables. A solution is an assignment of values to the variables satisfying the constraints. A Constraint Optimisation Problem (COP) is a CSP with given objective function over a subset of $\mathcal{X}$ that we wish to maximize or minimize.

Recall that a probabilistic CSP as introduced in [2] is defined as a 6-tuple $\langle \mathcal{X}, \mathcal{D}, \Lambda, \mathcal{W}, C, \Pr \rangle$ where:

- $\mathcal{X} = \{x_1, \ldots, x_n\}$ is a set of decision variables;
- $\mathcal{D} = D(x_1) \times \ldots \times D(x_n)$, where $D(x_i)$ is the domain of x_i;
- $\Lambda = \{\lambda_1, \ldots, \lambda_l\}$ is a set of uncertain parameters;
- $\mathcal{W} = W(\lambda_1) \times \ldots \times W(\lambda_l)$, where $W(\lambda_i)$ the domain of λ_i;
- C is a set of (probabilistic) constraints each involving at least one decision variable (and possibly some uncertain parameters);
- $\Pr : \mathcal{W} \to [0,1]$ is a probability distribution over uncertain parameters.

In [2] a complete assignment of the uncertain parameters (resp. of the decision variables) is called a *world* (resp. a *decision*). The probability that a decision is a solution is the probability of the set of the worlds in which it is a solution.

In the rest of this paper we will sometimes refer to classical constraints as deterministic constraints to distinguish them from the probabilistic ones. We will refer to the possible values of an uncertain parameter λ_i as $W(\lambda_i)$ and to the probability of λ_i taking a given value v in $W(\lambda_i)$ as $\Pr(\lambda_i = v)$. As in [2], we refer to a complete assignment of uncertain parameters as a *possible world* and denote by $\mathcal{W}$ the set of all possible worlds. We also assume that the probability of each possible world w is given by the probability function $\Pr : \mathcal{W} \to [0,1]$.

Definition 1 ([2]). *Given a probabilistic constraint c over decision variables and some uncertain parameters, the reduction of c by world $w \in \mathcal{W}$, denoted by $c_{\downarrow w}$, is the deterministic constraint obtained by setting all its uncertain parameters as in w.*

3.2   Modeling Framework

In EDP-CP some of the constraints can be designated by the user as *event constraints*. The user's objective is to maximize his/her chances of realizing these events. For instance, in our running example the user may consider the customer demand constraints as events. The objective is then to construct a plan satisfying customer demand constraints as far as possible. We now introduce a measure for event realization in a deterministic setting, and generalize it later to probabilistic events.

Definition 2. *Given a deterministic constraint c with $Scope(c) = \{x_1, \ldots, x_k\}$, an event realization measure $E\{c\}$ on c is a mapping M from $D(x_1) \times \ldots \times D(x_k)$ into $\{0, 1\}$ such that for all $t \in D(x_1) \times \ldots \times D(x_k)$, $M(t) = 1$ iff t satisfies c.*

Example 1. An event realization measure on event constraint $c : x_1 + x_2 = 8$, denoted by $E\{c\}$, takes value 1 only when the values v_1 and v_2 assigned to decision variables x_1 and x_2 (resp.) sum to 8, otherwise it takes value 0.

When the events are probabilistic constraints, the event realization measure is defined on the set of possible worlds as follows.

Definition 3. *Given a probabilistic constraint c over decision variables $Scope(c) = \{x_1, \ldots, x_k\}$ and uncertain parameters $\Lambda = \{\lambda_1, \ldots, \lambda_l\}$, a probabilistic event realization measure $E\{c\}$ on c is a mapping M from $D(x_1) \times \ldots \times D(x_k)$ into interval $[0, 1]$ such that*

$$E\{c\} = \sum_{w \in W(\lambda_1) \times \ldots \times W(\lambda_l)} Pr(w) E\{c_{\downarrow w}\}$$

Example 2. An event realization measure on probabilistic constraint $c : x_1 + x_2 \leq \xi$), where ξ is a discrete random variable assuming $\{6(0.2), 8(0.7), 11(0.1)\}$, is denoted by $E\{c\}$ and takes the value 0.8 when $x_1 = 4$ and $x_2 = 3$, and the value 0.1 when when $x_1 = 6$ and $x_2 = 3$.

For convenience we shall only consider the "expectation operator" in defining probabilistic event realization measure. However, any other relevant operator, such as the nth moment generator, can be used instead.

The feasibility of certain event constraints depends on the satisfaction of other constraints. For instance, having a plan that meets the customer demands depends on whether or not the supply constraints are met with such a plan. For this purpose we introduce a new meta-constraint useful for modeling such situations in our EDP-CP framework, which we refer to as a *dependency meta-constraint*. But again, we first introduce the dependency constraint in the deterministic setting.

Definition 4. DEPENDENCY(e, p, c) iff $Scope(e) \cap Scope(p) \neq \emptyset$ & $Scope(c) \subseteq Scope(e) \cap Scope(p)$ & $c \rightarrow (e \rightarrow p)$, *where e, p, and c are all deterministic constraints.*

The DEPENDENCY meta-constraint enforces that e is satisfied *only if* p is satisfied and *if* c holds. We refer to p as a *pre-requisite* constraint for event constraint e, and c as a *condition* constraint for e.

When either e or p or both are probabilistic, then the DEPENDENCY(e, p, c) constraint is defined as follows.

Definition 5. DEPENDENCY(e, p, c) iff $\forall w \in W \cdot$ DEPENDENCY$(e_{\downarrow w}, p_{\downarrow w}, c)$

Example 3. In Figure 1 the event e_1 is the demand constraint for the first customer $e_1 : x_{1,1} + x_{2,1} = 8$, while the pre-requisite constraints are the probabilistic supply constraints $p_1 : x_{1,1} + x_{1,2} \leq \xi_1$, $p_2 : x_{2,1} + x_{2,2} + x_{2,3} \leq \xi_2$, and $p_3 : x_{3,2} + x_{3,3} \leq \xi_3$. Now consider event e_1. From the constraint scopes we see that $Scope(e_1) \cap Scope(p_1) = \{x_{1,1}\}$, $Scope(e_1) \cap Scope(p_2) = \{x_{2,1}\}$ and $Scope(e_1) \cap Scope(p_3) = \emptyset$, so e_1 depends on p_1 and p_2, not p_3. From the problem semantics we should introduce the condition constraints $c_1 : x_{1,1} \neq 0$ and $c_2 : x_{2,1} \neq 0$, to express the fact that there is no dependency relation between e_1 and p_1 if $x_{1,1} = 0$, and that there is no dependency relation between e_1 and p_2 if $x_{2,1} = 0$. Thus we write the dependency meta-constraints DEPENDENCY$(e_1, p_1, x_{1,1} \neq 0)$ and DEPENDENCY$(e_1, p_2, x_{2,1} \neq 0)$.

Equipped with these concepts, we now define EDP-CP as follows.

Definition 6. *An EDP-CP is a 9-tuple* $\mathcal{P} = \langle \mathcal{X}, \mathcal{D}, \Lambda, \mathcal{W}, \mathcal{E}, \mathcal{C}, \mathcal{H}, \Psi, Pr \rangle$ *where:*

- $\mathcal{X} = \{x_1, \ldots, x_n\}$ *is a set of decision variables;*
- $\mathcal{D} = D(x_1) \times \ldots \times D(x_n)$, *where* $D(x_i)$ *is the domain of* X_i;
- $\Lambda = \{\lambda_1, \ldots, \lambda_l\}$ *is a set of uncertain parameters;*
- $\mathcal{W} = W(\lambda_1) \times \ldots \times W(\lambda_l)$, *where* $W(\lambda_i)$ *the domain of* λ_i;
- $\mathcal{E} = \{e_1, \ldots, e_m\}$ *is a set of event constraints. Each* e_i *may either be probabilistic (involving a subset of* $\mathcal{X}$ *and a subset of* Λ) *or deterministic (involving only a subset of* $\mathcal{X}$);
- $\mathcal{C} = \{c_1, \ldots, c_o\}$ *is a set of dependency meta-constraints. For each dependency meta-constraint* $c_i :$ DEPENDENCY(e, p, f) *we have* $e \in \mathcal{E}$, *where* p *may be either a probabilistic or a deterministic pre-requisite constraint, and* f *is a deterministic condition constraint;*
- $\mathcal{H} = \{h_1, \ldots, h_p\}$ *is a set of hard constraints. Each* h_i *may either be probabilistic (involving a subset of* $\mathcal{X}$ *and a subset of* Λ) *or deterministic (involving only a subset of* $\mathcal{X}$);
- Ψ *is any expression involving the event realization measures on the event constraints in* $\mathcal{E}$;
- $Pr : \mathcal{W} \to [0, 1]$ *is a probability distribution over uncertain parameters.*

Example 4. The motivational example of Section 2 can be expressed as an EDP-CP $\mathcal{P} = \langle \mathcal{X}, \mathcal{D}, \Lambda, \mathcal{W}, \mathcal{E}, \mathcal{C}, \mathcal{H}, \Psi, Pr \rangle$ where:

- $\mathcal{X} = \{x_{1,1}, x_{1,2}, x_{2,1}, x_{2,2}, x_{2,3}, x_{3,2}, x_{3,3}\}$;
- $\mathcal{D} = [0..99] \times [0..99] \times [0..99]$;
- $\Lambda = \{\xi_1, \xi_2, \xi_3\}$;

- $\mathcal{W} = \{3(0.3), 7(0.5), 12(0.2)\} \times \{6(0.4), 7(0.2), 10(0.4)\} \times \{3(0.3), 8(0.7)\}$;
- $\mathcal{E} = \{e_1 : x_{1,1} + x_{2,1} = 8, e_2 : x_{1,2} + x_{2,2} + x_{3,2} = 7, e_3 : x_{2,3} + x_{3,3} = 4\}$;
- $\mathcal{C} = \{c_1 : \textsc{Dependency}(e_1, p_1 : x_{1,1} + x_{1,2} \leq \xi_1, f_{1,1} : x_{1,1} \neq 0), \dots,$
  $c_7 : \textsc{Dependency}(e_3, p_3 : x_{3,2} + x_{3,3} \leq \xi_3, f_{3,3} : x_{3,3} \neq 0)\}$;
- $\mathcal{H} = \{x_{1,1} \geq 0, \dots, x_{3,3} \geq 0\}$;
- Ψ is $E(e_1) + E(e_2) + E(e_3)$;
- $\Pr(\langle \xi_1 = 3, \xi_2 = 6, \xi_3 = 3 \rangle) = 0.036, \dots, \Pr(\langle \xi_1 = 12, \xi_2 = 10, \xi_3 = 8 \rangle) = 0.056$.

Finally, we define optimal solutions to EDP-CPs as follows.

Definition 7. *An optimal solution to an EDP-CP $\mathcal{P} = \langle \mathcal{X}, \mathcal{D}, \Lambda, \mathcal{W}, \mathcal{E}, \mathcal{C}, \mathcal{H}, \Psi, Pr \rangle$ is any assignment S to the decision variables such that:*

1. *for each $h \in \mathcal{H}$, for each $w \in \mathcal{W}$, $h_{\downarrow w}$ is satisfied; and*
2. *all dependency constraints in $\mathcal{C}$ are satisfied; and*
3. *there exists no other assignment satisfying all hard and dependency constraints with a strictly better value for Ψ.*

Note that, in an optimal solution, event constraints and pre-requisite constraints may be violated when not also added explicitly as hard constraints. Note also that a *feasible* solution is any assignment that satisfies the first two conditions. Finally, note that when the total number of worlds is 1 with probability 1, the probabilistic event realization measure on c is the same as in the deterministic case.

4 Solution Methods for EDP-CP

We now show how to map an EDP-CP $\mathcal{P} = \langle \mathcal{X}, \mathcal{D}, \Lambda, \mathcal{W}, \mathcal{E}, \mathcal{C}, \mathcal{H}, \Psi, Pr \rangle$ into an equivalent classical COP $\mathcal{P}' = \langle \mathcal{X}', \mathcal{D}', \mathcal{C}', \Psi' \rangle$.

4.1 Mapping Variables and Domains

Algorithm 1 shows how to create the decision variables in $\mathcal{P}'$ starting from $\mathcal{P}$, in three steps. The first step (Line 3) duplicates the decision variables in $\mathcal{P}'$ along with their domains. The second step (Line 4) introduces a Boolean variable that is used later to represent the truth value of each event e in each possible world w. Similarly, the last step (Line 5) introduces a Boolean variable for each pre-requisite constraint in each possible world, used later these Boolean variables to represent the truth values of the pre-requisite constraints.

4.2 Mapping Constraints

Algorithm 2 also shows how to create the constraints in $\mathcal{P}'$, again in three steps. In step one (Line 2) we introduce a reification constraint between each event in each possible world ($e_{\downarrow w}$) and its corresponding Boolean variable (b_w^e). This ensures that $e_{\downarrow w}$ is satisfied iff b_w^e is assigned the value 1. In the second step

---

Algorithm 1. Variable-Mapping$(\mathcal{X}, \mathcal{D}, \Lambda, \mathcal{W}, \mathcal{E}, \mathcal{C})$:$\langle\, \mathcal{X}', \mathcal{D}'\, \rangle$

---

1 $\mathcal{X}' \leftarrow \emptyset$;
2 $\mathcal{D}' \leftarrow \emptyset$;
3 **foreach** $x \in \mathcal{X}$ **do**
  └ create x' with the same domain as x and add it to $\mathcal{X}'$;
4 **foreach** $e \in \mathcal{E}$ **do**
  │ **foreach** $w \in \mathcal{W}$ **do**
  │   └ create a Boolean b_w^e and add it to $\mathcal{X}'$;
5 **foreach** DEPENDENCY$(e, p, c) \in \mathcal{C}$ **do**
  │ **foreach** $w \in \mathcal{W}$ **do**
  │   │ **if** $b_w^p \notin \mathcal{X}'$ **then**
  │   │   └ create a Boolean b_w^p and add it to $\mathcal{X}'$;

Algorithm 2. Constraint-Mapping$(\mathcal{X}, \mathcal{D}, \Lambda, \mathcal{W}, \mathcal{E}, \mathcal{C}, \mathcal{H})$:$\mathcal{C}'$

---

1 $\mathcal{C}' \leftarrow \emptyset$;
2 **foreach** $e \in \mathcal{E}$ **do**
  │ **foreach** $w \in \mathcal{W}$ **do**
  │   └ add $b_w^e = 1 \leftrightarrow e_{\downarrow w}$ to $\mathcal{C}'$;
3 **foreach** DEPENDENCY$(e, p, c) \in \mathcal{C}$ **do**
  │ **foreach** $w \in W$ **do**
  │   │ add $b_w^p = 1 \leftrightarrow p_{\downarrow w}$ to $\mathcal{C}'$;
  │   └ add $c \rightarrow (b_w^e = 1 \rightarrow b_w^p = 1)$ to $\mathcal{C}'$
4 **foreach** $h \in \mathcal{H}$ **do**
  │ **foreach** $w \in \mathcal{W}$ **do**
  │   └ add $h_{\downarrow w}$ to $\mathcal{C}'$;

(Line 3) each dependency constraint DEPENDENCY(e, p, c) is transformed into two constraints: the first is a reification constraint similar to the previous case between each b_w^p and b_w^e, while the second one enforces the dependency constraints according to their definitions. In the final step (Line 4) each hard probabilistic constraint is transformed into a set of deterministic constraints in $\mathcal{C}'$.

4.3   Mapping the Objective Function

Finally, the objective function of $\mathcal{P}'$ is the same function Ψ as in $\mathcal{P}$, except that we replace each occurrence of an event measure $E\{e\}$ with

$$\sum_{w \in \mathcal{W}} Pr(w) b_w^e$$

as shown in Algorithm 3.

Algorithm 3. Objective-Function-Mapping($\mathcal{X}, \mathcal{D}, \Lambda, \mathcal{W}, \mathcal{E}, \mathcal{C}, \mathcal{H}, \Psi$):$\Psi'$

---

1 $\Psi' \leftarrow \Psi$;
2 **foreach** $E\{e\} \in \Psi'$ **do**
  $\lfloor$ replace $E\{e\}$ with $\sum_{w \in \mathcal{W}} Pr(w) b_w^e$;

---

5   Illustrative Examples

In this section we present two illustrative problems and model them using the EDP-CP framework. The first example is a probabilistic supply chain planning problem, which is an extended version of the example of Section 2. In this extended version, demand uncertainty, as well as supply uncertainty, is considered. The second example is a production planning problem with an emphasis on capital budgeting, and assumes that production rates, demands, prices and costs are all uncertain parameters.

5.1   An EDP-CP Model for Probabilistic Supply Chain Planning

There is a sizeable literature on supply chain modeling under uncertainty (see, for example, [4] and [8]). Recently, the authors of this work also experienced at first-hand the relevance of modeling supply and demand uncertainties during a research project carried out for a leading telecommunication company operating worldwide.

Here we adopt a simplified version of the problem, which was presented in Section 2.1 and Figure 1. The objective is to determine the most reliable plan that will meet customers' realised demands at $D_{1,2,3}$ by means of uncertain deliveries from suppliers denoted by $S_{1,2,3}$. It is assumed that (i) the order batch sizes $x_{i,j}$ from supplier i to customer j is not allowed to exceed 6 units, $x_i \leq 6$, (ii) D_3 requires that its order is supplied by only one supplier, $x_{2,3} x_{3,3} = 0$. Scenario

Table 2. Scenario Data

$Pr(w)$	0.036	0.084	0.018	0.042	0.036	0.084	0.060	0.140	0.030	0.070	0.060	0.140	0.024	0.056	0.012	0.028	0.024	0.056
w	1	2	3	4	5	6	7	8	9	10	11	12	13	14	15	16	17	18
S_1	3	3	3	3	3	3	7	7	7	7	7	7	12	12	12	12	12	12
S_2	6	6	7	7	10	10	6	6	7	7	10	10	6	6	7	7	10	10
S_3	3	8	3	8	3	8	3	8	3	8	3	8	3	8	3	8	3	8
D_1	8	8	8	7	7	7	8	8	9	9	9	9	8	8	8	7	7	7
D_2	7	7	7	5	5	5	5	7	7	5	5	5	5	3	3	3	5	5
D_3	4	6	6	4	4	6	6	4	4	6	6	4	4	6	6	4	4	6

parameters are given in Table 2. Excess supplies from suppliers are stored at customers with a negligible inventory carrying cost until the next order issue.

We consider two possible EDP-CP models for this probabilistic SC problem. In the first one we try to find a solution in which all events are realised, while in the second this condition is relaxed. The following EDP-CP model describes the first case:

$$\max E(e_1 : x_{1,1} + x_{2,1} \geq \zeta_1)+$$
$$E(e_2 : x_{1,2} + x_{2,2} + x_{3,2} \geq \zeta_2)+$$
$$E(e_3 : x_{2,3} + x_{3,3} \geq \zeta_3)$$

subject to $\text{DEPENDENCY}(e_1, p_1 : x_{1,1} + x_{1,2} \leq \xi_1, f_{1,1} : x_{1,1} \neq 0);$
$\quad\quad \text{DEPENDENCY}(e_1, p_2 : x_{2,1} + x_{2,2} + x_{2,3} \leq \xi_2, f_{2,1} : x_{2,1} \neq 0);$
$\quad\quad \text{DEPENDENCY}(e_2, p_1, f_{1,2} : x_{1,2} \neq 0);$
$\quad\quad \text{DEPENDENCY}(e_2, p_2, f_{2,2} : x_{2,2} \neq 0);$
$\quad\quad \text{DEPENDENCY}(e_2, p_3 : x_{3,1} + x_{3,2} \leq \xi_3, f_{3,2} : x_{3,2} \neq 0);$

$$\text{DEPENDENCY}(e_3, p_2, f_{2,3} : x_{2,3} \neq 0);$$
$$\text{DEPENDENCY}(e_3, p_3, f_{3,3} : x_{3,3} \neq 0);$$
$$0 \leq x_{i,j} \leq 6, \forall i, j \in \{1,2,3\};$$
$$x_{2,3}.x_{3,3} = 0;$$
$$e_i, \forall i \in \{1,2,3\}$$

The second case can be simply achieved by dropping e_1–e_3 from the set of hard constraints.

The EDP-CP model is compiled into a standard CP model using the algorithm presented in Section 4. The optimal solution is $x_{1,1} = 6$, $x_{1,2} = 1$, $x_{2,1} = 3$, $x_{2,2} = 4$, $x_{2,3} = 0$, $x_{3,2} = 2$, $x_{3,3} = 6$. In the optimal plan $E(e_1) = 0.420$, $E(e_2) = 0.294$ and $E(e_3) = 0.700$, giving an optimal objective function value of 1.414. In other words, this plan guarantees to meet customer demands at $D_{1,2,3}$ with probabilities 42.0%, 29.4% and 70.0%, respectively. This plan aims to satisfy customer demands completely.

In most circumstances it would be more realistic to assume that the event constraints e_1, e_2, and e_3 are not hard constraints and the expected plan should not aim for a complete demand satisfaction. When we drop these hard event constraint, the following plan is optimal under such a relaxation: $x_{1,1} = 6$, $x_{1,2} = 0$, $x_{2,1} = 3$, $x_{2,2} = 3$, $x_{2,3} = 0$, $x_{3,2} = 2$, $x_{3,3} = 6$. The event constraint satisfaction probabilities are now $E(e_1) = 0.700$, $E(e_2) = 0.476$ and $E(e_3) = 0.700$, giving a total of 1.876.

A comparison of two plans shows that there are differences between them at $x_{1,2}$ and $x_{2,2}$. It may not be immediately obvious why we change $x_{1,2}$ from 1 to 0, as in both plans the probability of acquiring the required capacity at S_1 (7 and 6, respectively) is 0.8. The explanation lies in the probability distribution of the uncertain capacity of S_2. Supplier S_2 can provide 6 units with a probability of 1.0, but not 7 units. The second plan exploits this situation and aims for a partial satisfaction at D_2 by providing only 5 units. Thus there is no need for any delivery from S_1 to D_2. The second plan has higher reliability at the expense of partial satisfaction at D_2. (There are alternative optimal solutions to this instance.)

5.2   An EDP-CP Model for Production Planning/Capital Budgeting

The production planning/capital budgeting problem assumes that there are $n = 7$ types of products to be produced, under uncertain demands d_i, $i = 1, ..., 7$. Each product can be produced on only one type of machine which is designated to this product only. The existing production floor space is $A = 50$ m², in which each machine type requires m_i ($m = [3, 6, 5, 3, 7, 8, 9]$) in m² per machine of type i. The cost of operating each machine involves two types of costs: fixed cost f_i ($f = [40, 75, 62, 39, 53, 19, 38]$) and variable production cost c_i. The total production budget is $B = \$670$. The variable production cost components $c_{1,...,7}$ are uncertain, taking different values in each world $w_{1,...,4}$ (see Table 3). The produced amount of each product depends on the number of machines used, x_i, and the uncertain machine production rate, r_i, is also given in Table 3. Table 3 shows two more uncertain problem parameters: demand d_i and selling price p_i.

Table 3. Problem Data

w	Pr	production cost							demand							selling price							production rate						
		c_1	c_2	c_3	c_4	c_5	c_6	c_7	d_1	d_2	d_3	d_4	d_5	d_6	d_7	p_1	p_2	p_3	p_4	p_5	p_6	p_7	r_1	r_2	r_3	r_4	r_5	r_6	r_7
1	0.16	3	6	1	1	6	10	2	4	7	2	8	3	5	2	8	14	4	16	14	10	4	2	3	2	1	2	1	2
2	0.19	4	4	7	2	4	7	7	7	9	9	9	4	7	4	10	16	18	18	10	14	14	4	4	5	2	4	3	6
3	0.38	5	3	5	8	7	6	10	9	11	12	10	7	8	7	18	22	14	18	14	16	24	5	5	6	3	5	5	4
4	0.27	5	6	8	5	5	3	6	11	13	17	11	13	16	13	22	26	26	22	16	24	18	9	6	8	4	7	7	7

Under these uncertainties, a realistic objective is to determine the most reliable plan (i.e. how many machines to purchase of each type) that maximizes our chances of meeting our demand constraints as much as possible, while achieving a specified target profit of $T = \$40$, not exceeding our budget B, and meeting all space and production constraints. It is assumed that meeting customer demands and the profit target are equally important events.

An EDP-CP model of the production planning/capital budgeting problem is as follows:

$$\max \frac{1}{2n} \sum_{i=1}^{n} E(e_i : \min(r_i x_i, d_i) = d_i) +$$
$$\frac{1}{2} E(\bar{e} : \sum_{i=1}^{n} p_i \min(r_i x_i, d_i) - f_i x_i - c_i r_i x_i \geq T)$$
subject to $\text{DEPENDENCY}(e_j, \sum_{i=1}^{n}(f_i + c_i r_i)x_i \leq B, \text{True}), \forall j \in \{1, \dots, n\};$
$$\text{DEPENDENCY}(\bar{e}, \sum_{i=1}^{n}(f_i + c_i r_i)x_i \leq B, \text{True});$$
$$\sum_{i=1}^{n} m_i x_i \leq A$$
$$x_i \in \mathbb{Z}^{0,+}$$

where $r_i x_i$ and $\min(r_i x_i, d_i)$ denote the amount produced and sold, respectively, of product type $i \in \{1, ..., n\}$, and x_i denotes the number of machine used in the production of type i product. There is only one pre-requisite constraint (the budget constraint) and no condition constraint.

The optimal solution found is $x^* = [2, 0, 2, 0, 0, 2, 2]$. This production plan gives $E(e_1) = 100\%$, $E(e_2) = 0$, $E(e_3) = 73\%$, $E(e_4) = 0$, $E(e_5) = 0$, $E(e_6) =$

38%, $E(e_7) = 100\%$, where event constraint e_i denotes the complete satisfaction of demand for product type i. In this plan the profit target is achieved $E(\bar{e}) = 65\%$ of the time.

6   Related Work

The EDP-CP framework is a generalization of the work of Liu [5] on dependent-chance programming. Firstly, our notion of constraint dependency introduces condition constraints in addition to the event and pre-requisite constraints. It should be noted that constraint dependency without condition constraints does not guarantee optimal plans since in certain instances common variables may take values which break the link between two dependent constraints. Secondly, while a feasible solution in Liu's framework satisfies all event constraints, in our framework such a requirement is relaxed, and this gives the decision-maker more flexibility in modeling. Finally, while Liu's work only considers Monte Carlo-based simulation methods, we propose a complete solution method.

EDP-CP is also related to the probabilistic CSPs framework [2]. However, probabilistic CSPs treat all probabilistic constraints uniformly, whereas EDP-CP distinguishes between event, pre-requisite, condition, and hard probabilistic constraints. For instance, in probabilistic CSPs, all customer and demand constraints will be treated in the same way. In a given world, either all constraints are satisfied or the problem is over-constrained. While finding a plan that has the highest probability of success is an interesting objective, our approach answers different questions and achieves different objectives. EDP-CP is also related to Soft CSPs [1], which can be viewed as a generalisation of probabilistic CSPs [2].

Another technique addressing constraint problems under uncertainty is Stochastic Constraint Programming (SCP) [6]. The SCP approach assumes that the constraints are stochastically independent (i.e., there are no DEPENDENCY constraints among them) and there is always at least one feasible solution which satisfies all constraints under all scenarios (i.e., worlds). Thus SCP addresses a completely different class of stochastic problems.

Another related work is that of [7], who extend the max-CSP framework with meta-constraints. However, their approach is defined for deterministic problems.

7   Conclusion

In this paper we propose EDP-CP as a novel modeling framework that helps decision makers in uncertain environments to realistically model their problems and find reliable solutions to such complex problems. The characteristic features of our modeling framework can be summarized as follows:

- To better model the uncertainties in real-world problems, we allow the set of constraints to be either deterministic or probabilistic;
- We move away from classical approaches that treat all constraints uniformly to one that distinguishes between event, pre-requisite, condition, and hard constraints;

- We introduce the DEPENDENCY meta-constraint that allows the modeler to state a problem by explicitly specifying dependency relationships between event, pre-requisite, and condition constraints;
- In an uncertain environment, it is quite unrealistic to assume that a solution is valid irrespective of the unfolding of the uncertain parameters. In fact, there is a certain degree of fuzziness associated with each candidate solution. Therefore, in our framework, we view the set of feasible solutions as probabilistic due to the inherent uncertainties;
- We introduce a probabilistic event realization measure, which can be used by the modeler to define solution reliability.

Our future work will extend the proposed framework in various directions, and provide efficient and effective solving methods. Our first steps will be:

- The development of specialized solution methods for EDC-CP. For instance a specialized global constraint for the DEPENDENCY meta-constraint can be designed.
- In large-scale uncertain problems, the number of worlds can be prohibitively large. We will investigate ways of reducing the number of world as well as employing effective decomposition techniques;
- We will look at ways of extending EDP-CP to deal with recourse actions.

Acknowledgements. Supported by Science Foundation Ireland under Grant-03/CE3/I405 as part of the Centre for Telecommunications Value-Chain-Driven Research and Grant-00/PI.1/C075.

References

1. Bistarelli, S., Montanari, U., Rossi, F.: Soft constraint logic programming and generalized shortest path problems, Journal of Heuristics, Kluwer, 2001.
2. Fargier, H., Lang, J., Martin-Clouaire, R., Schiex, T.: A constraint satisfaction framework for decision under uncertainty, Proc. of the 11th Int. Conf. on Uncertainty in Artificial Intelligence, Montreal, Canada, August 1995.
3. Kingsman, B. G.: Raw materials purchasing: an operational research approach, Pergamon Press, New York, 1985.
4. de Kok, A. G., Graves, S. C.: Supply chain management: design, coordination and operation, Handbook in OR/MS, vol. 11, Elsevier, Amsterdam, 2003.
5. Liu, B, Iwamura K.: Modelling stochastic decision systems using dependent-chance programming, European Journal of Operational Research, 101:193–203, 1997.
6. Manandhar, S., Tarim, S. A., Walsh, T.: Scenario-based stochastic constraint programming, Proc. of IJCAI-2003, Acapulco, Mexico, 257–262, 2003.
7. Petit, T., Régin, J.-C., Bessière, C.: Meta-constraints on violations for overconstrained problems. Proc. of ICTAI'2000, Vancouver, BC, Canada, November 2000, IEEE Computer Society, 358–365.
8. Porteus, E. L.: Foundations of stochastic inventory theory, Stanford University Press, Stanford, California, 2002.
9. Tarim, S. A., Manandhar, S., Walsh, T.: Stochastic constraint programming: a scenario based approach, Constraints, 11:53–80, 2006.

Online Stochastic Reservation Systems

Pascal Van Hentenryck, Russell Bent, and Yannis Vergados

Department of Computer Science, Brown University,
Providence, RI 02912, USA

Abstract. This paper considers online stochastic reservation problems, where requests come online and must be dynamically allocated to limited resources in order to maximize profit. Multi-knapsack problems with or without overbooking are examples of such online stochastic reservations. The paper studies how to adapt the online stochastic framework and the consensus and regret algorithms proposed earlier to online stochastic reservation systems. On the theoretical side, it presents a constant sub-optimality approximation of multi-knapsack problems, leading to a regret algorithm that evaluates each scenario with a single mathematical programming optimization followed by a small number of dynamic programs for one-dimensional knapsacks. On the experimental side, the paper demonstrates the effectiveness of the regret algorithm on multi-knapsack problems (with and without overloading) based on the benchmarks proposed earlier.

1 Introduction

In an increasingly interconnected and integrated world, online optimization problems are quickly becoming pervasive and raise new challenges for optimization software. Moreover, in most applications, historical data or statistical models are available, or can be learned, for sampling. This creates significant opportunities at the intersection of online algorithms, combinatorial and stochastic optimization, and machine learning. In fact, increasing attention has been devoted to these issues in a variety of communities (e.g., [10, 1, 6, 11, 9, 5, 8]).

This paper considers online stochastic reservation systems and, in particular, the stochastic multi-knapsack problems introduced in [1]. Typical applications include, for instance, reservation systems for holiday centers and advertisement placements in web browsers. These problems differ from the stochastic routing and scheduling considered in, say, [10, 6, 9, 5] in that online decisions are not about selecting the best request to serve but rather about how best to serve a request.

The paper shows how to adapt our online stochastic framework, and the consensus and regret algorithms, to online stochastic reservation systems. Moreover, in order to instantiate the regret algorithm, the paper presents a constant-factor suboptimality approximation for multi-knapsack problems using one-dimensional knapsack problems. As a result, on multi-knapsack problems with or without overbooking, each online decision involves solving a mathematical program and a series of dynamic programs. The algorithms were evaluated on the multi-knapsack problems proposed in [1] with and without overbooking. The results indicate that the regret algorithm is particularly effective, providing significant benefits over heuristic, consensus, and expectation approaches. It also dominates an earlier algorithm proposed in [1] (which applies the best-fit heuristic

J.C. Beck and B.M. Smith (Eds.): CPAIOR 2006, LNCS 3990, pp. 212–227, 2006.

within the expectation algorithm) as soon as the time constraints allows for 10 optimizations for each online decision or between each two online decisions. The results are particularly interesting in our opinion, because the consensus and regret algorithms have now been applied generically and successfully to online problems in scheduling, routing, and reservation using, at their core, either constraint programming, mathematical programming, or dedicated polynomial algorithms. The rest of the paper introduces online stochastic reservation problems in their simplest form, shows how to adapt our online stochastic algorithms for them, presents the sub-optimality approximation, and describes the experimental results.

2 Online Stochastic Reservation Problems

2.1 The Offline Problem

The offline problem is defined in terms of n bins B and each bin $b \in B$ has a capacity C_b. It receives as input a set R of requests. Each request is typically characterized by its capacity and its reward, which may or may not depend on which bin the request are allocated to. The goal is to find an assignment of a subset $T \subseteq R$ of requests to the bins satisfying the problem-specific constraints and maximizing the objective function.

The Multi-Knapsack Problem. The multi-knapsack problem is an example of a reservation problem. Here each request r is characterized by a reward $w(r)$ and a capacity $c(r)$. The goal is to allocate a subset T of the requests R to the bins B so that the capacities of the bins are not exceeded and the objective function $w(T) = \sum_{r \in T} w(r)$ is maximized. A mathematical programming formulation of the problem associates witch each request r and bin b a binary variable $x[r, b]$ whose value is 1 when the request is allocated to bin b and 0 otherwise. The integer program can be expressed as:

$$\max \quad \sum_{r \in R, b \in B} w(r)\, x[r, b]$$

such that

$$\sum_{b \in B} x[r, b] \leq 1 \quad (r \in R)$$
$$\sum_{r \in R} c(r)\, x[r, b] \leq C_b \quad (b \in B)$$
$$x[r, b] \in \{0, 1\} \quad (r \in R, b \in B)$$

The Multi-Knapsack Problem with Overbooking. In practice, many reservation systems allow for overbooking. The multi-knapsack problem with overbooking allows the bin capacities to be exceeded but overbooking is penalized in the objective function. To adapt the mathematical-programming formulation above, it suffices to introduce a nonnegative variable $y[b]$ representing the excess for each bin b and to introduce a penalty term $\alpha \times y[b]$ in the objective function. The integer programming model now becomes

$$\max \quad \sum_{r \in R, b \in B} w(r)\, x[r, b] - \sum_{b \in B} \alpha\, y[b]$$

such that $\sum_{b \in B} x[r, b] \leq 1 \quad (r \in R)$
$$\sum_{r \in R} c(r)\, x[r, b] \leq C_b + y[b] \quad (b \in B)$$
$$x[r, b] \in \{0, 1\} \quad (r \in R, b \in B)$$
$$y[b] \geq 0 \quad (b \in B)$$

This is the offline problem considered in [1].

Generic Formalization. To formalize the online algorithms precisely and generically, it is convenient to assume the existence of a dummy bin $\bot$ with infinite capacity to assign the non-selected requests and to use $B_\bot$ to denote $B \cup \{\bot\}$. A solution σ can then be seen as a function $R \rightarrow B_\bot$. The objective function can be specified by a function $\mathcal{W}$ over assignments and the problem-specific constraints can be specified as a relation over assignments giving us the problem $\max_{\sigma: \, C(\sigma)} \mathcal{W}(\sigma)$. We use $\sigma[r \leftarrow b]$ to denote the assignment where r is assigned to bin b, i.e.,

$$\sigma[r \leftarrow b](r) = b$$
$$\sigma[r \leftarrow b](r') = \sigma(r') \; \text{if } r' \neq r.$$

and $\sigma \downarrow R$ to denote the assignment where the requests in R are now unassigned, i.e.,

$$(\sigma \downarrow R)(r) = \bot \quad \text{if } r \in R$$
$$(\sigma \downarrow R)(r) = \sigma(r) \; \text{if } r \notin R.$$

Finally, we use $\sigma_\bot$ to denote the assignment satisfying $\forall r \in R : \sigma(r) = \bot$.

2.2 The Online Problem

In the online problem, the requests are not known a priori but are revealed online during the execution of the algorithm. For simplicity, we consider a time horizon $H = [1, h]$ and we assume that a single request arrives at each time $t \in H$. (It is easy to relax these assumptions). The algorithm thus receives a sequence of requests $\boldsymbol{\xi} = \langle \xi_1, \ldots, \xi_h \rangle$ over the course of the execution. At time i, the sequence $\boldsymbol{\xi}_i = \langle \xi_1, \ldots, \xi_i \rangle$ has been revealed, the requests $\xi_1, \ldots, \xi_{i-1}$ have been allocated in the assignment σ_{i-1} and the algorithm must decide how to serve request ξ_i. More precisely, step i produces an assignment $\sigma_i = \sigma_{i-1}[\xi_i \leftarrow b]$ that assigns a bin b to ξ_i keeping all other assignments fixed. The requests are assumed to be drawn from a distribution $\mathcal{I}$ and the goal is to maximize the expected value

$$\mathop{\mathbf{E}}_{\boldsymbol{\xi}}[\mathcal{W}(\sigma_\bot[\xi_1 \leftarrow b_1, \ldots, \xi_h \leftarrow b_h])$$

where the sequence $\boldsymbol{\xi} = \langle \xi_1, \ldots, \xi_h \rangle$ is drawn from $\mathcal{I}$.

The online algorithms have at their disposal a procedure to solve , or approximate, the offline problem, and the distribution $\mathcal{I}$. The distribution is a black-box available for sampling.[1] Practical applications often include severe time constraints on the decision time and/or on the time between decisions. To model this requirement, the algorithms may only use the optimization procedure $\mathcal{O}$ times at each time step.

It is interesting to contrast this online problem with those studied in [7, 5, 3]. In these applications, the key issue was to select which request to serve at each step. Moreover, in the stochastic vehicle routing applications, accepted requests did not have to be assigned a vehicle: the only constraint on the algorithm was the promise to serve every accepted request. The online stochastic reservation problem is different. The key issue

---

[1] Our algorithms only require sampling and do not exploit other properties of the distribution which makes them applicable to many applications. Additional information on the distribution could also be beneficial but is not considered here.

ONLINEOPTIMIZATION(ξ)
1 $\sigma_0 \leftarrow \sigma_\perp$;
2 **for** $t \in H$ **do**
3    $b \leftarrow$ CHOOSEALLOCATION(σ_{t-1}, ξ_t);
4    $\sigma_t \leftarrow \sigma_{t-1}[\xi_t \leftarrow b]$;
5 **return** σ_h;

Fig. 1. The Generic Online Algorithm

is not which request to serve but rather whether and how the incoming request must be served. Indeed, whenever a request is accepted, it must be assigned a specific bin and the algorithm is not allowed to reshuffle the assignments subsequently.

The Generic Online Algorithm. The algorithms in this paper share the same online optimization schema depicted in Figure 1. They differ only in the way they implement function CHOOSEALLOCATION. The online optimization schema receives a sequence of online requests ξ and starts with an empty allocation (line 1). At each decision time t, the online algorithm considers the current allocation σ_{t-1} and the current request ξ_t and chooses the bin b to allocate the request (line 3), which is then included in the new assignment σ_t (line 4). The algorithm returns the last assignment σ_h whose value is $\mathcal{W}(\sigma_h)$ (line 5). To implement function CHOOSEALLOCATION, the algorithms have at their disposal two black-boxes:

1. a function OPTSOL(σ, R) that, given an assignment σ and a R of requests, returns an optimal allocation of the requests in R given the past decisions in σ. In other words, OPTSOL(σ, R) solves an online problem where the decision variables for the requests in σ have fixed values.
2. a function GETSAMPLE(t) that returns a set of requests over the interval $[t, h]$ by sampling the arrival distribution.

To illustrate the framework, we specify a best-fit online algorithm as proposed in [1].

Best Fit (G). This algorithm assigns the request ξ to a bin that can accommodate ξ and has the smallest capacity given the assignment σ:

CHOOSEALLOCATION-G(σ, ξ)
1 **return** $argmin(b \in B_\perp : \mathcal{C}(\sigma[\xi \leftarrow b]))\ C_b(\sigma)$;

where $C_b(\sigma)$ denotes the remaining capacity of the bin $b \in B_\perp$ in σ, i.e.,

$$C_b(\sigma) = C_b - \sum_{r \in R: \sigma(r) = b} c(r).$$

3 Online Stochastic Algorithms

This section reviews the various online stochastic algorithms. It starts with the expectation algorithm and shows how it can be adapted to incorporate time constraints.

Expectation (E). Informally speaking, algorithm E generates future requests by sampling and evaluates each possible allocation against the samples. A simple implementation can be specified as follows:

```
CHOOSEALLOCATION-E(σ_{t-1}, ξ_t)
1  for b ∈ B_⊥ do
2      f(b) ← 0;
3  for i ← 1 . . . O/|B_⊥| do
4      R_{t+1} ← GETSAMPLE(t + 1);
5      for b ∈ B_⊥ : C(σ_{t-1}[ξ_t ← b]) do
6          σ* ← OPTSOL(σ_{t-1}[ξ_t ← b], R_{t+1});
7          f(b) ← f(b) + W(σ*);
8  return argmax(b ∈ B_⊥) f(b);
```

Lines 1-2 initialize the evaluation $f(b)$ of each request b. The algorithm then generates $O/|B_⊥|$ samples of future requests (lines 3–4). For each such sample, it successively considers each available bin b that can accommodate the request $ξ$ given the assignment $σ_{t-1}$ (line 5). For each such bin b, it schedules $ξ_t$ in bin b and applies the optimization algorithm using the sampled requests R_{t+1} (line 6). The evaluation of bin b is incremented in line 7 with the weight of the optimal assignment $σ*$. Once all the bin allocations are evaluated over all samples, the algorithm returns the bin b with the highest evaluation. Algorithm E performs O optimizations but uses only $O/|B_⊥|$ samples. When O is small (due to the time constraints), each request is only evaluated with respect to a small number of samples and algorithm E does not yield much information. To cope with tight time constraints, two approximations of E, consensus and regret, were proposed.

Consensus (C). The consensus algorithm C was introduced in [7] as an abstraction of the sampling method used in online vehicle routing [6]. Its key idea is to solve each sample once and thus to examine O samples instead of $O/|B_⊥|$. More precisely, instead of evaluating each possible bin at time t with respect to each sample, algorithm C executes the optimization algorithm once per sample. The bin to which request $ξ$ is allocated in optimal solution $σ*$ is credited $W(σ*)$ and all other bins receive no credit. Algorithm C can be specified as follows:

```
CHOOSEALLOCATION-C(σ_{t-1}, ξ_t)
1  for b ∈ B_⊥ do
2      f(b) ← 0;
3  for i ← 1 . . . O do
4      R_t ← {ξ_t} ∪ GETSAMPLE(t + 1);
5      σ* ← OPTSOL(σ_{t-1}, R_t);
6      f(σ*(ξ_t)) ← f(σ*(ξ_t)) + W(σ*);
7  return argmax(b ∈ B_⊥) f(b);
```

The core of the algorithm are once again lines 4–6. Line 4 defines the set R_t of requests that now includes $ξ_t$ in addition to the sampled requests. Line 5 calls the optimization algorithm with $σ_{t-1}$ and R_t. Line 6 increments only the bin $σ*(ξ_t)$ The main appeal of Algorithm C is its ability to avoid partitioning the available samples between the requests, which is a significant advantage when O is small and/or when the number of bins is large. Its main limitation is its *elitism*. Only the best allocatation is given some credit for a given sample, while other bins are simply ignored.

Regret (R). The regret algorithm R is the recognition that, in many applications, it is possible to estimate the loss of sub-optimal allocations (called regrets) quickly. In other words, once the optimal solution σ^* of a scenario is computed, algorithm E can be approximated with one optimization [5, 2].

Definition 1 (Regret). *Let σ be an assignment, R be a set of requests, r be a request in R, and b be a bin. The regret of a bin allocation $r \leftarrow b$ wrt σ and R, denoted by* DEVIATION$(\sigma, R, r \leftarrow b)$, *is defined as*

$$| \; \mathcal{W}(\text{OPTSOL}(\sigma, R)) - \mathcal{W}(\text{OPTSOL}(\sigma[r \leftarrow b], R \setminus \{r\}))) \; | \; .$$

Definition 2 (Sub-Optimality Approximation). *Let σ be an assignment, R be a set of requests, r be a request in R, and b be a bin. Assume that algorithm* OPTSOL(σ, R) *runs in time $O(f_o(R))$. A sub-optimatily approximation runs in time $O(f_o(R))$ and, given the solution $\sigma^* = optSol(\sigma, R)$, returns, for each bin $b \in B_\perp$, an approximation* SUBOPT$(\sigma^*, \sigma, R, r \leftarrow b)$ *to all regrets* REGRET$(\sigma, R, r \leftarrow b)$.

Intuitively, the $|B_\perp|$ regrets must not take more time than the optimization. We are ready to present the regret algorithm R:

```
CHOOSEALLOCATION-R(σ_{t-1}, ξ_t)
1   for b ∈ B_⊥ do
2       f(b) ← 0;
3   for i ← 1 ... O do
4       R_t ← {ξ_t} ∪ GETSAMPLE(t + 1);
5       σ* ← OPTSOL(σ_{t-1}, R_t);
6       f(σ*(ξ_t)) ← f(σ*(ξ_t)) + W(σ*);
7       for b ∈ B_⊥ \ {σ(ξ_t) : C(σ_{t-1}[ξ_t ← b])} do
8           f(b) ← f(b) + (W(σ*) − SUBOPT(σ*, σ_{t-1}, R_t, ξ_t ← b));
9   return argmax(b ∈ B_⊥) f(b);
```

Its basic organization follows algorithm C. However, instead of assigning some credit only to the bin selected by the optimal solution, algorithm R (lines 7-8) uses the sub-optimality approximation to compute, for each available allocation $\xi_t \leftarrow b$, an approximation of the best solution that allocates ξ_t to b. Hence every available bin is given an evaluation for every sample at time t for the cost of a single optimization (asymptotically). Observe that algorithm R performs $\mathcal{O}$ optimizations at time t.

4   Generalizations of the Framework

This section discusses two generalizations: precomputation and cancellations.

Precomputation. Many reservation systems require immediate responses to requests, giving only limited time to the online algorithm for decision making. However, as is the case in vehicle routing, there is time between decisions to generate scenarios and

ONLINEOPTIMIZATION(ξ, ζ)

1   $\sigma_0 \leftarrow \sigma_\perp$;
2   **for** $t \in H$ **do**
3       $\sigma_{t-1} \leftarrow \sigma_{t-1} \downarrow \zeta_t$;
4       $b \leftarrow$ SELECTALLOCATION(σ_{t-1}, ξ_t);
5       $\sigma_t \leftarrow \sigma_{t-1}[\xi_t \leftarrow b]$;
6   **return** σ_h;

Fig. 2. The Generic Online Algorithm with Cancellations

CHOOSEALLOCATION-C(σ_{t-1}, ξ_t)

1   **for** $b \in B_\perp$ **do**
2       $f(b) \leftarrow 0$;
3   **for** $i \leftarrow 1 \ldots \mathcal{O}$ **do**
4       $\langle R_{t+1}, Z_{t+1} \rangle \leftarrow$ GETSAMPLE($t+1$);
5       $\sigma^* \leftarrow$ OPTSOL($\sigma_{t-1} \downarrow Z_{t+1}, \{\xi_t\} \cup R_{t+1}$);
6       $f(\sigma^*(\xi_t)) \leftarrow f(\sigma^*(\xi_t)) + \mathcal{W}(\sigma^*)$;
7   **return** $argmax(b \in B_\perp) \, f(b)$;

Fig. 3. The Consensus Algorithm with Cancellations

optimize them. This idea can be accommodated in the framework by separating the optimization phase from the decision-making phase in the online algorithm. This is especially attractive for consensus and regret where each scenario is solved exactly once. Details on this separation can be found in [4] in the context of the original framework.

Cancellations. Most reservation systems allow requests to be cancelled after they are accepted. The online stochastic framework can accommodate cancellations by simple enhancements to the generic online algorithm and the sampling procedure. It suffices to assume that an (often empty) set of cancellations ζ_t is revealed at step t in addition to the request ξ_t and that the function GETSAMPLE return pairs $\langle R, Z \rangle$ of future requests R and cancellations Z. Figure 2 presents a revised version of the generic online algorithm: its main modification is in line 3 which removes the cancellations ζ_t from the current assignment σ_{t-1} before allocating a bin to the new request. Figure 3 shows the consensus algorithm with cancellations, illustrating the enhanced sampling procedure (line 4) and how cancellations are taken into account when calling the optimization.

5   The Suboptimality Approximation

This section describes a sub-optimality algorithm approximating multi-knapsack problems within a constant factor. Given a set of requests R, a request $r \in R$, and an optimal solution σ^* to the multi-knapsack problem, the sub-optimality algorithm must return approximations to the regrets of allocating r to bin $b \in B_\perp$. The sub-optimality algorithm must run within the time taken by a constant number of optimizations.

REGRET-SWAP$(i, 1, 2)$

1    $A \leftarrow bin(1, \sigma^*) \cup bin(2, \sigma^*) \cup U(\sigma^*) \setminus \{i\};$
2   **if** $C_1 - c(i) \geq C_2$ **then**
3        $bin(1, \sigma^a) \leftarrow knapsack(A, C_1 - c(i)) \cup \{i\};$
4        $bin(2, \sigma^a) \leftarrow knapsack(A \setminus bin(1, \sigma^a), C_2);$
5   **else**
6        $bin(2, \sigma^a) \leftarrow knapsack(A, C_2);$
7        $bin(1, \sigma^a) \leftarrow knapsack(A \setminus bin(2, \sigma^a), C_1 - c(i)) \cup \{i\};$
8    $e \leftarrow argmax(r \in bin(1, \sigma^*) \setminus bin(1..2, \sigma^a) : c(r) > \max(C_1 - c(i), C_2)) \; w(r);$
9   **if** e exists & $w(e) > \max(w(bin(1, \sigma^a)), w(bin(2, \sigma^a)))$ **then**
10        $j \leftarrow argmax(j \in 3..n) \; C_j;$
11        $bin(j, \sigma^a) \leftarrow knapsack(bin(j, \sigma^a) \cup \{e\}, C_j);$

Fig. 4. The Suboptimality Algorithm for the Knapsack Problem: Swapping i from Bin 2 to Bin 1

The key idea behind the suboptimality algorithm is to solve a small number of one-dimensional knapsack problems (which takes pseudo-polynomial time). There are two main cases to study: either request r is allocated to a bin in B in solution σ^* or it is not allocated (i.e., it is allocated to $\bot$. In the first case, the algorithm must approximate the optimal solutions in which r is allocated to other bins (procedure REGRET-SWAP) or not allocated (procedure REGRET-SWAP-OUT). In the second case, the request must be swapped in all the bins (procedure REGRET-SWAP-IN). The rest of this section presents algorithms for the non-overbooking case; they generalize to the overbooking case.

Since the names of the bins have no importance, we assume that they are numbered $1..n$. Moreover, without loss of generality, we formalize the algorithms to move request i from bin 2 to bin 1, to swap request i out of bin 1, and to swap request i into bin 1. We use σ^* to represent the optimal solution to the multi-knapsack problem, σ^s to denote the optimal solution in which request i is assigned to bin 1 (REGRET-SWAP and REGRET-SWAP-OUT) or is not allocated (REGRET-SWAP-IN), and σ^a to denote the sub-optimality approximation. We also use $bin(b, \sigma)$ to denote the requests allocated to bin b and generalize the notation to sets of bins. The solution to the one-dimensional knapsack problem on R for a bin with capacity C is denoted by $knapsack(R, C)$. We also use $c(R)$ to denote the sum of the capacities of the requests in R and $U(\sigma^*)$ the requests that are not allocated in the optimal solution σ^*.

Swapping a Request Between Two Bins. Figure 4 depicts the algorithm to swap request i from bin 1 to bin 2. The key idea is to consider all requests allocated to bins 1 and 2 in σ^* and to solve two one-dimensional problems for bin 1 (without the capacity taken by request i) and bin 2. The algorithm always starts with the bin whose remaining capacity is largest. After solving these two one-dimensional knapsacks, if there exists a request $e \in bin(1, \sigma^*)$ not allocated in $bin(1..2, \sigma^a)$ whose value is higher than the values of these two bins, the algorithm solves a third knapsack problem to place this request in another bin if appropriate. This is important if request e is of high value but cannot be allocated in bin 1 due to the capacity taken by request i.

Theorem 1. *Algorithm* REGRET-SWAP *is a constant-factor approximation.*

Proof. Let σ^s be the sub-optimal solution, σ^a be the regret solution, and σ^* be the optimal solution. Consider the following sets

$$
\begin{aligned}
I_1 &= \sigma^s \cap \sigma^a \\
I_2 &= (bin(1,\sigma^s) \setminus \sigma^a) \cap U(\sigma^*) \\
I_3 &= (bin(2,\sigma^s) \setminus \sigma^a) \cap U(\sigma^*) \\
I_4 &= (bin(3..n,\sigma^s) \setminus \sigma^a) \cap U(\sigma^*) \\
I_5 &= (bin(1,\sigma^s) \setminus \sigma^a) \cap bin(1,\sigma^*) \\
I_6 &= (bin(1,\sigma^s) \setminus \sigma^a) \cap bin(2,\sigma^*).
\end{aligned}
\qquad
\begin{aligned}
I_7 &= (bin(2,\sigma^s) \setminus \sigma^a) \cap bin(1,\sigma^*) \\
I_8 &= (bin(2,\sigma^s) \setminus \sigma^a) \cap bin(2,\sigma^*) \\
I_9 &= (bin(3..n,\sigma^s) \setminus \sigma^a) \cap bin(1,\sigma^*) \\
I_{10} &= (bin(3..n,\sigma^s) \setminus \sigma^a) \cap bin(2,\sigma^*) \\
I_{11} &= (bin(1..n,\sigma^s) \setminus \sigma^a) \cap bin(3..n,\sigma^*)
\end{aligned}
$$

The suboptimal solution σ^s can be partitioned into $\sigma^s = \bigcup_{k=1}^{11} I_k$ and the proof shows that $w(I_k) \leq c_k\, w(\sigma^a)$ $(1 \leq k \leq 13)$ which implies that $w(\sigma^s) \leq c\, w(\sigma^a)$ for some constant $c = c_1 + \ldots c_{11}$. The proof of each inequality typically separates two cases:

A: $C_1 - c(i) \geq C_2$;
B: $C_1 - c(i) < C_2$.

Observe also that the proof that $K(I_1) \leq K(\sigma^a)$ is immediate. We now give the proofs for the remaining sets. In the proofs, C_1' denotes $C_1 - c(i)$ and $K(E, C)$ is defined as follows:

$$
K(E, C) = w(knapsack(E, C)).
$$

$I_2.A$: By definition of I_2 and by definition of $bin(1, \sigma^a)$ in line 3,

$$
K(I_2, C_1') \leq K(U(\sigma^*), C_1') \leq K(bin(1, \sigma^a), C_1') \leq w(\sigma^a).
$$

$I_2.B$: By definition of I_2, $C_1' < C_2$, and by definition of $bin(2, \sigma^a)$ in line 6

$$
K(I_2, C_1') \leq K(U(\sigma^*), C_1') \leq K(U(\sigma^*), C_2) \leq K(bin(2, \sigma^a), C_2) \leq w(\sigma^a).
$$

$I_3.A$: By definition of I_3, $C_1' \geq C_2$, and by definition of $bin(1, \sigma^a)$ in line 3

$$
K(I_3, C_2) \leq K(U(\sigma^*), C_2) \leq K(U(\sigma^*), C_1') \leq K(bin(1, \sigma^a), C_1') \leq w(\sigma^a).
$$

$I_3.B$: By definition of I_3 and by definition of $bin(2, \sigma^a)$ in line 6

$$
K(I_3, C_2) \leq K(U(\sigma^*), C_2) \leq K(bin(2, \sigma^a), C_2) \leq w(\sigma^a).
$$

I_4 : Assume that $w(I_4) > w(\sigma^a)$. This implies

$$
\begin{aligned}
w(I_4) &> w(bin(1, \sigma^a)) + w(bin(2, \sigma^a)) + w(bin(3..n, \sigma^a)) \\
&> w(bin(3..n, \sigma^a)) > w(bin(3..n, \sigma^*))
\end{aligned}
$$

which contradicts the optimality of σ^* since $I_4 \subseteq U(\sigma^*)$.

$I_5.A$: By definition of I_5 and line 3 of the algorithm

$$
K(I_5, C_1') \leq K(bin(1, \sigma^*), C_1') \leq K(A, C_1') \leq w(bin(1, \sigma^a)) \leq w(\sigma^a).
$$

$I_5.B$: By definition of I_5, $C'_1 \geq C_2$, and line 6 of the algorithm

$$K(I_5, C'_1) \leq K(bin(1, \sigma^*), C'_1) \leq K(bin(1, \sigma^*), C_2) \leq K(A, C_2)$$
$$\leq K(bin(2, \sigma^a), C_2) \leq w(\sigma^a)$$

$I_6.A$: By definition of I_6 and line 3 of the algorithm

$$K(I_6, C'_1) \leq K(bin(2, \sigma^*) \setminus \{i\}, C'_1) \leq K(bin(1, \sigma^a), C'_1) \leq w(\sigma^a)$$

$I_6.B$: By definition of I_6 and line 6 of the algorithm.

$$K(I_6, C'_1) \leq K(bin(2, \sigma^*) \setminus \{i\}, C_2) \leq K(bin(2, \sigma^a), C_2) \leq w(\sigma^a)$$

$I_7.A$: by definition of I_7, $C_2 \leq C'_1$, and line 3 of the algorithm,

$$K(I_7, C_2) \leq K(I_7, C'_1) \leq K(bin(1, \sigma^*), C'_1) \leq K(bin(1, \sigma^a), C'_1) \leq w(\sigma^a).$$

$I_7.B$: By definition of I_7, $C_2 > C'_1$, and line 6 of the algorithm

$$K(I_7, C_2) \leq K(bin(1, \sigma^*), C_2) \leq K(bin(2, \sigma^a), C_2) \leq w(\sigma^a).$$

$I_8.A$: By definition of I_8, $C_2 \leq C'_1$, and line 3 of the algorithm

$$K(I_8, C_2) \leq K(I_8, C'_1) \leq K(bin(2, \sigma^*), C'_1) \leq K(bin(1, \sigma^a), C'_1) \leq w(\sigma^a)$$

$I_8.B$: by definition of I_8, $C_2 > C'_1$, and line 6 of the algorithm,

$$K(I_8, C_2) \leq K(bin(2, \sigma^*), C_2) \leq K(bin(2, \sigma^a), C_2) \leq w(\sigma^a).$$

$I_9.A$: Consider

$$T = knapsack(bin(1, \sigma^*), C'_1);$$
$$L = bin(1, \sigma^*) \setminus T$$

and let $e = \text{arg-max}_{e \in L} w(e)$. By optimality of T, we know that $c(T) + c(e) > C'_1$ and, since $bin(1, \sigma^*) = T \cup L$, we have that $c(L \setminus \{e\}) < c(i)$. If $w(e) \leq \max(w(bin(1, \sigma^a)), w(bin(2, \sigma^a)))$, then

$$w(I_9) \leq w(T) + w(L \setminus \{e\}) + w(e)$$
$$\leq w(bin(1, \sigma^a)) + w(bin(2, \sigma^a)) + w(e)$$
$$\leq 2(w(bin(1, \sigma^a)) + w(bin(2, \sigma^a))) \leq 2w(\sigma^a).$$

Otherwise, by optimality of $bin(1, \sigma^a)$ and $bin(2, \sigma^a)$, we have that

$$c(e) > C'_1 \;\&\; c(e) > C_2$$

and the algorithm executes lines 10–11. If $c(e) \leq C_j$, then

$$w(I_9) \leq w(T) + w(L \setminus \{e\}) + w(e)$$
$$\leq w(bin(1, \sigma^a)) + w(bin(2, \sigma^a)) + w(bin(j, \sigma^a)) \leq w(\sigma^a).$$

Otherwise, if $c(e) > C_j$, $e \notin \sigma^s$ and

$$w(I_9) \leq w(T) + w(L \setminus \{e\}) \leq w(bin(1, \sigma^a)) + w(bin(2, \sigma^a)) \leq w(\sigma^a).$$

$I_9.B$: Consider

$$T = knapsack(bin(1, \sigma^*), C_2);$$
$$L = bin(1, \sigma^*) \setminus T$$

and let $e = \text{arg-max}_{e \in L} \, w(e)$. If $w(T) \geq w(L)$, we have that

$$w(bin(1, \sigma^*)) \leq 2w(T) \leq 2w(bin(2, \sigma^a)) \leq 2w(\sigma^a).$$

Otherwise, $c(L) > C_2$ by optimality of T and thus $c(L) > c(i)$ since $C_2 \geq c(i)$. By optimality of T, $c(T \cup \{e\}) > C_2 > C_1'$ and, since $bin(1, \sigma^*) = T \cup L$, it follows that $c(L \setminus \{e\}) \leq c(i)$ Hence $w(L \setminus \{e\}) \leq w(T)$ by optimality of T and

$$w(I_9) \leq w(T) + w(L \setminus \{e\}) + w(e) \leq 2w(T) + w(e) \leq 2w(bin(2, \sigma^a)) + w(e).$$

If $w(e) \leq w(bin(2, \sigma^a))$, $w(I_9) \leq 3w(bin(2, \sigma^a)) \leq 3w(\sigma^a)$ and the result follows. Otherwise, by optimality of $bin(2, \sigma^a)$, $c(e) > C_2 \geq C_1'$ and the algorithm executes lines 10–11. If $c(e) \leq C_j$, then

$$w(I_9) \leq 2w(bin(1, \sigma^a)) + w(bin(j, \sigma^a)) \leq w(\sigma^a).$$

Otherwise, if $c(e) > C_j$, $e \notin \sigma^s$ and

$$w(I_9) \leq w(T) + w(L \setminus \{e\}) \leq 2w(bin(2, \sigma^a)) \leq 2w(\sigma^a).$$

$I_{10}.A$: By definition of I_{10}, $C_1' \geq C_2$, and line 3 of the algorithm

$$w(I_{10}) \leq w(bin(2, \sigma^*)) - w(i) \leq w(bin(1, \sigma^a)) \leq w(\sigma^a).$$

$I_{10}.B$: By definition of I_{10} and by line 6 of the algorithm

$$w(I_{10}) \leq w(bin(2, \sigma^*)) - w(i) \leq w(bin(2, \sigma^a)) \leq w(\sigma^a).$$

I_{11} : By definition of the algorithm, $K(bin(3..n, \sigma^*)) \leq K(3..n, \sigma^a)$.   □

Swapping a Request Out of a Bin. The algorithm to swap a request i out of bin 1 is depicted in Figure 5. It consists of solving a one-dimensional knapsack with the requests already in that bin and the unallocated requests. The proof is similar, but simpler, to the proof of Theorem 1.

Theorem 2. *Algorithm* REGRET-SWAP-OUT *is a constant-factor approximation.*

Swapping a Request Into a Bin. Figure 6 depicts the algorithm for swapping a request i in bin 1, which is essentially similar REGRET-SWAP but only uses one bin. It assumes that request i can be placed in at least two bins since otherwise a single additional optimization suffices to compute all the regrets. Once again, it solves a one-dimensional knapsack for bin 1 (after having allocated request i) with all the requests in $bin(1, \sigma^*)$ and the unallocated requests. If the resulting knapsack is of low quality (i.e., the remaining requests from $bin(1, \sigma^*)$ have a higher value than $bin(1, \sigma^a)$), REGRET-SWAP-IN solves an additional knapsack problem for the largest available bin. The proof is once again similar to the proof of Theorem 1.

REGRET-SWAP-OUT$(i, 1)$
1   $A \leftarrow bin(1, \sigma^*) \cup U(\sigma^*) \setminus \{i\}$;
2   $bin(1, \sigma^a) \leftarrow knapsack(A, C_1)$;

Fig. 5. The Suboptimality Algorithm for the Knapsack Problem: Swapping i out of Bin 1

REGRET-SWAP-IN$(i, 1)$
1   $A \leftarrow bin(1, \sigma^*) \cup U(\sigma^*)$;
2   $bin(1, R) \leftarrow knapsack(A, C_1 - c(i)) \cup \{i\}$;
3   $L \leftarrow bin(1, \sigma^*) \setminus bin(1, \sigma^a)$;
4   **if** $w(L) > w(bin(1, \sigma^a))$ **then**
5       $j \leftarrow argmax(j \in 2..n)\ C_j$;
6       $bin(j, \sigma^a) \leftarrow knapsack(bin(j, \sigma^a) \cup L, C_j)$;

Fig. 6. The Suboptimality Algorithm for the Knapsack Problem: Swapping i into Bin 1

Theorem 3. *Assuming that item i can be placed in at least two bins, Algorithm* REGRET-SWAP-IN *is a constant-factor approximation.*

6   Experimental Results

The Instances. The experimental results use the benchmarks proposed in [1]. Requests are classified in k types. Each type is characterized by a weight, a value, two exponential distributions indicating how frequently requests of that type arrive and are cancelled, and an overbooking penalty. We generated ten instances based on the master problem proposed in [1]. The goal was to try to produce a diverse set of problems revealing strengths and weaknesses of the various algorithms. The ten problems are named (A-J) here. Problem A scales the master problem by doubling the weight and value of the request types in the master problem, as well as halving the number of items that arrive. Problem B further scales problem A by increasing the weight and value of the types. Problem C considers 7 types of items whose cost ratio takes the form of a bell shape. Problem D looks at the master problem and doubles the number of bins while dividing their capacity by 2. Problem E considers a version of the master problem with bins of variable capacity. Problem F depicts a version of the master problem whose items arrive three times as often and cancel three times as often. Problem G considers a much larger problem with 35 requests types who cost ratio is also shaped in a bell. Problem H is like problem G, the main difference is that the cost ratio shape is reversed. Problem I is a version of G with an extra bin. Problem J is a version of H with fewer bins.

The mathematical programs are solved with CPLEX 9.0 with a time limit of 10 seconds. The optimal solutions can be found within the time limit for all instances but I and J. Every instance is executed under various time constraints, i.e., $\mathcal{O} = 1, 5, 10, 25, 50$, or 100, and the results are the average of 10 executions.

It is important to highlight that, on the master problem and its variations, the best-fit heuristic performs quite well. On the offline problems, it is 5% off the optimum in the

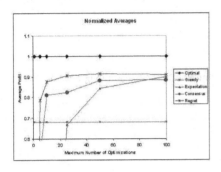

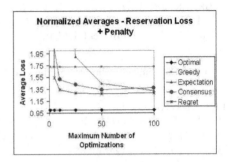

Fig. 7. Experimental Results over All Instances with Overbooking Allowed

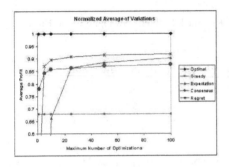

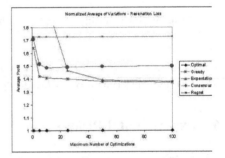

Fig. 8. Experimental Results over All Instances with Overbooking Disallowed

average and is never worse than 10% off. This will be discussed again when the regret algorithm is compared to earlier results.

Comparison of the Algorithms. Figure 7 describes the average profit (left) and loss (right) of the various online algorithms as a percentage of the optimal offline solution. The loss sums the weights of the rejected requests and the overbooking penalty (if any); it is often used in comparing online algorithms as it gives a sense of the "price" of uncertainty. The results clearly show the value of stochastic information as algorithms R, C, E recovers most of the gap between the online best-fit heuristic (G) and the offline optimum (which cannot typically be achieved in an online setting). Moreover, they show that algorithms R and C achieve excellent results even with small number of available optimizations (tight time constraints). In particular, algorithm R achieves about 89% of the offline optimum with only 10 samples and 91% with 50 optimizations. It also achieves a loss of 28% over the offline optimum for 25 optimizations and 34% for 10 optimizations. The regret algorithm clearly dominates the expectation algorithm E which performs poorly for tight time constraints. It becomes reasonable for 50 optimizations and reaches the quality of the regret algorithm for 100 optimizations.

Figure 8 shows the same results when no overbooking is allowed. These instances are easier in the sense that fewer optimizations are necessary for the algorithms to converge.

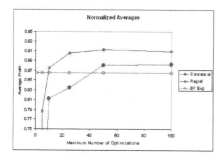 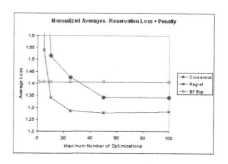

Fig. 9. Comparison with Earlier Results: Average Results for Instances with Overbooking

But they exhibit the same pattern as when overbooking is allowed. These results are quite interesting and shows that the benefits of the regret algorithm increase with the problem complexity but are significant even on easier instances.

Comparison with Earlier Results. As mentioned earlier, the best-fit algorithm is only 5% below the optimal offline solution in these problems. It is thus tempting to replace the IP solver in algorithm E by the best-fit heuristic to evaluate more samples. The algorithm, denoted by BF EXP, was proposed in [1] and was shown to be superior to several approaches including yield management and an hybridization with Markov Models [12]. Because the best-fit algorithm is so fast, BF EXP can easily be run with 10,000 samples and remedies the limitations of algorithm E under tight time constraints.

Figure 9 compares algorithms BF EXP, R, and C when overbooking is allowed. The results show that BF EXP indeed produces excellent results but is quickly dominated by R as time increases. In particular, the loss of BF EXP is above 40%, although it goes down to 34% for 10 optimizations and 28% for 25 optimizations in algorithm R. Similarly, the profit increases by 4% in the average starting at 25 optimizations. BF EXP is also dominated by algorithm C but only for 50 optimizations or more.

What is quite remarkable here is that the 5% difference in quality between the best-fit heuristic and the offline algorithm translates into a similar difference in quality in the online setting. Moreover, when looking at specific instances, one can see that BF EXP is often comparable to R but its loss (resp. profit) may be significantly higher (resp. lower) on instances that seem particularly difficult. This is the case for instances E and G, where the gap between the offline solutions and the solutions by algorithm R is larger. This seems to indicate that the harder the problems the more beneficial algorithm R becomes. This in fact confirms our earlier results on stochastic vehicle routing where the algorithms use a large neighborhood heuristic [3, 13]. Indeed, using a simpler, lower-quality, heuristic on more samples did not produce high-quality results in an online setting. The results presented here also show that the additional information produced by a more sophisticated solver quickly amortizes its computational cost, making algorithm R particularly effective and robust for many problems.

7 Conclusion

This paper adapted our online stochastic framework and algorithms to online stochastic reservations initially proposed in [1]. These problems, whose core can be modelled as multi-knapsacks, are significant in practice and are also different from the scheduling and routing applications we studied earlier. Indeed the main decision is not which request to select next but rather how best to serve a request given limited resources. The paper shows that the framework and its associated algorithms naturally apply to online reservation systems. It also presented a constant-factor sub-optimality approximation of multi-knapsack problems that only solves one-dimensional knapsack problems, leading to a regret algorithm that uses both mathematical programming and dynamic programming algorithms. The algorithms were evaluated on the multi-knapsack problems proposed in [1] with and without overbooking. The results indicate that the regret algorithm is particularly effective, providing significant benefits over heuristic, consensus, and expectation approaches. It also dominates an earlier algorithm proposed in [1] (which applies the best-fit heuristic with algorithm E) as soon as the time constraints allows for 10 optimizations at decision time or between decisions. Even more interesting perhaps, the regret algorithm has now been applied to online stochastic problems where the offline problem is solved by either constraint programming, integer programming, or (special-purpose) polynomial algorithms, indicating its versatility and benefits for a wide variety of applications.

References

1. T. Benoist, E. Bourreau, Y. Caseau, and B. Rottembourg. Towards stochastic constraint programming: A study of online multi-choice knapsack with deadlines. In *Proceedings of the Seventh International Conference on Principles and Practice of Constraint Programming (CP'01)*, pages 61–76, London, UK, 2001. Springer-Verlag.
2. R. Bent, I. Katriel, and P. Van Hentenryck. Sub-Optimality Approximation. In *Eleventh International Conference on Principles and Practice of Constraint Programming*, Stiges, Spain, 2005.
3. R. Bent and P. Van Hentenryck. A Two-Stage Hybrid Local Search for the Vehicle Routing Problem with Time Windows. *Transportation Science*, 8(4):515–530, 2004.
4. R. Bent and P. Van Hentenryck. Online Stochastic and Robust Optimization. In *Proceeding of the 9th Asian Computing Science Conference (ASIAN'04)*, Chiang Mai University, Thailand, December 2004.
5. R. Bent and P. Van Hentenryck. Regrets Only. Online Stochastic Optimization under Time Constraints. In *Proceedings of the 19th National Conference on Artificial Intelligence (AAAI'04)*, San Jose, CA, July 2004.
6. R. Bent and P. Van Hentenryck. Scenario Based Planning for Partially Dynamic Vehicle Routing Problems with Stochastic Customers. *Operations Research*, 52(6), 2004.
7. R. Bent and P. Van Hentenryck. The Value of Consensus in Online Stochastic Scheduling. In *Proceedings of the 14th International Conference on Automated Planning & Scheduling (ICAPS 2004)*, Whistler, British Columbia, Canada, 2004.
8. R. Bent and P. Van Hentenryck. Online Stochastic Optimization without Distributions . In *Proceedings of the 15th International Conference on Automated Planning & Scheduling (ICAPS 2005)*, Monterey, CA, 2005.

9. A. Campbell and M. Savelsbergh. Decision Support for Consumer Direct Grocery Initiatives. *Report TLI-02-09, Georgia Institute of Technology*, 2002.

10. H. Chang, R. Givan, and E. Chong. On-line Scheduling Via Sampling. *Artificial Intelligence Planning and Scheduling (AIPS'00)*, pages 62–71, 2000.

11. B. Dean, M.X. Goemans, and J. Vondrak. Approximating the Stochastic Knapsack Problem: The Benefit of Adaptivity. In *Proceedings of the 45th Annual IEEE Symposium on Foundations of Computer Science*, pages 208–217, Rome, Italy, 2004.

12. M. Puterman. *Markov Decision Processes*. John Wiley & Sons, New York, 1994.

13. P. Shaw. Using Constraint Programming and Local Search Methods to Solve Vehicle Routing Problems. In *Proceedings of Fourth International Conference on the Principles and Practice of Constraint Programming (CP'98)*, pages 417–431, Pisa, October 1998.

Traveling Tournament Scheduling:
A Systematic Evaluation of Simulated Annealling

Pascal Van Hentenryck and Yannis Vergados

Computer Science Department, Brown University,
Providence, RI 02912, USA
pvh@cs.brown.edu, vi@cs.brown.edu

Abstract. This paper considers all the variants of the traveling tournament problem (TTP) proposed in [17,7] to abstract the salient features of major league baseball (MLB) in the United States. The variants include different distance metrics and both mirrored and non-mirrored schedules. The paper shows that, with appropriate enhancements, simulated annealing is robust across the distance metrics and mirroring. In particular, the algorithm matches or improves most best-known solutions and produces numerous new best solutions spread over all classes of problems. The main technical contribution underlying these results is a number of compositive neighborhood moves that aggregate sequences of existing moves; these novel moves preserve the mirroring or distance structure of the candidate schedule, while performing interesting transformations.

1 Introduction

Sport league scheduling [7] has become an important class of combinatorial optimization applications as it represents significant sources of revenue for television networks and generates extremely challenging optimization problems. In 2001, Easton, Nemhauser, and Trick [7] proposed the traveling tournament problem (TTP) to abstract the salient features of Major League Baseball (MLB) in the United States. The key to the MLB schedule is a conflict between minimizing travel distances and feasibility constraints on the home/away patterns. Travel distances are a major issue in MLB because of the number of teams and the fact that teams go on "road trips" to visit several opponents before returning home. The feasibility constraints in the MLB restricts the number of successive games that can be played at home or away. The TTP is an abstraction of the MLB intended to stimulate research in sport scheduling. A solution to the TTP is a double round-robin tournament which satisfies sophisticated feasibility constraints (e.g., no more than three away games in a row) and minimizes the total travel distances of the teams. While both minimizing the total distance traveled, and satisfying the feasibility constraints, are easy problems when considered in isolation, the combination of the two (which is captured by the TTP) makes the problem very difficult. Moreover, [7] argues that, without an approach to the TTP, it is unlikely that suitable schedules can be obtained for the MLB.

The TTP has raised significant interest in recent years since the challenge instances were proposed. The work in [7] describes both constraint and integer programming

J.C. Beck and B.M. Smith (Eds.): CPAIOR 2006, LNCS 3990, pp. 228–243, 2006.

approaches to the TTP which generate high-quality solutions. The combination of Lagrangian relaxation and constraint programming proposed in [2] improves some of the results. In 2003, we proposed an simulated algorithm exploring a large neighborhood that produced most of the best-known solutions to the instances. Our neighborhood, or a subset of it, was reused in subsequent work (e.g., [8, 12, 16]) within tabu-search and GRASP approaches or, even, simulated annealing. Recently, Rasmussen and Trick also considered Benders decomposition approaches to the TTP [15].

On the original TTP instances, our original simulated algorithm has remained most effective in producing best-known solutions, but it was never applied to the variants proposed subsequently. However, these variants may fundamentally alter the combinatorial structure of the problem. For instance, with constant distances, the order of the games in a sequence of away games is irrelevant, which is obviously not the case with the original distances. Similarly, mirroring requires that the second half of the schedule be similar to the first half but with the home-away patterns of the games reversed. Mirroring imposes severe feasibility constraints on the schedule, making it harder to find feasible tournaments and to remain in the feasible region. As a result, it was not at all clear that the algorithm would scale and perform effectively on the entire spectrum of TTP instances.

This paper originated as an attempt to determine the effectiveness and limitations of our algorithm across all TTP instances. From a practical standpoint, its main contribution is to show that the original algorithm can be enhanced to be effective across all distance metrics and mirroring. More precisely, the enhanced algorithm matches or improves most of the best known solutions for all variants and it also produces numerous new best solutions for many of the variants. From a technical standpoint, the research led to new insights into the nature of the TTP and to the following contributions:

- It shows that the algorithm can smoothly handle mirroring constraints as soft constraints by including new neighborhood moves that preserve the mirroring structure of the candidate tournament;
- It shows that the algorithm can successfully accommodate all distance metrics by including new neighborhood moves that preserve the distance structure of the candidate tournament;
- It show how to refine the original strategic oscillation scheme to the instances where feasibility constraints are much stronger.

The novel neighborhood moves are in fact sequences of existing moves. As such, they do not improve the connectivity of the neighborhood for the TTP. Their significance comes from the fact that, in the original algorithm, these sequences have a low probability of taking place, although they capture fundamental aspects of the problem structure.

The rest of the paper is organized as follows. Section 2 describes the various travelling tournament problems. Section 3 reviews the original simulated sinnealing algorithm. Section 4 explains how to adapt the algorithm to accommodate mirroring, Section 5 presents new neighborhood moves for exploiting the distance structure, and Section 6 presents the remaining algorithmic enhancements. Section 7 presents the experimental results and Section 8 concludes the paper.

2 Problem Description

The traveling tournament problem, as originally introduced in [7], was first studied only for benchmarks based on the distances between cities of the 16 teams of National League Baseball. The solution space consisted of all double round-robin schedules and did not include mirroring. In recent years, increased attention to the problem has led to the study of a variety of variants. On one hand, the problem was studied under the additional constraint that the schedule be a *mirrored* double round-robin tournament. On the other hand, alternative sets of distances between cities were considered. Furthermore, the problem was been studied for larger number of teams (up to 24). This section describes the TTP and its various variants.

The Basic TTP Problem. A TTP input consists of n teams (n even) and an $n \times n$ symmetric matrix d, such that d_{ij} represents the distance between the homes of teams T_i and T_j. A solution is a schedule in which each team plays with each other twice, once in each team's home. Such a schedule is called a *double round-robin tournament*. It should be clear that a double round-robin tournament has $2n - 2$ rounds. For a given schedule S, the cost of a team as the total distance that it has to travel starting from its home, playing the scheduled games in S, and returning back home. The cost of a solution is defined as the sum of the cost of every team. The goal is to find a schedule with minimum cost satisfying the following two constraints:

1. **Atmost Constraints:** No more than three consecutive home or away games are allowed for any team.
2. **Norepeat Constraints:** A game of T_i at T_j's home cannot be followed by a game of T_j at T_i's home.

As a consequence, a double round-robin schedule is feasible if it satisfies the atmost and norepeat constraints and is infeasible otherwise.

In this paper, a schedule is represented by a table indicating the opponents of the teams. Each line corresponds to a team and each column corresponds to a round. The opponent of team T_i at round r_k is given by the absolute value of element (i, k). If (i, k) is positive, the game takes place at T_i's home, otherwise at T_i's opponent home. Consider for instance the schedule S for 6 teams (and thus 10 rounds).

T\R	1	2	3	4	5	6	7	8	9	10
1	6	-2	4	3	-5	-4	-3	5	2	-6
2	5	1	-3	-6	4	3	6	-4	-1	-5
3	-4	5	2	-1	6	-2	1	-6	-5	4
4	3	6	-1	-5	-2	1	5	2	-6	-3
5	-2	-3	6	4	1	-6	-4	-1	3	2
6	-1	-4	-5	2	-3	5	-2	3	4	1

Schedule S specifies that team T_1 has the following schedule. It successively plays against teams T_6 at home, T_2 away, T_4 at home, T_3 at home, T_5 away, T_4 away, T_3 away, T_5 at home, T_2 at home, T_6 away. The travel cost of team T_1 is

$$d_{12} + d_{21} + d_{15} + d_{54} + d_{43} + d_{31} + d_{16} + d_{61}.$$

Observe that long stretches of games at home do not contribute to the travel cost but are limited by the atmost constraints.

Mirroring. In some sports leagues (e.g., in most European soccer leagues), it is common practice to adopt a two-stage mirrored schedule. If n is the number of participating teams, each stage has $n - 1$ rounds and the $n - 1$ rounds of the second stage are simply a copy of the the first-stage rounds with a swap of home/away patterns of each game. In terms of the table representation, a schedule is *mirrored* if

$$\forall i, j : 1 \le j \le n - 1 : S[i, j + n - 1] = -S[i, j].$$

Distance Metrics. All the distance metrics studied in this paper can be found on the web site [17]. They are defined as follows:

- [NLn:] For $n = 4, 6, \ldots, 16$, NLn is the distances between the subset of the first n teams of National League Baseball.
- [Circular:] The n teams are labeled with numbers 0 through $n - 1$ and are placed on a circle in this order. Then, for any two teams, the distance is given by the length of the shortest arc connecting the teams. These distances satisfy

$$\forall i > j : d_{ij} = \min\{i - j, j - i + n\}.$$

- [Constant:] The n teams are at unit distance to one another, i.e., $d_{ij} = 1$, for all i, j.

For every distance metric, we are interested in solutions to both the mirrored and the non-mirrored version of the TTP.

3   The Original Simulated Annealing Algorithm

Our simulated algorithm TTSA in [1] is based on four main design decisions:

1. Constraints are separated into two groups: hard constraints, which are always satisfied by the configurations, and soft constraints, which may or may not be violated. The hard constraints are the round-robin constraints, while the soft constraints are the norepeat and atmost constraints. In other words, all configurations in the search represents a double round-robin tournament, which may or may not violate the norepeat and atmost constraints. To drive the search toward feasible solutions, TTSA modifies the original objective function to include a penalty term.
2. TTSA is based on a large neighborhood of size $O(n^3)$, where n is the number of teams. In addition, these moves may affect significant portions of the configurations. For instance, they may swap the schedule of two teams, which affects $4(n-2)$ entries in a configuration. In addition, some of these moves can be regarded as a form of ejection chains sometimes used in tabu search [11, 14].

3. TTSA dynamically adjusts the objective function to balance the time spent in the feasible and infeasible regions. This adjustment resembles the strategic oscillation idea [9] successfully in tabu search to solve generalized assignment problems [5], although the details differ since simulated annealing is used as the metaheuristics.
4. TTSA also uses reheats (e.g., [3]) to escape local minima at low temperatures. The "reheats" increase the temperature again and divide the search in several phases.

The rest of this section explore some of these aspects in more detail, as they are pertinent to the understanding of this paper. Since they are double round-robin tournaments, configurations are called schedules in the following. Observe that, contrary to simpler problems such as break minimization [18], we found it critical to explore the infeasible region in the TPP and to consider moves that significantly alter the schedules. In our experiments, neighborhoods consisting of simpler moves or considering only feasible schedules tend to produce local minima of low quality. Of course, this does not mean that such neighborhoods do not exist, simply that we have not be able to design them so far. This difference between the TTP and the break minimization stems from the fact that break minimization assumes a fixed schedule: only the home/away patterns must be determined. In contrast, the objective function in the TTP must not only determine the home/away patterns; it must also determine the schedule of each team to minimize the total travel distance. This additional difficulty and the tension it produces with the feasibility constraints is what makes the TTP particularly challenging.

3.1   The Neighborhood

The neighborhood of a schedule S is the set of the (possibly infeasible) schedules which can be obtained by applying one of five types of moves. The first three types of moves have a simple intuitive meaning, while the last two generalize them.

SwapHomes(S, T_i, T_j). This move swaps the home/away roles of teams T_i and T_j. In other words, if team T_i plays home against team T_j at round r_k, and away against T_j's home at round l, *SwapHomes*(S, T_i, T_j) is the same schedule as S, except that now team T_i plays away against team T_j at round r_k, and home against T_j at round r_l. There are $O(n^2)$ such moves. Consider the schedule S:

T\R	1	2	3	4	5	6	7	8	9	10
1	6	-2	4	3	-5	-4	-3	5	2	-6
2	5	1	-3	-6	4	■	6	-4	■	-5
3	-4	5	2	-1	6	-2	1	-6	-5	4
4	3	6	-1	-5	-2	■	5	2	■	-3
5	-2	-3	6	4	1	-6	-4	-1	3	2
6	-1	-4	-5	2	-3	5	-2	3	4	1

The move $SwapHomes(S, T_2, T_4)$ produces the schedule:

T\R	1	2	3	4	5	6	7	8	9	10
1	6	-2	4	3	-5	-4	-3	5	2	-6
2	5	1	-3	-6	-4		6	4		-5
3	-4	5	2	-1	6	-2	1	-6	-5	4
4	3	6	-1	-5	2		5	-2		-3
5	-2	-3	6	4	1	-6	-4	-1	3	2
6	-1	-4	-5	2	-3	5	-2	3	4	1

SwapRounds(S, r_k, r_l). The move simply swaps rounds r_k and r_l. There are also $O(n^2)$ such moves. Consider the schedule S:

T\R	1	2	3	4	5	6	7	8	9	10
1	6	-2	4		-5		-3	5	2	-6
2	5	1	-3		4		6	-4	-1	-5
3	-4	5	2		6		1	-6	-5	4
4	3	6	-1		-2		5	2	-6	-3
5	-2	-3	6		1		-4	-1	3	2
6	-1	-4	-5		-3		-2	3	4	1

The move $SwapRounds(S, r_3, r_5)$ produces the schedule

T\R	1	2	3	4	5	6	7	8	9	10
1	6	-2	-5		4		-3	5	2	-6
2	5	1	4		-3		6	-4	-1	-5
3	-4	5	6		2		1	-6	-5	4
4	3	6	-2		-1		5	2	-6	-3
5	-2	-3	1		6		-4	-1	3	2
6	-1	-4	-3		-5		-2	3	4	1

SwapTeams(S, T_i, T_j). This move swaps the schedule of Teams T_i and T_j (except, of course, when they play against each other). There are $O(n^2)$ such moves again. Consider the schedule S:

T\R	1	2	3	4	5	6	7	8	9	10
1	6	-2	4	3	-5	-4	-3	5	2	-6
2	5	1	-3	-6	4	3	6	-4	-1	
3	-4	5	2	-1	6	-2	1	-6	-5	4
4	3	6	-1	-5	-2	1	5	2	-6	-3
5	-2	-3	6	4	1	-6	-4	-1	3	
6	-1	-4	-5	2	-3	5	-2	3	4	1

The move $SwapTeams(S, T_2, T_5)$ produces the schedule

T\R	1	2	3	4	5	6	7	8	9	10
1	6	-5		3	-2		-3	2	5	
2	5	-3	6	4	1	-6	-4	-1	3	
3	-4	2	5		6	-5		-6	-2	
4	3	6	-1	-2	-5		2	5		-3
5	-2	1	-3	-6	4	3	6	-4	-1	
6	-1	-4	-2	5		2	-5		4	1

Note that, in addition to the changes in lines 2 and 5, the corresponding lines of the opponents of T_i and T_j must be changed as well. As a consequence, there are four values per round (column) that are changed (except when T_i and T_j meet).

It turns out that these three moves are not sufficient for exploring the entire search space and, as a consequence, they lead to suboptimal solutions for large instances. To improve these results, it is important to consider two, more general, moves. Although these moves do not have the apparent interpretation of the first three, they are similar in structure and they significantly enlarge the neighborhood, resulting to a more connected search space. More precisely, these moves are partial swaps: they swap a subset of the schedule in rounds r_i and r_j or a subset of the schedule for teams T_i and T_j. The benefits from these moves come from the fact that they are not as global as the "macro"-moves $SwapTeams$ and $SwapRounds$. As a consequence, they may achieve a better tradeoff between feasibility and optimality by improving feasibility in one part of the schedule, while not breaking feasibility in another one. They are also more "global" than the "micro"-moves $SwapHomes$.

PartialSwapRounds(S, T_i, r_k, r_l)**.** This move considers team T_i and swaps its games at rounds r_k and r_l. The rest of the schedule for rounds r_k and r_l is updated to produce a double round-robin tournament. Consider the schedule S

T\R	1	2	3	4	5	6	7	8	9	10
1	6	-2	2	3	-5	-4	-3	5	4	-6
2	5	1		-5	4	3	6	-4	-6	
3	-4	5	4	-1	6	-2	1	-6	-5	2
4	3	6	-3	-6	-2	1	5	2	-1	-5
5	-2	-3	6	2	1	-6	-4	-1	3	4
6	-1	-4	-5	4	-3	5	-2	3	2	1

and the move $PartialSwapRounds(S, T_2, r_2, r_9)$. Obviously swapping the game in rounds r_2 and r_9 would not lead to a round-robin tournament. It is also necessary to swap the games of team 1, 4, and 6 in order to obtain:

T\R	1	2	3	4	5	6	7	8	9	10
1	6	4		3	-5	-4	-3	5	-2	
2	5	-6		-5	4	3	6	-4	1	
3	-4	5	4	-1	6	-2	1	-6	-5	2
4	3	-1		-6	-2	1	5	2	6	
5	-2	-3	6	2	1	-6	-4	-1	3	4
6	-1	2		4	-3	5	-2	3	-4	

This move, and the next one, can thus be regarded as a form of ejection chain [11, 14].

Finding which games to swap is not difficult: it suffices to find the connected component which contains the games of T_i in rounds r_k and r_l in the graph where the vertices are the teams and where an edge contains two teams if they play against each other in rounds r_k and r_l. All the teams in this component must have their games swapped. Note that there are $O(n^3)$ such moves.

PartialSwapTeams(S, T_i, T_j, r_k). This move considers round r_k and swaps the games of teams T_i and T_j. Then, the rest of the schedule for teams T_i and T_j (and their opponents) is updated to produce a double round-robin tournament. Note that, as was the case with *SwapTeams*, four entries per considered round are affected. There are also $O(n^3)$ such moves. Consider the schedule S

T\R	1	2	3	4	5	6	7	8	9	10
1	6	-2	4	3	-5	-4	-3	5	2	-6
2	5	1	-3	-6	4	3	6	-4	-1	
3	-4	5	2	-1	6	-2	1	-6	-5	4
4	3	6	-1	-5	-2	1	5	2	-6	
5	-2	-3	6	4	1	-6	-4	-1	3	2
6	-1	-4	-5	2	-3	5	-2	3	4	1

The move *PartialSwapRounds*(S, T_2, T_4, r_9) produces the schedule

T\R	1	2	3	4	5	6	7	8	9	10
1	6	-2	2		-5	-4	-3	5	4	
2	5	1	-1	-5		3	6	-4	6	-3
3	-4	5	4		6	-2	1	-6	-5	2
4	3	6	-3	-6		1	5	2	-1	-5
5	-2	-3	6	2		-6	-4	-1	3	4
6	-1	-4	-5	4		5	-2	3	2	

3.2 The Objective Function

As mentioned already, the configurations in algorithm TTSA are schedules which may or may not satisfy the norepeat and atmost constraints. To drive toward feasible solution, the standard objective function function *cost* is replaced by a more complex objective function which combines travel distances and the number of violations. The new objective function C is defined as follows:

$$C(S) = \begin{cases} cost(S) \text{ if } S \text{ is feasible,} \\ \sqrt{cost(S)^2 + [w \cdot f(nbv(S))]^2} \text{otherwise,} \end{cases}$$

where $nbv(S)$ denotes the number of violations of the norepeat and atmost constraints, w is a weight, and $f : N \to N$ is a sublinear function such that $f(1) = 1$.

It is interesting to give the rationale behind the choice of f. The intuition is that the first violation costs more than subsequent ones, since adding 1 violation to a schedule with 6 existing ones does not make much difference. More precisely, crossing the feasible/infeasible boundary costs w, while v violations only cost $wf(v)$, where $f(v)$ is sublinear in v. In our experiments, we chose $f(v) = 1 + (\sqrt{v} \ln v)/\lambda$ with λ equal to 2 on small instances and 1 on larger ones. This choice makes sure that f does not grow too slowly to avoid solutions with very many violations.

3.3 Strategic Oscillation

TTSA also includes a strategic oscillation strategy often used in tabu search when the search explores both the feasible and infeasible regions (e.g., [9, 5]). The key idea is to vary the weight parameter w during the search. In advanced tabu-search applications (e.g., [5]), the penalty is updated according to the frequencies of feasible and infeasible configurations in the last iterations. Such a strategy is meaningful in that context, but is not particularly appropriate for simulated annealing since very few moves may be selected. TTSA uses a very simple scheme. Each time it generates a new best solution TTSA multiplies w by some constant $\delta > 1$ if the new solution is infeasible or divide w by some constant $\theta > 1$ if the new solution is feasible.

The rationale here is to keep a balance between the time spent exploring the feasible region and the time spent exploring infeasible schedule. After having spent a long time in the infeasible region, the weight w, and thus the penalty for violations, will become large and it will drive the search toward feasible solutions. Similarly, after having spent a long time in the feasible region, the weight w, and thus the penalty for violations, will become small and it will drive the search toward infeasible solutions. In our experiments, we chose $\delta = \theta$ for simplicity.

4 Mirroring

This section reviews the enhancements of the algorithm to find mirrored tournaments.

4.1 Mirroring Constraints

Mirrored constraints, like atmost and norepeat constraints, are considered soft constraints in the algorithm, since restricting the neighborhood graph to only mirrored schedules was not found effective. In other words, the neighborhood graph consists of nodes representing both mirrored and non-mirrored schedules and it is the role of the objective function to drive the search toward mirrored schedules. More precisely, the algorithm associates a soft mirrored constraint with each entry $S[i,j]$ $(1 \leq i \leq n$ & $1 \leq j < n-1)$ and the constraint holds when

$$S[i,j] = -S[i,j+n-1]$$

It is thus possible to include the violations of these constraints in the objective function as was the case for the atmost and norepeat constraints.

Unfortunately, this simple modeling is not directly effective given the large number of mirroring constraints that can be violated in candidate schedules. As a result, the algorithm weights the mirroring constraints appropriately and rewrite the function $nbv(S)$ of the objective function as the sum

$$nbv_a(S) + nbv_r(S) + \frac{nbv_m(S)}{\mu}$$

where nbv_a, nbv_r, and nbv_m represent the violations of the atmost, norepeat, and mirroring constraints respectively. In particular, the violations of the mirroring constraints is simply defined as

$$nbv_m(S) = |\{(i,j) : S[i,j] \neq -S[i,j+n-1] \ \& \ 1 \leq i \leq n \ \& \ 1 \leq j \leq n-1\}|.$$

4.2 Mirroring Moves

Once the algorithm reaches a (possibly infeasible) mirrored schedule, it is beneficial to let the search explored neighboring *mirrored* schedules with fewer violations of the remaining constraints or smaller distances. The simulated-annealing algorithm, with its present moves, has a low probability of exploring such moves. It is thus important to design mirrored versions of the moves that affect the mirroring constraints: the *SwapRounds*, *PartialSwapRounds*, and *PartialSwapTeams* moves. The basic idea behind the aggregate moves is to apply the original moves to both parts of the schedule simultaneously. For instance, the new move $SwapRoundsMirrored(S, r_k, r_l)$ for $1 \leq r_k < r_l \leq n-1$ is the aggregate

$$(SwapRounds(S, r_k, r_l), SwapRounds(S, T_i, r_k+n-1, r_l+n-1)).$$

Mirroring constraints are invariant with respect to mirrored moves: in particular, if the schedule is mirrored, it remains so. As a result, the algorithm is able to preserve the structure of the schedule with respect to mirroring constraints, while performing transformations that affect the remaining constraints or the distances.

5 Distance Metrics

This section presents two additional composite moves that affect the distances in interesting ways. Once again, the novel moves aggregate sequences of existing moves that have a low probability of taking place in the original algorithm. Hence, they also preserve some significant structure in the schedule, while performing some interesting transformations.

The key idea underlying the novel moves is to reverse subsequences of away moves. Recall that travel only occurs for successive pairs of away games and for successive pairs of (home,away) and (away,home) games. So, by reversing a subsequence of away games, we preserve a significant part of the distance structure, while modifying it in a way that is difficult to achieve by sequences of original moves. In particular, the distances in the reversed subsequence, as well as the distances in the sequence of the other teams that must also be reversed to maintain a double round-robin tournament, remain the same. It is only at the beginning and at the end of the subsequences that distances are changing. In fact, moves similar in spirit are also used in car sequencing [4] but they are simpler since they do not have to account for the round-robin constraints and the distance structure.

The algorithm thus considers moves of the form $ReverseAwayRun(S, T_i, r_k, m)$ where team T_i plays an away sequence from round r_k to round r_{k+m}. The effect of the move is similar to the sequence of $p = (m+1)/2$ moves

$PartialSwapRounds(S, T_i, r_k, r_{k+m})$
$PartialSwapRounds(S, T_i, r_{k+1}, r_{k+m-1})$

$\ldots$

$PartialSwapRounds(S, T_i, r_{k+p-1}, r_{k+m+1-p})$

For instance, consider the schedule S

T\R	1	2	3	4	5	6	7	8	9	10
1	6	2	4	3	-5	-4	-3	-6	-2	
2	5	-1	-3	-6	4	3	6	-5	1	-4
3	-4	5	2	-1	6	-2	1	4	-5	-6
4	3	6	-1	-5	-2	1	5	-3	-6	2
5	-2	-3	6	4	1	-6	-4	2	3	-1
6	-1	-4	-5	2	-3	5	-2	1	4	3

The move $ReverseAwayRun(S, T_1, r_6, 4)$ produces the schedule S':

T\R	1	2	3	4	5	6	7	8	9	10
1	6	2	4	3	-5	-2	-6	-3	-4	
2	5	-1	-3	-6	4	1	-5	6	3	
3	-4	5	2	-1	6	-5	4	1	-2	
4	3	6	-1	-5	-2	-6	-3	5	1	
5	-2	-3	6	4	1	3	2	-4	-6	
6	-1	-4	-5	2	-3	4	1	-2	5	

When $d_{52} + d_{41} < d_{54} + d_{21}$, the move improves the distance with respect to team T_1, without affecting the atmost violations of team T_1 and the distance structure inside the subsequence. There are several points worth highlighting here. First, reversing entire sequences of away games does not change the distance for the team considered and should not be considered. Second, the value m is never very large in the TTP instances, since the atmost constraints drive the search toward small subsequences of away games. Finally, the algorithm must include a mirrored version *ReverseAwayRunMirrored* of the moves since they affect the mirroring constraints.

6   Algorithmic Refinements

Strategic Oscillation. The mirroring constraints make it harder to find feasible tournaments and the search may spend considerable time in the infeasible region before finding a first feasible solution. As a result, even small values for μ and λ, the strategic oscillation scheme will overly inflate the violation weight w, leading the search to stagnation. To alleviate this pathological case, the algorithm now takes a two-step approach. In a first phase, which lasts until the first feasible tournament is found, no oscillation takes place. In the second phase, the strategic oscillation scheme is activated as before and balances the time spent in the infeasible and feasible regions. Note also the synergy between this scheme and the new neighborhood moves. By including mirrored moves, the algorithm is able to better balance the time it spends in the feasible and infeasible regions in presence of mirroring constraints.

Initial Schedules. The simple backtrack search used in [1] to find initial schedules does not scale well when the number of teams increases, which is the case in the constant and circular variants. As a result, like in [16], the algorithm now uses a randomized version of the hill-climbing algorithm for generating 1-factorizations [6]. The initial schedules generated by this randomized algorithm are more diversified than perturbations of schedules obtained by the polygon algorithm. The algorithm in [6] works for single round-robin schedules but a double round-robin schedule can be obtained by a simple mirroring.

7   Experimental Results

The enhanced version of the algorithm was run on all the mirrored and non-mirrored instances given on Michael Trick's webpage [17].[1] For each instance, 20 experiments were carried out from randomly chosen schedules on an AMD Athlon 64 at 2Ghz. The results are reported in two tables for each variant. The first table reports the best, mean, and worst solutions found by the algorithm, as well as the standard deviation and the best known solution value at the time of writing. The second table reports the time to reach the best solution, the mean time of each experiment, and the standard deviation. Bold face indicates improved results. It is also important to mention that many authors (e.g., [8, 12, 16]) now use the neighborhood we originally proposed in [1] which makes it much harder to improve the results (since, in a sense, we are also competing with ourselves). Our implementation is also slightly more incremental than in 2003, but this is not seen as a major factor in these results in contrast to the new moves proposed herein.

---

[1] They do not include the NFL instances just posted in December.

Table 1. Solution Quality and Solution Times for the NLn Distances with Mirroring

n	Old Best	min(D)	max(D)	mean(D)	std(D)
8	41928	41928	43025	42037.65	291.98
10	63832	63832	64409	63860.85	125.75
12	120665	**119608**	120774	**120121.55**	417.07
14	208086	**199363**	210599	**202400.50**	2883.39
16	279618	**279077**	297173	284036.95	4770.61

n	T for min	mean(T)	std(T)
8	0.1	1555.55	1880.94
10	477.2	8511.29	17132.49
12	15428.1	49373.31	32834.88
14	34152.3	70898.90	48551.27
16	55640.8	47922.16	36948.40

Table 2. Solution Quality and Solution Times for the Constant Distances with Mirroring

n	Old Best	min(D)	max(D)	mean(D)	std(D)
8	80	80	80	80	0
10	130	130	130	130	0
12	192	192	192	192	0
14	253	253	253	253	0
16	342	342	342	342	0
18	432	432	432	432	0
20	524	**522**	**522**	**522**	0
22	650	650	650	650	0
24	768	768	768	768	0

n	T for min	mean(T)	std(T)
8	0.1	0.06	0
10	0.1	0.10	0
12	0.3	0.56	0.38
14	6.0	154.26	147.95
16	2.7	3.29	1.53
18	8.1	24.60	19.20
20	1106.3	12556.20	10347.58
22	24.3	45.42	22.90
24	813.3	1791.77	983.47

Table 3. Solution Quality and Solution Times for the Circular Distances with Mirroring

n	Old Best	min(D)	max(D)	mean(D)	std(D)
8	140	140	140	140	0
10	272	272	276	273.60	1.01
12	456	**432**	**444**	**434.90**	3.12
14	714	**696**	726	**708.90**	7.05
16	978	**968**	1072	1001.60	28.55
18	1306	1352	1364	1357.80	3.40
20	1882	**1852**	2198	2017.60	60.64

n	T for min	mean(T)	std(T)
8	0.2	74.18	55.13
10	28160.0	12527.18	12208.25
12	93.1	4658.58	3560.27
14	53053.5	23549.14	16311.15
16	38982.7	23360.81	14451.53
18	178997.5	106139.77	57175.01
20	59097.9	43137.13	22515.46

Mirrored Instances. Tables 1, 2, and 3 report the results for mirrored instances, which are particularly impressive. The algorithm matches or improves all best-known solutions (but one). It produces 8 new best solutions and the improvements essentially occur for larger instances. This was a surprising result for us, since we thought that mirroring instances would be significantly more challenging for the algorithm. Some of the improvements may also be quite large and reach more than 4%.

Non-mirroring Instances. Tables 4, 5, and 6 report the results for the non-mirrored instances. On the NLn and constant distance metrics, the algorithm is once again impressive, matches or improves all the best-known solutions, and improves 6 instances. Once again, the gains are obtained on the larger instances. The results on the circular

Table 4. Solution Quality and Solution Times for the NLn Distances without Mirroring

n	Old Best	min(D)	max(D)	mean(D)	std(D)
8	39721	39721	39721	39721	0
10	59436	59436	59583	59561.63	48.33
12	111483	**111248**	116018	112663.32	738.55
14	190056	**189156**	195742	193187.85	1432.99
16	270794	**267194**	282005	273552.64	3461.49

n	T for min	mean(T)	std(T)
8	1169.0	1639.33	332.38
10	2079.6	27818.24	64873.91
12	202756.2	150328.30	92385.48
14	90861.4	77587.86	40346.49
16	344633.4	476191.65	389371.71

Table 5. Solution Quality and Solution Times for the Constant Distances without Mirroring

n	Old Best	min(D)	max(D)	mean(D)	std(D)
8	80	80	80	80	0
10	124	124	124	124	0
12	181	181	181	181	0
14	252	252	252	252	0
16	327	327	329	328	0.31
18	418	**417**	418	**417.65**	0.47
20	521	**520**	522	**520.90**	0.53
22	632	**628**	630	**629.40**	0.58
24	757	**750**	753	**750.65**	0.91

n	min(T)	mean(T)	std(T)
8	0.2	0.14	0.14
10	4.6	3.96	2.43
12	128.7	1126.85	1480.45
14	26.1	95.32	59.42
16	82884.1	16042.20	22332.36
18	10362.8	7091.27	6614.78
20	7781.7	22850.72	25094.76
22	39380.3	22618.46	20364.85
24	16356.6	28941.04	22987.96

Table 6. Solution Quality and Solution Times for the Circular Distances without Mirroring

n	Old Best	min(D)	max(D)	mean(D)	std(D)
8	132	132	132	132.00	0
10	242	242	256	252.70	3.24
12	408	420	432	427.13	3.43
14	654	666	690	679.70	5.14
16	928	968	1072	1001.60	28.55
18	1306	1352	1364	1357.80	3.40
20	1842	1852	2198	2017.60	60.64

n	T for min	mean(T)	std(T)
8	3.2	589.23	590.74
10	19261.6	14491.27	7937.21
12	151459.1	96717.13	52788.38
14	12908.9	86766.67	68408.73
16	38982.7	23360.81	14451.53
18	178997.5	106139.77	57175.01
20	59097.9	43137.13	22515.46

instances are somewhat disappointing. The algorithm cannot match the best-known results on the larger instances, although it is often very close to the best-known solutions. This may be due to the fact that the algorithm only uses mirrored starting schedules, which may bias the search. In fact, the best solutions found by our algorithm for 16 and 20 teams are mirrored schedules. These instances need to be investigated more carefully, since very little time was spent tuning the parameters.

8 Conclusion

This paper reconsidered our original simulated algorithm for the travelling tournament problem (TTP) and studied its effectiveness across all TTP variants. The variants include various distance metrics, as well as mirroring constraints. From a practical standpoint, its main contribution is to show that the original algorithm can be enhanced to be effective across all distance metrics and mirroring. The main technical novelty in the algorithm is the introduction of novel neighborhood moves that capture sequences of earlier moves. As such, these novel moves do not improve the connectivity of the neighborhood for the TTP. Their significance comes from the fact that, in the original algorithm, these sequences have a low probability, although they capture fundamental aspects of the mirroring or distance structure. The resulting algorithm matches or improves most best-known solutions and it also produces numerous new best solutions for many of the variants. It is thus quite remarkable that a single algorithm be so robust in producing high-quality solutions to all instances.

An important area of future research is to find high-quality solutions more quickly. For instance, Gaspero and Schaerf [8] embedded a subset of our neighborhood in a tabu-search algorithm and obtained high-quality solutions quickly, although their best solutions still do not match our best found solutions. Whether it is possible to find the same solution quality with the speed of their algorithm is an important issue to address.

References

1. A. Anagnostopoulos, L. Michel, P. Van Hentenryck, and Y. Vergados. A Simulated Annealing Approach to the Traveling Tournament Problem In *CP-AI-OR'2003*, Montreal, Canada, May 2003.
2. T. Benoist, F. Laburthe, and B. Rottembourg. Lagrange Relaxation and Constraint Programming Collaborative Schemes for Travelling Tournament Problems. In *CP-AI-OR'2001*, Wye College (Imperial College), Ashford, Kent UK, April 2001.
3. D.T. Connelly. General Purpose Simulated Annealing. *European Journal of Operations Research*, 43, 1992.
4. A. Davenport and E. Tsang. Solving Constraint Satisfaction Sequencing Problems by Iterative Repair. In *Procceedings of the First International Conference on the Practical Applications of Constraint Technologies and Logic Programming (PACLP-99)*, pages 345–357, London, April 1999.
5. Juan A. Díaz and Elena Fernández. A tabu search heuristic for the generalized assignment problem. *European Journal of Operational Research*, 132(1):22–38, July 2001.
6. J. H. Dinitz, and D. R. Stinson. A Hill-climbing Algorithm for the Construction of One-Factorizations and Room Squares. *SIAM J. Alg. Disc. Meth.*, 8(3):430–438, July 1987.
7. K. Easton, G. Nemhauser, and M. Trick. The traveling tournament problem description and benchmarks. In *Seventh International Conference on the Principles and Practice of Constraint Programming (CP'01)*, pages 580–589, Paphos, Cyprus, 2001. Springer-Verlag, LNCS 2239.
8. L. Di Gaspero, and A. Schaerf. A Tabu Search Approach to the Traveling Tournament Problem. In *Proceedings of RCRA 2005, Associazione Italiana per l'Intelligenza Artificiale (AI*IA)*, pages 23–27, Ferrara, Italy, June 2005
9. F. Glover and M. Laguna. *Tabu Search*. Kluwer Academic Publishers, 1997.
10. S. Kirkpatrick, C. Gelatt, and M. Vecchi. Optimization by Simulated Annealing. *Science*, 220:671–680, 1983.
11. M. Laguna, J.P. Kelly, Gonzalez-Velarde, and F. Glover. Tabu search for the multilevel generalized assignment problems. *European Journal of Operational Research*, 42:677–687, 1995.

12. A. Lim, B. Rodrigues and X. Zhang. Scheduling Sports Competitions at Multiple Venues Revisited. European Journal of Operational Research, 2005 (Accepted for Publication).
13. I. H. Osman. Metastrategy simulated annealing and tabu search algorithms for the vehicle routing problem. *Annals of Operations Research*, 41:421–451, 1993.
14. E. Pesch and F. Glover. TSP Ejection Chains. *Discrete Applied Mathematics*, 76:165–181, 1997.
15. R. Rasmussen and M. Trick. A Benders Approach to the Constrained Minimum Break Problem. European Journal of Operational Research, 2005 (Accepted for Publication)
16. C. C. Ribeiro, and S. Urrutia. Heuristics for the Mirrored Traveling Tournament Problem. *Proceedings of the 5th International Conference on the Practice and Theory of Automated Timetabling (PATAT'04)*, 323-342, 2004.
17. M. Trick. http://mat.gsia.cmu.edu/TOURN/, 2002.
18. P. Van Hentenryck and Y. Vergados Minimizing Breaks in Sport Scheduling with Local Search In *Proceedings of the 15th International Conference on Automated Planning and Scheduling* Monterey, CA, June 2005.

Open Constraints in a Closed World

Willem-Jan van Hoeve[1] and Jean-Charles Régin[2,*]

[1] Department of Computer Science, Cornell University, Ithaca, NY 14853, USA
vanhoeve@cs.cornell.edu
[2] ILOG Sophia Antipolis, Les Taissounières HB2, 1681 route des Dolines,
06560 Valbonne, France
regin@ilog.fr

Abstract. We study domain filtering algorithms for *open constraints*, i.e., constraints that are not a priori defined on specific sets of variables. We present an efficient filtering algorithm, achieving set-domain consistency, for open global cardinality constraints. We extend this result to conjunctions of them, in case they are defined on disjoint sets of variables. We also analyze the case when the sets of variables may overlap. As establishing set-domain consistency is NP-complete in that case, we propose a weaker, though efficient, filtering algorithm instead. Finally, we extend our results to conjunctions of similar open constraints.

1 Introduction

Traditionally, constraint programming has focused on solving problems in closed-world scenarios: all variables and constraints are fixed from the beginning. In many real-life applications, however, the scope of a constraint may not be defined a priori. Instead, the variables on which the constraints are defined may only be revealed during the solution process. This happens very often in scheduling applications and other distributed settings.

For example, consider a set of activities and suppose that each activity can be processed either on the factory line 1 formed by the set of unary resources R_1, or on the factory line 2 formed by the set of unary resources R_2. Thus, at the beginning, the set of resources that will be used by an activity is not known. Also the set of activities that will be processed by a resource is not known. However, it is useful to express that the activities that will be processed on each line must be pairwise different. This can be done by defining two `alldifferent` constraints, involving the start variables of each activity, and by stating that a start variable will be involved in exactly one `alldifferent` constraint. Initially, each `alldifferent` constraint is defined on a set of variables formed by all start variables. Then this set will be modified when it can be proved that a variable cannot be a member of an `alldifferent` constraint (i.e., the corresponding activity cannot be processed on the corresponding factory line), or that a start variable (activity) will be processed on the specific factory line.

---

* A large part of this work was carried out while the author was at Cornell University.

J.C. Beck and B.M. Smith (Eds.): CPAIOR 2006, LNCS 3990, pp. 244–257, 2006.

Constraints of this nature are called *dynamic constraints* [2] or *open constraints* [4, 5]. For instance, the above `alldifferent` constraints are examples of open constraints. In this case, they live in a *closed world*, because the set of possible variables is explicitly known. The extension of constraint programming with open constraints is called *open constraint programming* [4, 5].

The use of efficient domain filtering algorithms is a key element in solving problems with constraint programming. This is particularly true when the filtering is based on a *global constraint*, i.e., a constraint that encapsulates and exploits a substructure of the problem. Efficient filtering algorithms for *open global constraints* therefore have high potential to improve the solution process of open constraint programming, together with its rich application area. Nevertheless, such filtering algorithms have not yet been proposed, until now.

In this work we study the problem of filtering open global constraints in a closed world. We focus in particular on conjunctions of open *global cardinality constraints*, or `gccs`, because of their practical applicability and generality. We present an efficient filtering algorithm, obtaining "set-domain" consistency, when the scopes of the `gccs` are restricted to non-overlapping sets of variables. We also analyze the case when the scopes of the `gccs` are allowed to share variables. In that case obtaining domain consistency is NP-complete. Hence we propose a weaker, though efficient, filtering algorithm.

Our filtering algorithms are based on techniques from flow theory. In fact, we are able to generalize the techniques used in the filtering algorithm for the original (closed) global cardinality constraint to conjunctions of open global cardinality constraints. It furthermore allows us to filter the domains of the set variables that underly the open global cardinality constraints. Finally, we extend our results to conjunctions of arbitrary flow-based open global constraints. Due to the application of efficient flow theoretic techniques, our algorithms are also efficient.

The outline of this paper is as follows. In Section 2, we present definitions and other preliminaries. In Section 3 we outline the general problem and give a motivating example. Section 4 describes our main result, the filtering algorithm for conjunctions of open global cardinality constraints on non-overlapping sets of variables. In Section 5 we consider the case where the sets of variables may overlap. In Section 6 we present a filtering algorithm for the set variables that underly the open constraints. In Section 7 we extend our results to conjunctions of similar open global constraints. Finally, we conclude in Section 8.

2   Background

2.1   Constraint Programming

Let x be a variable. The *domain* of x is a finite set of elements (also called domain values) that can be assigned to x. It is denoted by $D(x)$. For a set of variables X we define $D(X) = \cup_{x \in X} D(x)$.

A *set variable* is a variable whose domain values are sets. We often represent the domain of a set variable S by an "interval" $[L, U]$, where L and U are sets, such that $D(S) = \{s \mid L \subseteq s \subseteq U\}$. For example, let V be a set, and let S be a set variable with domain $D(S) = [\varnothing, V]$. Then $D(S)$ is the power set of V, i.e., it contains all possible subsets of V.

Let $X = \{x_1, x_2, \ldots, x_k\}$ be a set of variables. A *constraint* C on X is defined as a subset of the Cartesian product of the domains of the variables in X, i.e., $C \subseteq D(x_1) \times D(x_2) \times \cdots \times D(x_k)$. We say that X is the *scope* of C. A tuple $(d_1, \ldots, d_k) \in C$ is called a *solution* to C. We also say that the tuple *satisfies* C. C is *inconsistent* if it does not contain a solution. Otherwise, C is called *consistent*.

Sometimes a constraint C is defined on variables X together with a certain set of parameters p. In such cases, we denote the constraint as $C(X, p)$ for syntactical convenience, while admissible tuples are still of size $|X|$.

Next we introduce open constraints. For the purpose of this paper, we define a constraint to be *open* when its scope is the domain of a set variable whose domain values are sets of variables. The explicit representation of the domain of this set variable reflects that the constraint lives in a closed world. For example, let X be a set of variables and let S be a set variable with domain $D(S) = [\varnothing, X]$. The constraint $C(D(S))$ is an open constraint, whose scope depends on the actual instantiation of S. We also write $C(S)$ as a shorthand for $C(D(S))$, if there is no confusion.

A *constraint satisfaction problem*, or a *CSP*, is defined by a finite set of variables X, together with a finite set of constraints C, each on a subset of X. The goal is to find an assignment $x = d$ with $d \in D(x)$ for all $x \in X$, such that all constraints are satisfied simultaneously. This assignment is called a *solution to the CSP*. Note that by using set variables to define the scope of open constraints, we maintain this common definition of a CSP.

The solution process of constraint programming interleaves *constraint propagation*, and *search*. The search process essentially consists in enumerating all possible combinations of variable domain values, until we find a solution to the CSP or prove that none exists. We say that this process constructs a *search tree*. To reduce the exponential number of combinations, we *filter* the domains of the variables and *propagate* this information through all constraints:

> Given the current domains and a constraint C, remove domain values that do not belong to a solution to C. This is repeated for all constraints until no more domain values can be removed.

We typically apply constraint propagation at each node in the search tree. In order to be effective, filtering algorithms should be efficient, because they are applied many times during the solution process. Furthermore, they should remove as many domain values that are not part of a solution as possible. If a filtering algorithm for a constraint C removes *all* such values from the domains with respect to C, we say that it makes C *domain consistent*:

Definition 1 (domain consistency). *A constraint C on the variables $x_1, x_2,$ $\ldots, x_k$ is called* domain consistent *if for each variable x_i and each domain value*

$d_i \in D(x_i)$ $(i = 1, \ldots, k)$, there exists a domain value $d_j \in D(x_j)$ for all $j \neq i$ such that $(d_1, \ldots, d_k) \in C$.

In the literature, domain consistency is also referred to as *hyper-arc consistency* or *generalized-arc consistency*. Note that domain consistency does only guarantee that each individual constraint contains a solution; it does *not* guarantee that the CSP has a solution.

If we make an open constraint $C(S)$ domain consistent, we should remove from the domain of S all sets s of variables for which $C(s)$ has no solution. As set variables are only represented by an interval, we use *bounds consistency* for this purpose instead:

Definition 2 (bounds consistency). *An open constraint C on the set variables $S_1, S_2, \ldots, S_k$ is called* bounds consistent *if for each S_i and each $s_i \in \{\min S_i, \max S_i\}$ $(i = 1, \ldots, k)$, there exist sets $s_j \in [\min S_j, \max S_j]$ for all $j \neq i$ such that $C(s_1, \ldots, s_k)$ has a solution.*

Rather than filtering the domain of the set variable S, however, we would like to filter the domain of the actual variables that appear in any instantiation of S. Hence, we next introduce a slight variant of domain consistency for open constraints that captures exactly this:

Definition 3 (set-domain consistency). *An open constraint $C(S)$ is called* set-domain consistent *if for each variable $x \in \{y \mid y \in s, s \in D(S)\}$ and all domain values $d \in D(x)$ there exists a set $s' \in D(S)$ with $x \in s'$ such that $x = d$ belongs to a solution of $C(s')$.*

By introducing this notion of set-domain consistency, we separate the filtering of the set variable S and the variables that appear in its domain. An open constraint $C(S)$ can hence be made set-domain consistent, while $C(S)$ itself may not be bounds consistent.

2.2   Flow Theory

In this section we present some concepts of flow theory that are necessary to understand this paper. For more information on flow theory we refer to [1].

Let $G = (V, A)$ be a directed graph, or *digraph*, with vertex set V and arc set A. Let $s, t \in V$. A function $f : A \to \mathbb{R}$ is called a *flow from s to t*, or an *s-t flow*, if

$$(i) \quad f(a) \geq 0 \qquad \text{for each } a \in A, \text{ and}$$
$$(ii) \quad f(\delta^{\text{out}}(v)) = f(\delta^{\text{in}}(v)) \text{ for each } v \in V \setminus \{s, t\},$$

where $\delta^{\text{in}}(v)$ and $\delta^{\text{out}}(v)$ denote the multiset of arcs entering and leaving v, respectively. Here $f(S) = \sum_{a \in S} f(a)$ for all $S \subseteq A$. Property (ii) ensures *flow conservation*, i.e., for a vertex $v \neq s, t$, the amount of flow entering v is equal to the amount of flow leaving v.

As we will always consider s-t flows in this paper, we will often speak of a *flow* instead of *s-t* flow, for convenience. Furthermore, we say that an arc a *belongs* to a flow if $f(a) > 0$.

Let $d : A \to \mathbb{R}_+$ and $c : A \to \mathbb{R}_+$ be a "demand" function and a "capacity" function, respectively[1]. We say that a flow f is *feasible* if $d(a) \leq f(a) \leq c(a)$ for each $a \in A$.

Let f be an *s-t* flow in G. The *residual graph* of G with respect to f, c and d is defined as $G_f = (V, A_f)$ where for each $(u, v) \in A$,

if $f(u, v) < c(u, v)$ then $(u, v) \in A_f$ with residual demand $\max\{d(u, v) - f(u, v), 0\}$ and residual capacity $c(u, v) - f(u, v)$, and

if $f(u, v) > d(u, v)$ then $(v, u) \in A_f$ with residual demand 0 and residual capacity $f(u, v) - d(u, v)$.

Finally, a digraph $G = (V, A)$ is *strongly connected* if for any two vertices $u, v \in V$ there is a directed path from u to v. A maximally strongly connected non-empty subgraph of a digraph G is called a *strongly connected component* of G.

3   Open Global Cardinality Constraints

A *global cardinality constraint* (gcc) on a set of variables specifies for each domain value in the union of their domains an upper and lower bound to the number of variables that are assigned to this value:

Definition 4 (global cardinality constraint). *Let $X = \{x_1, \ldots, x_n\}$ be a set of variables, and let $l_d, u_d \in \mathbb{N}$ for all $d \in D(X)$. Then*

$$\mathbf{gcc}(X, l, u) = \{(d_1, \ldots, d_n) \mid \forall i \in \{1, \ldots, n\}\ d_i \in D(x_i), \\ \forall d \in D(X)\ l_d \leq |\{d_i \mid d_i = d\}| \leq u_d\}.$$

A special case of the gcc is the alldifferent constraint, which specifies that all variables should be pairwise different. If we set $l_d = 0$ and $u_d = 1$ for all $d \in D(X)$, the gcc is equal to the alldifferent constraint. A filtering algorithm for the gcc, establishing domain consistency, was developed in [7], making use of network flows.

3.1   A Single Open Global Cardinality Constraint

We first consider the case of a single open gcc. In order to filter this constraint, we compute a flow in a particular graph, similar to the filtering of the original (closed) gcc [7].

Let X be a set of variables, and let S be a set variable with domain $[L, U]$, such that $L \subseteq U \subseteq X$. Let $\mathbf{gcc}(S, l, u)$ be the open gcc under consideration. We build the following graph. The vertex set of the graph is composed of U, $D(U)$, a source s, and a sink t. The arc set is composed of:

- Arcs (s, x) for all $x \in U$ with capacity 1. If $x \in L$, then this arc has demand 1, otherwise its demand is 0.
- Arcs (x, d) for all $x \in U$, $d \in D(x)$ with demand 0 and capacity 1,
- Arcs (d, t) for all $d \in D(U)$, with demand l_d and capacity u_d.

---

[1] Here $\mathbb{R}_+$ denotes $\{x \in \mathbb{R} \mid x \geq 0\}$.

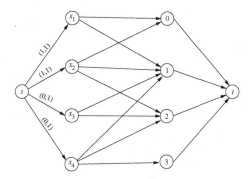

Fig. 1. Graph representation for the **gcc** of Example 1. For the (s, x_i) arcs, the demand d and capacity c is explicitly given as (d, c). All other arcs have demand 0 and capacity 1.

Call the resulting graph $\mathcal{G}_{\text{single}}$. We have the following result:

Theorem 1. *A feasible integer flow in* $\mathcal{G}_{\text{single}}$ *corresponds to a solution to* **gcc**(S, l, u) *and vice versa.*

Proof. Let f be a feasible integer flow in $\mathcal{G}_{\text{single}}$. We construct a solution to **gcc**(S, l, u) by defining $S = \{x \mid f(s, x) = 1\}$ and $x = d$ for all $x \in S, d \in D(x)$ with $f(x, d) = 1$.

Conversely, given a solution to **gcc**(S, l, u) we construct a feasible integer flow f in $\mathcal{G}_{\text{single}}$ by defining

for all $x \in U$: $f(s, x) = 1$ if $x \in S$, and $f(s, x) = 0$ otherwise,
for all $x \in U$ and $d \in D(x)$: $f(x, d) = 1$ if $x = d$, and $f(x, d) = 0$ otherwise,
for all $d \in D(U)$: $f(d, t) = |\{x \mid x \in S, x = d\}|$.   □

Example 1. Let $X = \{x_1, x_2, x_3, x_4\}$ be a set of variables with integer domains: $x_1 \in \{0, 1\}$, $x_2 \in \{0, 1, 2\}$, $x_3 \in \{1, 2\}$, and $x_4 \in \{1, 2, 3\}$. Let $S \in [\{x_1, x_2\}, X]$ be a set variable. Furthermore, let $l_d = 0$ and $u_d = 1$ for all $d \in \{0, 1, 2, 3\}$.

The graph representation for the constraint **gcc**(S, l, u) is presented in Figure 1. In fact, this **gcc** corresponds to an **alldifferent** constraint for this choice of l and d. Note that the demand of the arcs (s, x_1) and (s, x_2) is 1, because S must include $\{x_1, x_2\}$.

By applying Theorem 1, we get the following result:

Corollary 1. *The constraint* **gcc**(S, l, u)*, where* $S \in [L, U]$*, is set-domain consistent if and only if for all* $x \in U$ *and* $d \in D(x)$*, there exists an arc* (x, d) *with* $d \in D(x)$ *that belongs to a feasible integer flow in* $\mathcal{G}_{\text{single}}$*.*

The proof is immediate because there is a one to one correspondence between a solution of **gcc**(S, l, u) and a feasible flow in $\mathcal{G}_{\text{single}}$.

Corollary 1 gives rise to a set-domain consistency algorithm, similar to the original (closed) **gcc**. First, we compute an initial feasible integer flow f in $\mathcal{G}_{\text{single}}$.

This can be done in $O(nm)$ time, where $n = |U|$ and $m = \sum_{x \in U} D(x)$. If f does not exist, the constraint is not consistent. Otherwise, we identify and remove all arcs (x, d) that do not belong to any feasible integer flow. As indicated by [7], inconsistent arcs are exactly those that do not belong to a strongly connected component in the residual graph $(\mathcal{G}_{single})_f$. Computing the strongly connected components of $(\mathcal{G}_{single})_f$ can be done in $O(n + m)$ time [9], where n and m are defined as above. Moreover, our algorithm is incremental. When k variables have changed their value, we can recompute the flow in $O(km)$ time and re-establish domain consistency in $O(n + m)$ time.

3.2   The Conjunction of Open Global Cardinality Constraints

Next we consider a more general problem; the conjunction of several open gccs. First, consider the following motivating example.

Example 2. Let $X = \{x_1, x_2, x_3, x_4, x_5\}$ be a set of variables with integer domains: $x_1 \in \{0, 1\}$, $x_2 \in \{0, 1\}$, $x_3 \in \{0, 1\}$, $x_4 \in \{0, 1\}$ and $x_5 \in \{0, 1, 2, 3, 4, 5\}$. Let $S_1 \in [\varnothing, X]$ and $S_2 \in [\varnothing, X]$ be set variables. We define the following conjunction of constraints:

$$\begin{aligned}
&\texttt{alldifferent}(S_1) \wedge \\
&\texttt{alldifferent}(S_2) \wedge \\
&(S_1 \cup S_2) = X \wedge \\
&(S_1 \cap S_2) = \varnothing.
\end{aligned} \tag{1}$$

Here $\texttt{alldifferent}(S_1)$ and $\texttt{alldifferent}(S_2)$ are open constraints, as they are defined on the domain of set variables whose domain values are sets of variables.

From conjunction (1) we are able to deduce that

$$\begin{aligned}
2 &\leq |S_1| \leq 3, \\
2 &\leq |S_2| \leq 3, \\
x_5 &\in \{2, 3, 4, 5\}.
\end{aligned}$$

Namely, as x_1, x_2, x_3 and x_4 all have domain $\{0, 1\}$, no more than two of them can appear in one $\texttt{alldifferent}$ constraint. Since we need to include all variables into both constraints, we have that $2 \leq |S_1| \leq 3$ and $2 \leq |S_2| \leq 3$. Moreover, each $\texttt{alldifferent}$ constraint will involve exactly two variables from $\{x_1, x_2, x_3, x_4\}$, and the variables x_1, x_2, x_3 and x_4 will saturate the values 0 and 1 in both $\texttt{alldifferent}$ constraints. Hence those values are removed from the domain of x_5.

Our general problem is the conjunction of k open gccs. Let X be a set of variables, and let $S_1, S_2, \ldots, S_k$ be set variables, with respective domains $[L_i, U_i]$, such that $L_i \subseteq U_i \subseteq X$ ($i = 1, \ldots, k$). The conjunction of k open gccs is defined as:

$$\bigcap_{1 \leq i \leq k} \texttt{gcc}(S_i, l^i, u^i), \tag{2}$$

where $l_d^i, u_d^i \in \mathbb{N}$ for all $d \in D(X)$ and $i = 1, \ldots, k$.

If the set variables $S_1, S_2, \ldots, S_k$ are not constrained, there is not much that can be deduced. We know that, for $1 \le i \le k$,

$$\sum_{d \in D(X)} l_d^i \le |S_i| \le \sum_{d \in D(X)} u_d^i,$$

but in general this is not sufficient to make further deductions with respect to the domains of the variables. Hence, we impose additional constraints on the set variables. In particular we distinguish the following four cases and combinations thereof:

$$(\bigcup_{1 \le i \le k} S_i) = X, \tag{3}$$

$$(\bigcup_{1 \le i \le k} S_i) \subset X, \tag{4}$$

$$\forall_{1 \le i < j \le k} \; S_i \cap S_j = \varnothing, \tag{5}$$

$$\exists_{1 \le i < j \le k} \; S_i \cap S_j \ne \varnothing. \tag{6}$$

For example, the combination of (3) and (5) restricts the set variables to be a partition of X. In the remainder of this paper we will study filtering algorithms for the conjunction of k open gccs, in combination with one or more of the constraints (3) up to (6).

4 Disjoint Set Variables

In this section, we present an efficient set-domain consistency filtering algorithm for k open gccs together with restriction (5), i.e., all set variables should be pairwise disjoint. Our work is based on the domain consistency filtering algorithm for the single gcc as developed in [7], and an extension of the algorithm presented above for a single open gcc.

Again, we base our algorithm on finding a flow in a particular graph. The key observation is that for each open gcc, one duplicates the corresponding variables and domain values, and associates the corresponding lower and upper bounds to each domain value. This allows us to build a graph similar to the graph of a single gcc, and also to apply similar efficient flow algorithms to establish set-domain consistency.

We build our graph as follows; see Figure 2 for a schematic representation. In order to distinguish variables in different open gccs, we duplicate every variable $x \in U_i$ as x^i, for $i \in \{1, \ldots, k\}$, and denote the respective set of variables by X^i. We also duplicate the domain values $D(X)$ as $D^1(X), \ldots, D^k(X)$. Then the vertex set of the graph is composed of X, $X^1, \ldots, X^k$, $D^1(X), \ldots, D^k(X)$, a source s, "intermediate sinks" $t_1, \ldots t_k$ and a sink t. The arc set is composed of:

- arcs (s, x) for all $x \in X$, with demand 0 and capacity 1,
- arcs (x, x^i) for all $i \in \{1, \ldots, k\}$ and $x \in L_i$, with demand 1 and capacity 1,

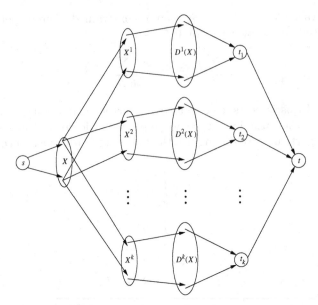

Fig. 2. Schematic graph representation for the conjunction of k open **gccs**

- arcs (x, x^i) for all $i \in \{1, \ldots, k\}$ and $x \in U_i \setminus L_i$, with demand 0 and capacity 1,
- arcs (x^i, d) for all $i \in \{1, \ldots, k\}$, $x \in U_i$ and $d \in D^i(x)$, with demand 0 and capacity 1,
- arcs (d, t_i) for all $i \in \{1, \ldots, k\}$ and $d \in D^i(X)$, with demand l_d^i and capacity u_d^i,
- arcs (t_i, t) for all $i \in \{1, \ldots, k\}$, with demand $|L_i|$ and capacity $|U_i|$.

Call the resulting graph $\mathcal{G}$. Note that we may actually omit arcs (x, x^i) if $x \in L_j$ for some $j \neq i$. This follows from the disjointness of the set variables. We nevertheless prefer the above description, because it can be easily extended to the non-disjoint case, as we will see in Section 5.

We have the following result:

Theorem 2. *A feasible integer flow in $\mathcal{G}$ corresponds to a solution to the conjunction of (2) and (5) and vice versa.*

Proof. Let f be a feasible integer flow in $\mathcal{G}$. We construct a solution to the conjunction of (2) and (5) by defining $S_i = \{x \mid f(x, x^i) = 1\}$ $(1 \leq i \leq k)$ and $x = d$ for all $x \in X, d \in D(x)$ with $f(x^i, d) = 1$ for some $i \in \{1, \ldots, k\}$.

Conversely, given a solution to the conjunction of (2) and (5), we construct a feasible integer flow f in $\mathcal{G}$ by defining

for all $x \in X$: $f(s, x) = 1$ if $x \in S_i$ for some $i \in \{1, \ldots, k\}$, and $f(s, x) = 0$ otherwise,

for all $x \in X$ and $i \in \{1, \ldots, k\}$: $f(x, x^i) = 1$ if $x \in S_i$, and $f(x, x^i) = 0$ otherwise,

for all $x \in X$, $i \in \{1, \ldots, k\}$ and $d \in D^i(x)$: $f(x^i, d) = 1$ if $(x = d) \wedge (x \in S_i)$,
    and $f(x^i, d) = 0$ otherwise,
for all $i \in \{1, \ldots, k\}$ and $d \in D^i(X)$: $f(d, t_i) = |\{x \mid x \in S_i, x = d\}|$,
for all $i \in \{1, \ldots, k\}$: $f(t_i, t) = |\{x \mid x \in S_i\}|$.   □

Notice that if $L_i \cap L_j \neq \varnothing$ for some $i \neq j$, there is no feasible flow in $\mathcal{G}$, because the demand requirements on the arcs involving S_i and S_j cannot be fulfilled.

An illustration applied to Example 2 is given in Figure 3. In Figure 3.a we present the graph $\mathcal{G}$ corresponding to this example. In Figure 3.b we present a

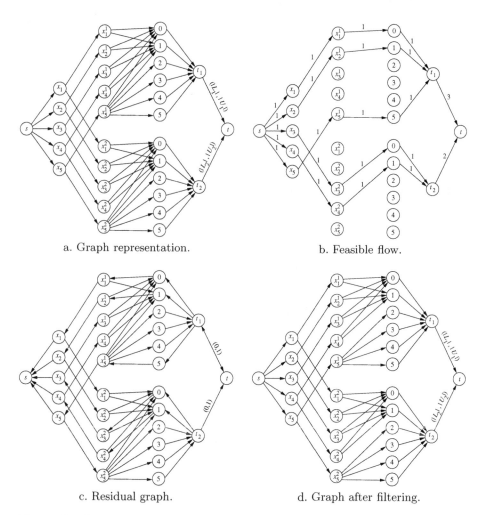

a. Graph representation.   b. Feasible flow.

c. Residual graph.   d. Graph after filtering.

Fig. 3. Graph representation for the conjunction of $\mathtt{alldifferent}(S_1)$ and $\mathtt{alldifferent}(S_2)$ on disjoint set variables S_1 and S_2, where $S_1 \cup S_2 = \{x_1, x_2, x_3, x_4, x_5\}$ (following Example 2). Arcs (t_1, t) and (t_2, t) have demand and capacity determined by the cardinality of the lower bounds L_1, L_2 and upper bounds U_1, U_2 of the respective set variables. All other arcs have demand 0 and capacity 1.

feasible flow in $\mathcal{G}$, corresponding to the solution $S_1 = \{x_1, x_2, x_5\}$, $S_2 = \{x_3, x_4\}$, $x_1 = 0$, $x_2 = 1$, $x_3 = 0$, $x_4 = 1$ and $x_5 = 5$.

By applying Theorem 2, we get the following result.

Corollary 2. *The conjunction of (2) and (5) is set-domain consistent if and only if for all $x \in X$ and $d \in D(x)$, there exists an arc (x^i, d) with $d \in D^i(x)$ for some $1 \leq i \leq k$, that belongs to a feasible integer flow in $\mathcal{G}$.*

The proof follows from the one to one correspondence between a solution of the conjunction of (2) and (5), and a feasible flow in $\mathcal{G}$. Note that Theorem 2 does not consider the set variables on which the gccs are defined. We will deal with them in Section 6.

Similar to the above single open gcc, we apply Corollary 2 to design a set-domain consistency algorithm. First, we compute an initial feasible integer flow f in $\mathcal{G}$. This can be done in $O(nm)$ time, where $n = |X|$ and $m = k \cdot \sum_{x \in X} D(x)$. If f does not exist, the conjunction is not consistent. Otherwise, we identify and remove all arcs (x^i, d) that do not belong to any feasible integer flow. Note however, that one arc (x^i, d) with $d \in D^i(x)$ for some $1 \leq i \leq k$ is already sufficient to make $d \in D(X)$ consistent. Again, inconsistent arcs are exactly those that do not belong to a strongly connected component in the residual graph $\mathcal{G}_f$. Computing the strongly connected components of $\mathcal{G}_f$ can again be done in $O(n+m)$ time, where n and m are defined as above. Also this algorithm is incremental. When l variables have changed their value, we can recompute the flow in $O(lm)$ time and re-establish domain consistency in $O(n + m)$ time.

As an example, consider again Figure 3. In Figure 3.c we show the residual graph with respect to the flow given in Figure 3.b. Figure 3.d shows the graph after filtering, i.e., all inconsistent arcs are removed.

5   Non-disjoint Set Variables

We next study the conjunction of k open gccs together with restriction (6), i.e., the set variables are allowed to be non-disjoint.

Unfortunately, establishing set-domain consistency in this case is an NP-complete problem. Namely, in [3], it is proved that establishing domain consistency on the conjunction of alldifferent constraints on overlapping sets of variables is NP-complete. As the alldifferent constraint is a special case of the gcc, their result immediately implies that our task is NP-complete. To overcome the NP-completeness, we propose to relax the requirement of establishing set-domain consistency.

We use the same graph representation as in the previous section, with one modification. In the previous section, the graph representation does not allow a variable to appear in several gccs. Namely, because the capacity of the arcs (s, x) for $x \in X$ is 1, each variable can be assigned to at most one value in $D^1(X), \ldots, D^k(X)$. This means that each variable can only occur in a single open gcc. We relax this by defining the capacity of the arcs (s, x) to be k. On the positive side, this allows a variable to occur in several open gccs at the same time, which yields a filtering method for the non-disjoint case. On the negative side, a variable may take different values in different gccs, which is likely to weaken the filtering. With this modification, however, we can apply the same efficient algorithm as in the previous section.

6   Filtering the Set Variables

In this section we consider the filtering of the set variables on which the open constraints are defined. As stated earlier, we would like to establish bounds consistency with respect to these variables.

Consider again the conjunction of k open gccs (2) and restriction (5), i.e., the constraints are defined on disjoint sets of variables. We can use the graph $\mathcal{G}$ to establish bounds consistency on the set variables $S_1, \ldots, S_k$ as well. To this end, we apply the following three rules for all $1 \leq i \leq k$:

i) when there is no arc between x^i and $D^i(X)$, then x is removed from S_i, i.e., $U_i := U_i - x$,

ii) when there are only arcs between x^i and $D^i(X)$, and $f(x, x^i) = 1$ for all feasible flows f, then x is added to S_i, i.e., $L_i := L_i + x$,

iii) $|L_i| \geq \min\{f(t_i, t) \mid f \text{ feasible flow}\}$ and $|U_i| \leq \max\{f(t_i, t) \mid f \text{ feasible flow}\}$.

When we apply these rules, we know by Theorem 2 that we have established bounds consistency with respect to the set variables. The application of the above three rules can be done in $O(k^2 nm)$ time, by subsequently computing minimum and maximum flows.

7   Extension

In the above, we have focused on conjunctions of gccs because of their generality and applicability to real-life problems with open constraints. The results can easily be extended to similar cases, however.

7.1   Optimization Constraints

A first extension is to apply our technique to optimization constraints. For example, consider a conjunction of open *weighted* global cardinality constraints [8]. In that case, a weight is assigned to each pair (x, d), for all $x \in X$ and $d \in D(X)$. Then a solution to the problem induces a weight, defined by the sum of the weight of the pairs (x, d) if $x = d$ is in the solution. The aim is to find a solution with minimum total weight.

We can handle this case similar to the original filtering algorithm for weighted gccs [8]. With each arc (x^i, d), for all $x \in X^i$ $(i = 1, \ldots, k)$ and $d \in D(x)$ in $\mathcal{G}$, we associate a cost that is equal to the weight of this pair. Then a solution to the conjunction of open weighted gccs corresponds to a minimum-cost feasible flow in the graph. Hence, the cost-based version of our filtering algorithm is immediate.

7.2 Soft Constraints

Soft constraints can be viewed as special optimization constraints. A number of soft global constraints can be represented by a flow in a graph, similar to the gcc, see [6]. In this case, rather than associating a cost to an arc (x, d), for all $x \in X^i$ $(i = 1, \ldots, k)$ and $d \in D(x)$, costs may appear on "any" arc in the graph. We can again apply the same machinery as for the open weighted gccs to open soft global constraints.

7.3 Mixture

Finally, it is also possible to apply our results to a mixture of open constraints, provided that they are reasonably compatible. For example, we can group together open `alldifferent` constraints and open gccs in one conjunction. Another example is to join together open soft gccs and open weighted `alldifferent` constraints.

8 Conclusion

For the first time, we have proposed filtering algorithms for open global constraints. We have in particular studied open global cardinality constraints and conjunctions of them. We have proposed an efficient filtering algorithm, based on techniques from flow theory, establishing set-domain consistency, when the constraints are defined on disjoint sets of variables. In case the constraints are defined on non-disjoint sets of variables, this task becomes NP-complete. For that case we have proposed a weaker, but efficient, filtering algorithm. We have also presented a bounds consistency filtering algorithm for the set variables that underly these open constraints. Finally, we have shown how to extend our results to other conjunctions of open constraints, for example optimization constraints and soft global constraints.

References

1. R.K. Ahuja, T.L. Magnanti, and J.B. Orlin. *Network Flows*. Prentice Hall, 1993.
2. R. Barták. Dynamic Global Constraints in Backtracking Based Environments. *Annals of Operations Research*, 118(1–4):101–119, 2003.
3. K. Elbassioni, I. Katriel, M. Kutz, and M. Mahajan. Simultaneous matchings. In X. Deng and D. Du, editors, *Proceedings of the 16th Annual International Symposium on Algorithms and Computation (ISAAC 2005)*, volume 3827 of *LNCS*, pages 106–115. Springer, 2005.
4. B. Faltings and S. Macho-Gonzalez. Open Constraint Satisfaction. In P. Van Hentenryck, editor, *Proceedings of the 8th International Conference on Principles and Practice of Constraint Programming (CP 2002)*, volume 2470 of *LNCS*, pages 356–370. Springer, 2002.
5. B. Faltings and S. Macho-Gonzalez. Open Constraint Programming. *Artificial Intelligence*, 161(1–2):181–208, 2005.

6. W.-J. van Hoeve, G. Pesant, and L.-M. Rousseau. On Global Warming: Flow-Based Soft Global Constraints. *Journal of Heuristics*, 2006. To appear.

7. J.-C. Régin. Generalized Arc Consistency for Global Cardinality Constraint. In *Proceedings of AAAI/IAAI*, volume 1, pages 209–215. AAAI Press/The MIT Press, 1996.

8. J.-C. Régin. Cost-Based Arc Consistency for Global Cardinality Constraints. *Constraints*, 7:387–405, 2002.

9. R. Tarjan. Depth-first search and linear graph algorithms. *SIAM Journal on Computing*, 1:146–160, 1972.

Conditional Lexicographic Orders in Constraint Satisfaction Problems*

Richard J. Wallace and Nic Wilson

Cork Constraint Computation Center and Department of Computer Science
University College Cork, Cork, Ireland
{r.wallace, n.wilson}@4c.ucc.ie

Abstract. The lexicographically-ordered CSP ("lexicographic CSP" for short) combines a simple representation of preferences with the feasibility constraints of ordinary CSPs. Preferences are defined by a total ordering across all assignments, such that a change in assignment to variable k is more important than any change in assignment to any variable that comes after it in the ordering. In this paper, we show how this representation can be extended to handle conditional preferences. This can be done in two ways. In the first, for each conditional preference relation, the parents have higher priority than the children in the original lexicographic ordering. In the second, the relation between parents and children need not correspond to the basic ordering of variables. For problems of the first type, any of the algorithms originally devised for ordinary lexicographic CSPs can also be used when some of the domain orderings are dependent on the assignments to "parent" variables. For problems of the second type, algorithms based on lexical orders can be used if the representation is augmented by variables and constraints that link preference orders to assignments. In addition, the branch-and-bound algorithm originally devised for ordinary lexicographic CSPs can be extended to handle CSPs with conditional domain orderings.

1 Introduction

An important contribution of artificial intelligence to the study of preferences has been the development of methods for representing and handling conditional preferences. This work assumes that preference orderings are often context-dependent. Once one considers preferences in this way, many examples spring to mind. To take one such: what I prefer to eat may depend on the country I am in, especially if I am inclined to 'go native'. So in Spain I may prefer paella and tortillas, while in Germany I may prefer bratwurst and sauerkraut.

The most widely discussed representation of conditional preferences is the "CP-net", which is characterized by *ceteris paribus* conditions on preferences between the different values of a feature [1] [2]. In the present work we describe an alternative representation based on lexicographic orders. On the one hand, this representation is rather rigid in that it requires a strict priority ordering on the variables (which is not essential for CP-nets), and requires total orders on variable domains. On the other hand, in

---

* This work received support from Science Foundation Ireland under Grant 00/PI.1/C075.

J.C. Beck and B.M. Smith (Eds.): CPAIOR 2006, LNCS 3990, pp. 258–272, 2006.
© Springer-Verlag Berlin Heidelberg 2006

many cases a user may be willing to supply such inputs, and it leads to a much more decisive system, and one which is computationally simpler. Moreover, as we show here, conditional lexicographic orders can support conditionalities that oppose the priority ordering. This is in contrast to CP-nets or to to TCP-nets, an extension of CP-nets which allows some priority ordering on the variables. These alternative representations, therefore, have different strengths and limitations, and each may have applications where it is more useful.

In this work, conditional preferences are studied in the context of constraint satisfaction problems (CSPs). This means that outcomes are "framed" in relation to domains of values associated with distinct variables (cf. [3]). As with CP-nets, this allows us to specify conditions of preferential independence and conditional preferential independence between values in different domains.

In earlier work, we investigated lexicographic orderings incorporated into a standard CSP representation, which we termed the *lexicographically ordered CSP* [4]. This is a special kind of soft constraint system in which a lexicographic ordering is imposed on complete assignments, based on orderings of variables and domain values. In this case, variable selection is the primary factor and value assignment is secondary. This means that a good assignment for a more-preferred variable is more important than a good assignment for a less-preferred variable in deciding the overall ranking of solutions. The preference ordering is assumed to be independent of any constraints that hold among these variables. The latter, therefore, restrict the alternatives given by an ideal preference ordering to those that can be realized.

Lexicographic CSPs represent problems in which preferences involve multiple objectives and attributes and where feasibility constraints impose restrictions on assignments that are actually possible. From the point of view of representation as well as computation they offer significant benefits. This is partly because of the radical decoupling of the preference structure from the feasibility conditions [4], which allows users to concentrate on their preferences without regard to feasibility constraints, which they may not know or understand. A similar argument has been made by [5] in connection with CP-nets.

When this form of lexicographic ordering is extended to *conditional lexicographic orders*, the same type of ordering holds as in ordinary lexicographic CSPs, but domain orderings are conditional on assignments chosen from other domain. We consider two important classes of conditional lexicographic CSPs. In the first, conditionalities respect the priority ordering of the variables; in the second, they do not. Despite complications engendered in the latter case, the basic algorithmic strategies devised for ordinary (unconditional) lexicographic CSPs can be extended to handle these problems.

As a motivating (and clarifying) example, consider a situation in which a customer is deciding among vacations. There are two seasons when he can travel: spring and summer. For simplicity, we consider only two locations: Naples and Helsinki. In the first scenario (first type of conditional lexicographic ordering), location is more important than time of travel and the preferred season depends on the location chosen. This is shown in Figure 1a, where following [1], the conditional preference is represented as a conditional preference table. The preference statement Naples: spring $\succ$ summer, for example, means that if Naples is chosen, then spring is preferred to summer. The associated preference ordering is:

$$\langle \text{ Naples, spring } \rangle \succ \langle \text{ Naples, summer } \rangle \succ \langle \text{ Helsinki, summer } \rangle \succ \langle \text{ Helsinki, spring } \rangle$$

In the second scenario, location is again the primary feature, but the preference for location depends on the city chosen. Thus, our customer prefers Naples in the spring and Helsinki in the summer, and a vacation in spring is preferred over summer. The customer's input preferences do not tell us whether the location is better in scenario $\langle$ Naples, spring $\rangle$ or in scenario $\langle$ Helsinki, summer $\rangle$. We choose in this example to interpret this as the scenarios being equally good as far as location is concerned, while the former scores better on the season criterion; more generally, the customer could be asked to give these comparisons explicitly. In this case, the preference ordering is:

$$\langle \text{ Naples, spring } \rangle \succ \langle \text{ Helsinki, summer } \rangle \succ \langle \text{ Helsinki, spring } \rangle \succ \langle \text{ Naples, summer } \rangle$$

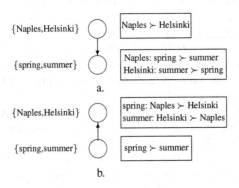

Fig. 1. Examples of conditional lexicographic preference orderings. As with CP-nets, a directed graph represents conditional preference relations among variables. Conditional preference tables are on the right. In both cases, the more important variable is above the less important one. In (a) the conditionality is consistent with relative importance; in (b) the two are opposed.

To summarise, lexicographically ordered CSPs have some appealing features: the simplicity of the inputs, the decisiveness of the ordering, and the fact that they support efficient solving methods. The decoupling of preferences and feasibility constraints is also quite intuitive. The present paper shows that this formulation can be extended to conditional preferences, which are natural in many situations. This combines the expressiveness of CP-nets (and related formalisms) with the simplicity and relative efficiency of lexicographically ordered CSPs.

In the remainder of the paper, Section 2 defines lexicographic CSPs and CSPs with conditional lexicographic orders, and discusses relations with general soft constraint representations. Section 3 discusses relations with CP-nets and TCP-nets. Section 4 considers algorithms for simple conditional lexicographic CSPs, Sections 5 and 6 for extended conditional lexicographic CSPs. Section 7 gives some experimental results. Section 8 gives conclusions.

2   Background and Definitions

2.1   Lexicographic and Conditional Lexicographic CSPs

Definition 1. Lexicographic CSP. A finite CSP is defined in the usual way as a triple $\langle V, D, C \rangle$, where V is a set of variables, D is a set of domains each of which is associated with a member of V, and C is a set of constraints, or relations holding between subsets of variables.

To specify a CSP as lexicographic, we introduce the following definitions. A labelling of set V is a bijection between $\{1, \ldots, |V|\}$ and V. A *lexicographic structure L* over V is a pair $\langle \lambda, \{>_X : X \in V\} \rangle$, where the second component is a family of total orders, with $>_X$ being a total order on the domain of X, and λ is a labelling of V. We write $\lambda(i)$ as X_i, so that the ordering of the variables is $X_1, \ldots, X_n$. The associated *lexicographic order* $>_L$ on (complete) assignments is defined as follows: $\alpha >_L \beta$ if and only if $\alpha \neq \beta$ and $\alpha(X_i) >_{X_i} \beta(X_i)$, where X_i is the first variable (i.e., with minimum i) such that α and β differ.

A *lexicographic CSP* is a tuple $\langle V, D, C, \lambda, \{>_X : X \in V\} \rangle$, where $\langle V, D, C \rangle$ is a finite CSP and $\langle \lambda, \{>_X : X \in V\} \rangle$ is a lexicographic structure over V.

A solution to a lexicographic CSP is the unique assignment α such that

(i) α is a satisfying assignment, that is, it is consistent with, or satisfies, all constraints in C, and

(ii) $\alpha >_L \beta$ for any other satisfying assignment β.

The lexicographically-ordered CSP is a special case of the "lexicographic CSP" or "lex-VCSP" as defined in [6]. As these authors show, lex-VCSPs are in turn equivalent to a kind of weighted CSP. However, because of the character of the ordering in our case, we do not need to represent preferences numerically, and we can build up partial solutions correctly without reference to numerical operations such as addition. So, while we follow [6] in using the term "lexicographic CSP", we are designating a very special case of the class they describe, with implications both for its usefulness as a representation in the context of preferences and its ability to support efficient algorithms. For this reason, we also retain the term "lex-VCSP", using it to refer to the more general category of CSPs whose evaluations can be ordered lexicographically.

We can embed a lexicographic ordering within a weighted CSP framework:

Lexicographic CSP as a weighted CSP. For each $i = 1, \ldots, n$ we define a unary weighted constraint W_i over variable X_i, given by $W_i(x) = k b^{n-i}$, where x is the kth best value in the domain of X_i and b is the largest domain size. Then for assignments α and β, the sum of weights associated to α is less than the sum associated to β if and only if $\alpha >_L \beta$.

Similar embeddings can be used for the conditional lexicographic and extended conditional lexicographic cases.

Definition 2. Conditional lexicographic CSP. A *conditional lexicographic network* over V is defined as a tuple $K = \langle \lambda, G, CPT \rangle$, where λ is a labelling of V, with $\lambda(i)$ being written X_i, and G is a directed acyclic graph on V which is compatible with λ, i.e., $(X_i, X_j) \in G$ implies $i < j$. If $(X_i, X_j) \in G$ then X_i is said to be a parent of X_j (and X_j is a child of X_i)). CPT is a function which associates a conditional preference table $CPT(X)$ to each $X \in V$. As in CP-nets [2], each conditional preference table $CPT(X_i)$ associates a total order $>_u^{X_i}$ with each instantiation u of the parents U_i of X_i (with respect to G). The associated *conditional lexicographic order* $\succ_K$ on assignments is defined as follows: $\alpha \succ_K \beta$ if and only if $\alpha \neq \beta$ and $\alpha(X_i) >_u^{X_i} \beta(X_i)$, where X_i is the first variable (i.e., with smallest i) such that $\alpha(X_i) \neq \beta(X_i)$, and $u = \alpha(U_i)$ (which also equals $\beta(U_i)$). It is easily seen that $\succ_K$ is a total order on assignments. In this definition, it is essential that the graphical structure G is compatible with the importance ordering of the variables λ, as this ensures that $\alpha(U_i) = \beta(U_i)$ in the above definition.

2.2   Extended Conditional Lexicographic Orders

Definition 3. Conditional lexicographic CSP with extended conditional preference orders. An *extended conditional preference network* involves, like a conditional preference network, a directed graph G and a labelling λ which represents an ordering $X_1, \ldots, X_n$ of the variables. It also involves, for each variable X_i, a function Q_i which assigns a number $Q_i(x, u)$ for every value x of X_i and assignment u to U_i, where U_i is defined to be the set of parents of X_i with respect to graph G. If α is a complete assignment, we write also $Q_i(\alpha)$ for $Q_i(\alpha(X_i), \alpha(U_i))$, where e.g., $\alpha(X_i)$ is the value that α assigns to X_i.

The associated ordering, *the extended conditional preference order* is then defined as follows: to compare complete assignments α and β we find the first X_i (i.e., with smallest i) such that $Q_i(\alpha)$ is not equal to $Q_i(\beta)$; if $Q_i(\alpha)$ is less than $Q_i(\beta)$, we prefer α to β; else we prefer β to α. If, on the other hand, there exists no such X_i—so that for all $i = 1, \ldots, n$, $Q_i(\alpha) = Q_i(\beta)$—then α and β are equivalent with respect to the ordering; neither is (strictly) preferred to the other. We write $\alpha \succ_E \beta$ if α is strictly preferred to β, and we write $\alpha \succeq_E \beta$ if α is either strictly preferred to β, or α and β are equivalent. So $\alpha \succ_E \beta$ holds if and only if $\beta \nsucceq_E \alpha$.

Another way to view the construction of this ordering is that we are converting each assignment $\alpha = (x_1, \ldots, x_n)$ to an n-tuple of numbers $\alpha^* = (Q_1(\alpha_1), \ldots, Q_n(\alpha_n))$, i.e., $(Q_1(x_1, u_1), \ldots, Q_n(x_n, u_n))$, where u_i is the assignment α makes to U_i. The extended conditional lexicographic order $\succeq_E$ is then essentially the standard lexicographic order on these n-tuples of numbers: α is strictly preferred to β if and only if α^* is lexicographically better than β^*, and they are equivalent in the order if and only if $\alpha^* = \beta^*$. This implies that $\succeq_E$ is a total pre-order, i.e., it is reflexive, transitive and complete. It is not necessarily a total order since we can have $\alpha \succeq_E \beta \succeq_E \alpha$ for $\alpha \neq \beta$, which happens when $\alpha^* = \beta^*$.

This definition leaves the question open of where the values of each function Q_i come from, and what they mean. One interpretation is as follows (another approach is based on generating the functions Q_i from conditional preference tables, as in the introductory example in Section 1). The idea is that we have n criteria, where the first is much more important than the second, and so on; these are in one-to-one correspondence with the variables X_i. If, for example, variable X_i gives the location of a holiday, the corresponding criterion says how good that location is. The function Q_i tells us how well the criterion is satisfied, i.e., how good the value of X_i is. This depends not just on X_i, but on other variables (the parents of X_i); for example, how good a location is depends on the season. Complete assignments are judged primarily on the first criterion i.e., on the value of Q_1, and then on the second etc.

The user is required to order, for each i, the values of Q_i; we represent this by assigning numbers consistent with this ordering. The overall order on complete assignments depends only on the ordering of the values of each Q_i. For example, suppose X_5 with domain $\{a_5, b_5\}$ has a single parent X_7, with domain $\{a_7, b_7\}$. The user has to judge how good X_5 is in the four different scenarios: a_5a_7, a_5b_7, b_5a_7 and b_5b_7. For example, she might judge (i) that X_5 is best if $X_5 = a_5$ and its parent, X_7, equals a_7; (ii) that the worst case is b_5b_7, and (iii) that X_5 is equally good in scenarios a_5b_7 and b_5a_7. We could express this by setting e.g., $Q_5(a_5, a_7) = 1$, $Q_5(a_5, b_7) = Q_5(b_5, a_7) = 2$, and $Q_5(b_5, b_7) = 3$. However we could equally well set $Q_5(a_5, a_7) = -1$, $Q_5(a_5, b_7) = Q_5(b_5, a_7) = 0$, and $Q_5(b_5, b_7) = 8$, since only the order of the values of Q_5 matters to the final ordering of complete assignments. Because of this, each function Q_i in the definition of an extended conditional preference network can be replaced by a pre-order $\geq_i$ on the set of assignments to $\{X_i\} \cup U_i$, where $xu \geq_i x'u'$ if and only if $Q_i(x, u) \geq Q_i(x', u')$; this equivalent representation emphasizes the ordinal nature of extended conditional lexicographic orders.

3   Conditional Lexicographic CSPs and CP-Nets

In recent years, the most widely-discussed method for representing conditional preferences within the AI community has been the conditional preference network with *ceteris paribus* assumptions, or CP-net [1]. A more recent variant, the TCP-net [7], includes elaborations to handle relations of importance between the features of user-selections (this relates to, but is weaker than, the ranking of variables in lexicographic CSPs [8]).

CP-net structures are based on assignments of values to variables, or "features". A conditional preference is encoded in a "conditional preference table" (CPT) associated with a particular variable X_i, examples of which are shown in Figure 1. TCP-nets also encode importance relations between variables in terms of an ordering with lexicographic features.

A critical feature of (T)CP-nets is that preferences are only defined under "*ceteris paribus*" conditions. If, for example, features A and B each have two values, a_1, a_2 and b_1, b_2, respectively, and $a_1 >_{X_A} a_2$ and $b_1 >_{X_B} b_2$, then we can deduce from *ceteris paribus* assumptions that $a_1b_1 >_N a_2b_1$, $a_2b_1 >_N a_2b_2$, etc, but we cannot order a_1b_2 and a_2b_1 on this basis. As a result of this feature, preference orders can be established on the basis of "flipping sequences"; e.g., the sequence of two "flips" $a_1b_1 \succ a_2b_1 \succ a_2b_2$ enables us to deduce the preference $a_1b_1 \succ a_2b_2$. This is still true of TCP-nets, although in this case adjacent outcomes in a sequence can be separated by a "double flip" of two variables.

It was shown in [9] that, except in some trivial cases, the order on assignments generated by a CP-net, or by a TCP-net, is never a lexicographic order. The reason for this is that flipping sequences require that consecutive elements in the ordering differ by at most one (CP-nets) or two (TCP-nets) elements. However, consecutive elements in a lexicographic ordering can differ by up to $|V|$ elements.

Perhaps the most important implication of these differences is that, while determining whether solution α is preferred to solution β is easy for lexicographic orderings, since it is based on successive comparisons of values of a variable, this is more problematic with (T)CP-nets, since it depends on finding flipping sequences that transform one alternative into another. Although there are special cases where this problem is polynomial [10] [2], it appears, in general, to be a very hard problem [11]. On the other hand, CP-nets allow a weaker form of comparison, which indicates for two solutions α and β that the preference of the latter over the former is *not* entailed by the CP-net structure. This comparison can be carried out in low order polynomial time [2].

In considering the algorithmics of CP-nets, the emphasis has been on cases where the set of dependencies forms an acyclic graph. In particular, algorithms developed for solving constrained optimisation problems are based on this assumption [5] [7]. In many cases, this seems natural, but as we have seen, when importance relations are mixed with conditional preference relations it is sometimes reasonable to consider cases where the two do not always correspond. As we will see, for conditional lexicographic orderings this does not appear to be as important as for CP-nets, since the basic algorithms can be extended, in some cases without marked effects on performance.

For acyclic networks in which conditional preferences correspond to the priority ordering, a CP-net ordering can be extended to a conditional lexicographic ordering [12]: we choose a conditional lexicographic network with the same conditional preference tables and the same directed acyclic graph G as the CP-net. Let $\succ_K$ be the associated conditional preference order. It is obvious that if there is a worsening flip from α to β then $\alpha \succ_K \beta$. Transitivity of $\succ_K$ then implies that if α is better than β according to the CP-net, then it is better according to the conditional lexicographic order ($\alpha \succ_K \beta$).[1] Similarly, the preference ordering corresponding to an acyclic conditional preference theory [9] can be extended to a conditional lexicographic ordering. This means that if one wants to find a solution of a set of constraints C which is optimal with respect to a CP-net, one can generate an associated conditional preference order, as described above, and find the optimal solution with respect to this. The above result implies that this solution will also be optimal with respect to the CP-net. Hence, *constrained optimisation algorithms for conditional lexicographic orders can be used for finding a single optimal solution of a constrained optimisation problem for an acyclic CP-net.*

4   Constrained Optimisation Algorithms for Conditional Lexicographic CSPs

4.1   Methods for Solving Ordinary Lexicographic CSPs

In earlier work, we showed that this form of lexicographic representation of preferences for CSPs offers wide scope for developing optimization algorithms (see [4] [12] for detailed descriptions). The most successful algorithms were,

---

[1] Similar techniques are used in [2], for ordering queries, and proving consistency of an acyclic CP-net.

1. *Lexical search*: ordinary CSP algorithms that follow the importance ordering of the variables. These work very well when problems are not too strongly constrained, but are inefficient for problem in the critical complexity region.
2. A *branch and bound* algorithm that works well for problems in the critical complexity region.
3. A specialised restart algorithm termed the *"staged lexical"* algorithm in which on the kth restart, the kth variable is instantiated in lexical order while better ordering heuristics are used for the remaining variables. This works almost as well as branch and bound for problems in the phase transition region and tends to be more effective than the latter as the number of solutions increases.

For the branch and bound procedure, the cost function (based on the representation in terms of weighted constraints) gives large values for any but very small problems; however we do not need to calculate it directly. Instead, we simply compare successive values following the lexical variable ordering until we encounter a difference. Suppose that variable X_k is the variable currently being considered for instantiation, i.e., the kth most important variable in the ordering. To evaluate the current partial solution, we start from the first variable X_1 in the lexical ordering. If X_1 has been assigned, we check this against its instantiation in the best assignment found so far; if it does not yet have an assignment, we check the best remaining value in its domain against the best assignment. If this favors the best assignment found so far, then search can back up. Otherwise, if they are equal on X_1 we perform a similar check on X_2 (and so on).

In staged lexical search, search is done repeatedly, in each case until the first solution is found, and for each repetition, or stage, one more variable is chosen in lexical order, beginning with X_1 at stage 1. Values are always chosen according to the lexical ordering. On the kth repetition, when we have found a feasible solution, we know that the assignment for X_k is optimal, so we retain it for the remainder of search. Although developed independently, this algorithm is, in fact, a special case of preference-based search [13], where the criteria on which search is based form a total order.

4.2   Algorithms for Simple Conditional Lexicographic CSPs

When the parent-child order is compatible with the importance order of the variables, any of our methods for constrained optimization can be used to return a solution that is optimal for the conditional lexicographic CSP. In particular, the staged lexical algorithm can be applied in exactly the same way as before to the conditional lexicographic case, since at stage i we know the ordering of the values of X_i, as its parents have been instantiated already. For branch and bound, if a child is chosen for instantiation before its parents, bounding can be done provided parent values are available for comparison: in this case, any differences naturally override differences for the child, while if these values are identical then the orderings on the child variable will be the same for the candidate and current best values.

5   Lexical Search Algorithms for Extended Conditional Lexicographic Orders

By adding extra variables, an extended conditional lexicographic order can be related to an ordinary lexicographic order; this enables one to generalise the algorithms for lexicographic orders to the conditional case.

Auxiliary Variables Representation. Let R_i be the set of all values taken by Q_i, i.e., the set of numbers $Q_i(x, u)$ over all values x of X_i and assignments u to the parents of X. For each variable X_i, create a new variable Y_i with domain R_i. Variable Y_i can be considered as telling us how well the ith criterion is satisfied. We create a constraint with scope $U_i \cup \{X_i, Y_i\}$ consisting of all tuples uxq with $Q_i(x, u) = q$, where x is a value of X_i, u is an assignment to U_i and $q \in R_i$. Let B be the set of these extra constraints. Let $V' = V \cup \{Y_1, \ldots, Y_n\}$. Each assignment α to V clearly extends uniquely to an assignment α' to V' satisfying these extra constraints B: we define, for each $i = 1, \ldots, n$, $\alpha'(X_i) = \alpha(X_i)$ and $\alpha'(Y_i) = Q_i(\alpha)$. Hence α' is essentially α extended with α^*.

There is a natural lexical order on assignments to V' defined by variables $Y_1, \ldots, Y_n$ in that order of importance. $\alpha \succ_E \beta$ if and only if α' is lexically better than β' according to this lexical order (which is if and only if α^* is lexically better than β^*). In particular, α is an optimal solution of C with respect to the extended conditional lexicographic network if and only if α' is a lexically optimal solution of constraints $C \cup B$. Therefore, the strategies mentioned earlier for finding optimal solutions with respect to a lexical order can be elaborated to produce algorithms for finding optimal solutions for extended conditional lexicographic orders.

5.1   Extending Lexical Search

We can adapt lexical search for extended conditional lexicographic optimisation by using auxiliary variables to represent Q-values and adding constraints to represent acceptable parent-child assignments given a particular value for the child. The key idea is that the Q-values can serve as the basis for a lexically-ordered search. In other words, search can be done in a way that is lexicographic on the Q-values rather than on the decision variables themselves, even though the domains of the latter have no *a priori* preference order.

Recall that R_i is the set of all values taken by Q_i. For any $q \in R_i$, we define constraint c_i^q on variables $U_i \cup \{X_i\}$ to be the constraint $Q_i = q$, i.e., xu is a tuple in c_i^q if and only if $Q_i(x, u) = q$.

The search tree for the lexical search can be defined as follows: A node N at level j, for $j = 0, \ldots, n$ has an associated set of constraints C^N of the form $C \cup \{c_1^{q_1}, \ldots, c_j^{q_j}\}$, where for each $i \leq j$, q_i is an element of R_i. The initial node, which is the root node of the search tree, is at level 0. A node at level n is said to be a *complete node*; the other nodes are said to be *partial nodes*. At each partial node we will need to maintain some form of partial consistency. If we deduce that the associated set of constraints C^N is inconsistent then we can backtrack at this point. Otherwise, we branch on constraints c_{j+1}^q, for $q \in R_{j+1}$; that is, for each $q \in R_{j+1}$ we generate a child node of N with associated set of constraints $C^N \cup \{c^q_{j+1}\}$. The child corresponding to the smallest element q of R_{j+1} is explored first, in a depth-first manner. When we reach a complete node N, we determine if C^N has a solution; if it does, we return the solution and stop; otherwise we backtrack. This algorithm will return an optimal solution, given that the initial set of constraints C has a solution.

If the Q_i's are one-to-one functions, each constraint of the form $c^{q_i}_i$ just contains a single tuple which is an instantiation of X_i and the parents of X_i. The algorithm then behaves in a fairly similar way to standard lexical search.

On the other hand, if the Q_i's have many 'ties' so that added constraints of the form c^q_i include several tuples, then we may find that, e.g., maintaining arc consistency is not sufficiently strong to prune the search effectively, since we are not directly instantiating the variables X_i. For example, it could happen that an initial constraint $c^{q_1}_1$ is inconsistent with constraints C, but we might only discover this at complete nodes, when we have generated values of all the other q_i, and we test the consistency of the associated set of constraints $C \cup \{c^{q_1}_1, \ldots, c^{q_n}_n\}$. This will tend to make the algorithm extremely slow in such situations. For this reason, it is natural to consider stronger forms of consistency checking at each node. In particular, we can use a search to check global consistency at a node, which is essentially what the staged lexical algorithm does.

5.2   Extending Staged Lexical Algorithm

Unlike the lexical search algorithm, the staged lexical algorithm can be done without backtracking over Y_i variables. These are instantiated in order over successive stages, in each stage before any X_i variables are instantiated. The latter can be instantiated using any heuristic ordering.

Let C_0 be the original set of constraints C, which we assume to be satisfiable. Let α be an optimal solution, and let α^* be the corresponding n-tuple of numbers, as defined above. The fact that C_0 is satisfiable implies that there exists some $q \in R_1$ such that $C_0 \cup \{c^q_1\}$ is satisfiable, since R_1 includes all possible values of Q_1. We find minimal value $q_1 \in R_1$ such that $C_0 \cup \{c^{q_1}_1\}$ is satisfiable. (By definition of α^*, we have $q_1 = \alpha^*(X_1)$, since q_1 is the best feasible value of Q_1.) This can be found by starting with the lowest (i.e., best) value in R_1 and continuing until we find q_1 with $C_0 \cup \{c^{q_1}_1\}$ satisfiable.

We then add this constraint $c^{q_1}_1$ to the set of constraints, setting $C_1 = C_0 \cup \{c^{q_1}_1\}$ (and we will not backtrack over this decision). (With the auxillary variable representation this amounts to setting $Y_1 = q_1$.) We move on to optimising Q_2: we find minimal value $q_2 \in R_2$ such that $C_1 \cup \{c^{q_2}_2\}$ is satisfiable. We set $C_2 = C_1 \cup \{c^{q_2}_2\}$. We continue this until we have generated minimal $q_n \in R_n$ such that $C_{n-1} \cup \{c^{q_n}_n\}$ is satisfiable; we let $C_n = C_{n-1} \cup \{c^{q_n}_n\}$, which, by construction, is satisfiable.

It is easy to see that any optimal solution α is a solution of C_n (or else α would be worse than a solution of C_n). Also, if β is any other solution of C_n then α and β have exactly the same Q_i-values, so β is also optimal. This leads to the following result which shows that a solution of the set of constraints C_n (in particular, the one found when checking that C_n is satisfiable) is an optimal solution of the constraints C.

Proposition 1. *With the above notation, complete assignment α is an optimal solution of C if and only if α is a solution of the set of constraints C_n.*

```
conditional-bnb (partial-solution, remaining-variables)
        if remaining-variables ≡ nil
                save new best-solution
                and continue            //backtrack
        else
                select next-variable and remove from remaining-variables
                for each value in its ordered domain
                        if new instantiation gives an arc consistent problem
                                and
                                bounds-check(next-variable, next-value) returns true //under bound
                                        conditional-bnb (new-partial-solution, remaining-variables)
                        continue                //backtrack

bounds-check (candidate-var, candidate-value)
        while variables remain to be compared
                select next-variable in order
                get value next-best for this variable from current best-solution
                if next-variable == candidate-var
                        curr-assign = candidate-value
                else if next-variable is instantiated
                        curr-assign = current assignment of next-variable
                                                //perform comparisons
                if next-variable ∉ any child-set
                        compare curr-assign or best value in default-current-domain with next-best
                else if next-variable has current-preference order
                        compare curr-assign or best value in default-current-domain with next-best
                else if candidate-var is a remaining uninstantiated parent
                        get domain-order associated with parent values
                        compare curr-assign or best value according to domain-order with next-best
                else
                        set comparison to succeeded and bound to not-exceeded
                if comparison has succeeded break
        if bound was exceeded
                return false
        else
                return true
```

Fig. 2. Pseudocode for branch and bound for CSP with extended lexicographic ordering

6   Branch and Bound Algorithms for Extended Conditional Lexicographic Orders

Like the lexically-based search algorithms, the present branch-and-bound procedure (Figure 2) relies on the fact that for lexicographic orderings, value orderings can be indexed by the Q-values. This allows it to check bounds in terms of Q-values, thereby comparing a candidate assignment with previous assignments even when the preference ordering for the past assignment is different from the present ordering.

Regardless of the search order, bounds testing always proceeds according to the priority ordering of the variables (until the current variable is reached), and the decision to bound depends on the first difference found between the current partial assignment and the best solution α. Note that for the latter all $Q_i(x, u)$ are known. If testing is restricted to the following conditions:

- variable k has no parents, in which case either its current assignment or the best assignment available can be used for comparing with the current best assignment for this variable (In this case, the $Q_k(x, u)$ are independent of other assignments),
- variable k has a current ordering (because its parents have assignments), in which case the $Q_k(x, u)$ are known,

- variable k has one uninstantiated parent, which is the current variable, in which case the $Q_k(x, u)$ can be determined for a candidate value of variable k,

and bounding is not done otherwise, then if the first difference found favors the current best solution, since this is also the first difference in the lexicographic order, no extension of the current assignment can produce a solution β such that $\beta \succ_E \alpha$. This guarantees the correctness of the present algorithm.

The present branch and bound algorithm has one difference from that used for CSPs with simple conditional lexicographic orderings. In the latter case, comparisons can also be made during bounds testing if the preference order of the domain of the remaining uninstantiated parents of the variable whose assignments are being compared can be ordered. In this case, assignments for the parents will already have been tested, and so they must have been equal since a bounding decision could not be made. In this case, for the variable currently under consideration the preference ordering over the domain is the same for the best solution and the current assignment. For extended conditional lexicographic orders, where the parent is not necessarily more important than the child, this situation does not always hold.

7 Experimental Tests

We present some results of experimental tests with random binary CSPs to show comparative performance of the different algorithms described above. Since in previous work [4] [12] a MAC-based algorithm proved to be much more effective than forward checking, the former is used in all tests reported here. Algorithms were coded in lisp and run using Xlisp on a Dell Work Station PWS 330 running at 1800 MHz. For each condition, solutions were compared for the different algorithms to verify that the implementations were correct.

7.1 Problem Generator

For these tests, CSPs with conditional lexicographic orders were generated with a program written by the first author. This program starts with an existing CSP and transforms it into a conditional lexicographic CSP by selecting variables for conditional preferences and building a CPT for each relation. Q-values are derived from successive positions of successive domain elements within an ordering, so they do not have to be generated explicitly. The user specifies the following parameters: (i) number of preference relations, (ii) maximum number of parents per relation, (iii) maximum number of children per relation, (iv) maximum number of attempts to make a relation with p parents and c children, since at some point in generation it may not be possible to do this under the given restrictions. (If this number is ever exceeded, the program writes a message to standard-output, but continues with problem generation.) In addition, the following restrictions are made during generation:

1. A child-variable only appears as such in one preference relation (otherwise the CPT is ill-defined).
2. The graph of conditional relations is directed-acyclic, so there is no *directed* path from a node back to itself.

3. A variable occurs in no more than one single-parent relation. This is not a required restriction, but it prevents selection from undermining the maximum-child specification since k singleton-parent relations involving the same parent variable are indistinguishable from a single relation with one variable and k children.

There are two further restrictions that the user can specify optionally:

1. That parent-child relations always correspond to the priority ordering of the variables. (This specifies that the conditional lexicographic CSP is of the simpler type.)
2. That the parents and children in a relation do not have parents in common. (This option was not used in the experiments reported here.)

For problems used in the present tests, there was a maximum of two parents and two children per relation. The number of relations per problem was set to be 3, 7 or 11 in different experiments. With 7 relations, 70-80% of the variables were included in at least one conditional relation (i.e. were either parents or children in at least one relation); with 11 it was approximately 100%.

7.2   Experimental Results

Performance comparisons are given in Table 1. (Note that the number of values per domain is large in comparison with problems typically considered in this context and that there are numerous hard constraints. In addition, since the same Q-values were used for all domains, these constitute particularly difficult problems for algorithms like the staged lexical method where constraint size is related to number of different Q-values [cf. Sect. 5.1].) There are some striking differences due to problem difficulty and to the character of the conditional lexicographic ordering, and no single algorithm is superior overall.

Results for simple conditional lexicographic orderings were quite similar to those found for similar problems for the unconditional case. In addition, increasing the number of conditional relations had very little effect on search efficiency. For hard problems of this type, branch and bound and staged lexical were comparable, and both were better than straight lexical as domain size increased. For easy problems, either of the lexically-based search strategies outperformed branch and bound.

For the extended lexicographic orderings, performance differences depended on the number of constraints and number of conditional preference relations, as well as the heuristic used to order decision variables. For hard problems, the difference was decisive with branch and bound outperforming the staged lexical search by about an order of magnitude. For easy problems, staged lexical search was sometimes more efficient than branch and bound. This depended on the search order heuristic as well as the number of conditional preference relations (Table 1). (Branch and bound does worse on these problems if a lexical variable ordering is used, in contrast to the results for staged lexical search shown in the table.)

These preliminary results suggest that it is important to have some form of lexically-based search for the extended conditional lexicographic case - because branch and bound is adversely affected when the number of feasible solutions becomes large, while the effects on staged lexical search are not always as severe. With our suite of

Table 1. Search Efficiency Comparisons

domain/tight	hard problems						easy problems					
	10/.45			20/.55			10/.50			20/.60		
	simple conditional lexicographic											
	lex	stg	bb	lex	stg	bb	lex	stg	bb	lex	stg	bb
P3 nodes	348	390	217	9083	2334	2020	36	237	394	276	363	1346
ccks(000)	161	128	115	6505	2260	2124	12	38	112	118	148	949
P7 nodes	349	390	219	9063	2329	2020	36	238	386	269	365	1297
ccks(000)	161	128	116	6492	2255	2123	12	39	111	115	149	927
P11 nodes	348	391	218	9098	2330	2022	36	237	416	269	362	1408
ccks(000)	161	128	116	6514	2256	2125	12	38	116	117	148	985
	extended conditional lexicographic											
	lex	stg/dm	bb	lex	stg/dm	bb	stg/lex	stg/dm	bb	stg/lex	stg/dm	bb
P3 nodes	-	3547	439	-	27743	3496	2659	8559	20215	12085	73652	47428
ccks(000)	-	1645	203	-	30294	3660	795	2330	985	6833	47508	17219
P7 nodes	-	4397	587	-	43790	5078	8645	8777	29880	98031	106701	87209
ccks(000)	-	2043	267	-	46330	5241	1829	2508	1544	36832	66849	29578
P11 nodes	-	6038	957	-	38344	6801	17835	11726	70261	213311	97095	123614
ccks(000)	-	2927	416	-	40415	7050	3799	3644	3217	59348	62312	44385

Notes. 20-variable problems with density = 0.5. Data are means for 100 problems. "hard problems" are near the critical complexity peak. "easy problems" are near the edge of the hard region. Branch and bound employed min domain variable ordering; this was also used for staged lexical search for simple conditional lexicographic orderings. (For the extended case, decision variables, X_i, were either ordered in this way or ordered lexically, as indicated.) "Pk" is number of conditional preference relations per problem.

algorithms, we are, therefore, able to accomodate the reduced restrictions on conditional relations under a wider range of conditions so as to solve these problems efficiently.

8   Conclusions

Lexicographic orders allow a very simple and basic representation of preferences in combinatorial problems. The assumptions are strong, but the user inputs are of an easily-understandable form, and there are powerful algorithmic approaches for constrained optimisation.

In many situations, preferences are naturally conditional, i.e., context-dependent, and there has been a good deal of recent work in the AI literature on qualitative frameworks for conditional preferences, especially, CP-nets and their extensions. In this paper we define conditional lexicographic orders, which gives a simple approach for reasoning with conditional preferences, which has computational advantages over more sophisticated methods. We also show how algorithms for constrained optimisation for ordinary lexicographic CSPs can be extended to handle conditional orderings. Somewhat unexpectedly, this can be done even when conditional preference relations do not correspond to the relative importance of the variables. Therefore, efficiency of search in combinatorial optimisation can be maintained despite the additional complexity of this form of representation, in some cases to a surprising degree. These algorithms can also be used for finding a single optimal solution of a constrained optimisation problem for an acyclic CP-net.

References

1. Boutilier, C., Brafman, R.I., Hoos, H.H., Poole, D.: Reasoning with conditional *ceteris paribus* preference statements. In: Proc. Fifteenth Annual Conf. on Uncertainty in Artif. Intell. (1999) 71–80
2. Boutilier, C., Brafman, R.I., Domshlak, C., Hoos, H.H., Poole, D.: CP-nets: A tool for representing and reasoning with conditional *ceteris paribus* preference statements. Journal of Artificial Intelligence Research (2004) 135–191
3. Wellman, M.P., Doyle, J.: Preferential semantics for goals. In: Proc. Nineth Nat. Conf. on Artif. Intell. (1991) 698–703
4. Freuder, E.C., Wallace, R.J., Heffernan, R.: Ordinal constraint satisfaction. In: Fifth Internat. Workshop on Soft Constraints - SOFT'02. (2003)
5. Boutilier, C., Brafman, R.I., Domshlak, C., Hoos, H., Poole, D.: Preference-based constrained optimization with CP-nets. Computational Intelligence (2004) 137–157
6. Schiex, T., Fargier, H., Verfaillie, G.: Valued constraint satisfaction problems: Hard and easy problems. In: Proc. Fourteenth Internat. Joint Conf. on Artif. Intell. (1995) 631–637
7. Brafman, R.I., Domshlak, C.: Introducing variable importance tradeoffs into CP-nets. In: Proc. Eighteenth Annual Conf. on Uncertainty in Artif. Intell. (2002)
8. Wilson, N.: Consistency and constrained optimisation for conditional preferences. In: Proc. Sixteenth Europ. Conf. on Artific. Intell. (2004) 888–892
9. Wilson, N.: Extending CP-nets with stronger conditional preference statements. In: Proc. Nineteenth Nat. Conf. on Artif. Intell. (2004) 735–741
10. Domshlak, C., Brafman, R.I.: CP-nets - reasoning and consistency testing. In: Proc. Eighth Conf. on Principles of Knowledge Representation and Reasoning. (2002) 121–132
11. Goldsmith, J., Lang, J., Truszczynski, M., Wilson, N.: The computational complexity of dominance and consistency in CP-nets. In: Proc. Nineteenth Int. Jt. Conf. on Artif. Intell. (IJCAI-05). (2005) 144–149
12. Freuder, E.C., Heffernan, R., Prestwich, S., Wallace, R.J., Wilson, N.: Lexicographically-ordered constraint satisfaction problems. unpublished (2005)
13. Junker, U.: Preference-based search and multi-criteria optimization. In: Proc. Eighteenth Nat. Conf. on Artif. Intell. (2002) 34–40

An Efficient Hybrid Strategy for Temporal Planning

Zhao Xing, Yixin Chen, and Weixiong Zhang

Department of Computer Science and Engineering
Washington University in Saint Louis, Saint Louis, MO, USA
{zx2, chen, zhang}@cse.wustl.edu

Abstract. Temporal planning (TP) is notoriously difficult because it requires to solve a propositional STRIPS planning problem with temporal constraints. In this paper, we propose an efficient strategy for solving TP, which combines, in an innovative way, several well established and studied techniques in AI, OR and constraint programming. Our approach integrates graph planning (a well studied planning paradigm), max-SAT (a constraint optimization technique), and the Program Evaluation and Review Technique (PERT), a well established technique in OR. Our method first separates the logical and temporal constraints of a TP problem and solves it in two phases. In the first phase, we apply our new STRIPS planner to generate a parallel STRIPS plan with a minimum number of parallel steps. Our new STRIPS planner is based on a new max-SAT formulation, which leads to an effective incremental learning scheme and a goal-oriented variable selection heuristic. The new STRIPS planner can generate optimal parallel plans more efficiently than the well-known SATPLAN approach. In the second phase, we apply PERT to schedule the activities in a parallel plan to create a shortest temporal plan given the STRIPS plan. When applied to the first optimal parallel STRIPS plan, this simple strategy produces optimal temporal plans on most benchmarks we have tested. This strategy can also be applied to optimal STRIPS plans of different parallel steps in an anytime fashion to find an optimal temporal plan. Our experimental results show that the new strategy is effective and the resulting algorithm outperforms many existing temporal planners.

1 Introduction

In this paper, we are concerned with propositional STRIPS planning and temporal planning (TP) problems, two major classes of planning problems. STRIPS planning [9] refers to those problems whose actions are instantaneous and without durations, transition constraints are discrete by nature, and solution quality is usually measured by the number of parallel actions or time steps. TP refers to problems in which actions have continuous durations and can overlap with one another in time. The most popular quality metric for temporal plans is the makespan (duration of execution) of a plan.

Since STRIPS planning problems are discrete by nature, they are often solved by traditional AI techniques such as heuristic search [13]. On the other hand, because actions in TP have continuous durations, the discrete state-space representation used by AI techniques is inadequate, and TP is more difficult than STRIPS planning. As a result, many existing TP methods rely on techniques from operations research (OR), such as linear programming [28] and constraint programming [27].

J.C. Beck and B.M. Smith (Eds.): CPAIOR 2006, LNCS 3990, pp. 273–287, 2006.

In this research, we develop an efficient approach for TP that combines many well established AI and OR techniques. Our contribution is twofold. First, we propose an efficient STRIPS planning algorithm that is based on the planning-as-satisfiability (SATPlan) paradigm [25, 17] but improves SATPlan by incorporating an objective function to accommodate an effective goal-oriented variable selection heuristic. Second, we apply the Program Evaluation and Review Technique (PERT) [12], a widely applied technique in OR, to extend STRIPS planning to TP.

STRIPS Planning by Maximum Satisfiability. Over the past decade, SATPlan [25, 17] has emerged as one of the most effective formulations of STRIPS planning in AI. SATPlan was proposed to take advantage of much celebrated progress made over the years in the research of satisfiability (SAT), an extensively studied problem in AI. The SATPlan method first transforms a STRIPS planning problem into a SAT problem, and then solves it using a generic SAT solver. By transforming a planning problem into a SAT, many powerful methods developed for SAT can be utilized for STRIPS planning.

However, despite its success, the potential of the SATPlan paradigm has been neither fully explored nor exploited. A critical limitation of the current best realization of SATPlan is its inability to handle TP. In general, modelling continuous duration constraints in a SAT problem is costly and often problematic, as SAT variables only represent binary values.

In this paper, we first propose a novel max-SAT formulation for STRIPS planning. In this formulation, the original SAT formulation is maintained to encode hard constraints that cannot be violated, and furthermore, an objective function is introduced to specify the objective of a given problem. Using this formulation, we develop a new general variable ordering heuristic to improve search efficiency. Furthermore, we develop an accumulative learning scheme to collect and utilize the knowledge learnt from multiple SAT problems during the incremental planning process.

Temporal Planning Combining max-SAT and PERT. Many techniques have recently been proposed for TP by combining constraint programming methods and methods in OR, such as linear programming [28] and constraint programming [27]. However, these methods only have achieved a limited success. They are either suboptimal methods with inferior solution quality such as SGPlan [29], or optimal but expensive, such as CPT [6]. In contrast to the advances in STRIPS planning, progresses on TP have been slower. We believe that combining the strengths of AI and OR techniques can lead to more efficient TP methods.

Program Evaluation and Review Technique (PERT) is a well established and broadly applied OR technique for evaluating a network of events. Edelkamp [8] first applied PERT to TP by combining PERT with a forward chaining heuristic search-based planner. Nevertheless, the idea of using PERT for TP has not been well studied. In this paper, we apply PERT to the SATPlan paradigm.

2   Temporal Planning and a Two-Phase Approach: An Overview

The temporal planning (TP) problems that we consider in this paper can be represented by a tuple $T = (F, O, I, G)$, where F is a set of facts, $I \subseteq F$ the initial state, $G \subseteq F$

the goal state, and O a set of actions. An action has preconditions pre(o), add effects add(o), delete effects del(o), and duration dur(o). In addition, the preconditions of each action should be held either at the beginning of the action or duration the overall action period; the add/delete effects of each action can take effect either right after the action starts or after the action finishes.

The key idea of our planning system is to separate propositional and temporal constraints, resulting in a two-phase algorithm. In the first phase, we first simplify a TP problem to a STRIPS problem. Specifically, we treat each action to have unit duration, each precondition of an action to be held satisfied during the execution of the action, and each add/delet effect of an action to take effect after the action execution. We then represent the STRIPS problem by a maximum SAT (max-SAT) formulation, to be discussed in detail in Section 3, and solve the max-SAT problem with a max-SAT solver. In the second phase, we translate the solution from max-SAT solving back into a STRIPS plan, and apply PERT to it to generate a temporal plan.

We have several ideas to integrate many existing techniques to develop our TP algorithm. First, we follow the SATPlan strategy [25], which falls into the paradigm of graph planning [4]. However, instead of using SAT, we encode a planning problem in a hybrid max-SAT formulation, which consists of an objective function to be maximized and a set of SAT constraints.

This hybrid max-SAT representation naturally captures some problem structures in that most critical variables are included in the objective function. This leads to a *goal-oriented variable selection rule*, i.e., the variables involved in the objective functions are critical (or independent) variables and thus have higher priorities to be chosen in the search than those not in the objective function.

Another key idea that significantly improves the SATPlan strategy is accumulative learning. This strategy requires an incremental process to look for a plan with k parallel steps in one iteration; if it fails, it continues on to finding a plan with $k + 1$ steps in the next iteration, and so on. During the SAT solving in one iteration, many existing SAT solvers can learn a substantial number of no-goods clauses to speed up search. It is important to notice that such clauses learnt in one iteration can be reused in the next iteration if the planning problem is incrementally encoded over iterations, i.e., the max-SAT for one iteration is enclosed in the max-SAT for the next iteration. Our max-SAT formulation supports such incremental encoding.

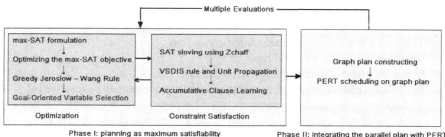

Fig. 1. An illustration of the two-phase constraint optimization search of our TP algorithm

We also adopt several other techniques in solving a max-SAT. The most significant is the Zchaff algorithm [30] as the core SAT solver. Other techniques include the Jeroslow-Wang greedy variable selection rule [14] and VSIDS variable selection [20] for goal and non-goal variables, respectively. Figure 1 illustrates the overall structure of our new strategy and algorithm for TP.

3   Phase I: Planning as Maximum Satisfiabilty

The objective of this phase is to create a network of actions or activities to satisfy all the logical constraints of a planning problem. A parallel plan is in essence a network of actions, to which many efficient algorithms exist, such as Graphplan [4]. In this research, we use SATPlan, a general and efficient graph planning strategy. The latest implementation SATPlan is SATPLAN04 [16]. Starting with a discussion on SATPlan, we describe our new approach based on planning as max-SAT in the rest of the section.

3.1   Review of Planning as Satisfiability and SATPLAN04

SATPlan is an incremental strategy for finding shortest parallel plans. For a planning problem, starting with a small target number k of parallel steps, it tries to find a parallel plan with k steps. If such a plan is found, the algorithm terminates, otherwise, it extends to $k + 1$ steps and repeats. In the iteration for finding a k-step plan, the algorithm uses two ideas [25, 17]. The first is to convert a planning problem in the STRIPS formulation [9] into a SAT formulation. The second idea is to solve the SAT problem using a generic SAT solver. The front-end of this approach contains a plan generator to construct a graphplan-style planning graph [4] and encode the graph into a SAT formulation. The back-end includes a SAT engine using any existing SAT solver.

Several planning systems have been developed under the SATplan paradigm, including SATPLAN, BLACKBOX [25, 17], and the latest SATPLAN04 [16]. These systems, SATPLAN04 in particular, have enjoyed a great deal of success, mainly due to the sophistication of generic SAT algorithms used. SATPLAN04 supports action-based encoding. We leave the detail of this encoding to the original papers [25, 17], while simply mention its main features here. The central idea of this encoding is to convert all fact variables to action variables. Any fact at certain time step is represented as a disjunction of the actions that add this fact at previous time steps. This encoding includes three classes of clauses based on action variables as follows:

1. *A clauses* – from actions' precondition constraints
   An action implies its preconditions.
2. *E clauses* – from mutually exclusive constraints
   Two actions cannot be true at the same time if they are mutually exclusive [4].
3. *G clauses* – from goal constraints
   A subgoal at step k implies disjunction of the actions that add this fact at step $k - 1$.

We call the variables in G clauses *G variables* and others *non-G variables*.

Despite its success, SATPLAN04 has several limitations. The first is the scope that it can apply; it cannot handle TP problems. The second limitation is due to the way it applies clause learning. SATPLAN04 uses an incremental process to search for an optimal plan. However, it only uses clause learning by its underlying SAT solver within each iteration, but does not share learnt clauses across iterations. The third limitation is due to its "blackbox" nature. SATPLAN04 does not utilize any structure information of a problem, which can be acquired during the encoding stage, to speed up its search process. SAT problems derived from real-world problems are highly structured and contain a large portion of variables that are dependent of other variables.

3.2 Planning as Maximum Satisfiability

To better handle STRIPS planning, we propose the following hybrid max-SAT formulation to replace the SAT formulation in SATPlan.

$$\begin{cases} \text{objective : maximize } f(G) \\ \text{subject to : SATPLAN constraints,} \end{cases} \tag{1}$$

where $f(G)$ is the objective function, G is the set of goals, and the constraints are from the SATPLAN formulation. Furthermore, we have $G = \{g_{t,i}\}\, k_0 \leq t \leq k_m, i = 1 \ldots n$ where each subgoal $g_{t,i}$ takes value 1 if the i-th subgoal is fulfilled at step t or 0, otherwise; k_0 and k_m are, respectively, lower and upper bounds of the shortest achievable step; and n is the number of subgoals. Note that k_0 can be obtained by applying a reachability analysis, e.g., using GraphPlan, and k_m can be estimated or a solution of a fast approximation planning method.

This max-SAT formulation highlights the optimization nature of a planning problem, i.e., minimizing the number of parallel steps. Therefore, the objective function can be represented as $f(G) = \sum_{k_0 \leq t \leq k_m} w_t * (\prod_{i=1 \ldots n} g_{t,i})$, where we use $w_t = n^{k_m - k_o - t}$ to force to satisfy all the subgoals at a shallower step prior to satisfy any subgoal at a longer step. Focusing on an objective function leads to effective variable selection methods that can better exploit the intrinsic structures of planning problems. By including the goal variables in the objective function, we can isolate the core variables from dependent variables and derived effective variable selection methods to utilize the structures of planning problems.

3.3 Goal-Oriented Variable Selection (GOVS) Heuristic

To make the max-SAT formulation effective, we introduce what we call *goal oriented variable selection* (GOVS) heuristic. Since the objective is to maximize the objective function $f(G)$ in (1), we select a variable, from the set of unassigned goal-related variables, i.e. G variables, to instantiate. In other words, the G variables are preferred over the variables only in SAT constraints. Assigning a G variable can often lead to a chain of unit propagations, which further result in satisfying some SAT constraints. We depend on the existing SAT solvers for such unit propagations.

The selection and assignment of G variables are based on a greedy heuristic; we adopted the *Jeroslow-Wang rule* [14] to select G clauses. Let $\{C_1, C_2, \ldots, C_m\}$ be the set of G clauses to be satisfied. We selects a variable v from the set of G clauses to maximize $J(v) + J(\overline{v})$ over all un-instantiated variables, where $J(v) = \sum_{v \in C_i} 2^{-n_i}$ and n_i is the number of literals in the i-th clause.

SATPLAN04 uses Siege [24] as its SAT engine. Unfortunately, the source code of Siege is not available. We thus chose Zchaff [30], one of the best SAT solvers, and integrate the GOVS heuristic with its original VSIDS [20] variable selection heuristic. Integrating GOVS with VSIDS, we have the following algorithm for max-SAT:

Zchaff-GOVS for solving max-SAT:

1. Have a counter, initialize to 0, for each of the two literals of a non-G variable.
2. If Select an uninstantiated G variable, if any, according to the Jeroslow-Wang rule.
3. Otherwise (i.e., all G variables have been assigned), choose a non-G variable whose literal (either positive or negative) has the highest counter value.
4. Assign the selected variable (from step 2 or 3) to True or False, simplify the CNF formula according to this variable assignment, and apply unit propagation to the simplified formula.
5. When a conflict occurs, apply the resolution rule to conflicting clauses, record the new generated learnt clause, increase the counters of the literals in the learnt clause, and backtrack to an earlier decision level that causes the conflict.
6. Periodically decay the counters of all non-G variables as implemented in VSIDS [20].
7. If an uninstantiated variable remains, goto step 2, Otherwise, terminate.

The GOVS heuristic and max-SAT strategy have several advantages. Our preliminary experimental analysis (data not shown) has indicated that after all the G variables have been fixed, the SAT constraints can be satisfied quickly by unit propagation. This means that the G variables are the most critically constrained in a planning problem. By focusing on such critical variables, the overall search procedure is geared toward the regions of the search space where high quality solutions locate. In other words, GOVS leads to a focused search. Furthermore, GOVS can also be viewed as a heuristic for backward search because those variables directly related to goal (or objective) in the G clauses are considered first. Moreover, the number of goal-related variables is relatively small comparing to the total number of variables in a planning problem, which in many cases contributes to a significant speed up of the search procedure. Finally, the GOVS heuristic can be easily integrated with a SAT solver for max-SAT.

3.4 Accumulative Learning

The problem structures over two consecutive iterations resemble each other except some G clauses because of the increase in time steps. We take advantage of such structural similarities and use what we called *Accumulative Learning* (AL) scheme. Instead of re-encoding the whole problem from scratch after each iteration, we modify and patch the previous encoding to meet the new constraint requirements for the next iteration. Therefore, the time for encoding can be significantly reduced; such saving can be significant for large planning problems. More importantly, we can retain all the learnt clauses, which are not related to goal clauses, in all the previous iterations and use them in the next iteration. As a result, most learnt clauses only need to be learnt once, which gives a dramatic reduction in running time. More specifically, the accumulative encoding and learning for time steps k can be described as follows:

Accumulative encoding and learning:

1. Delete all G clauses for steps $k - 1$, and all learnt clauses related to these G clauses.
2. Add new G clauses for steps k.
3. Add all additional A and E clauses required for steps k.

Note that the learnt clauses that are retained may be encountered in the next iteration. Most existing efficient SAT solvers have mechanisms for clause learning that support managing and deleting learnt clauses intelligently. Therefore, the accumulative learning scheme incurs a limited overhead, if any, over the SAT solver used.

4   Phase II: Integrating STRIPS Planning with PERT

The maxSAT solution obtained from Phase I corresponds to a STRIPS plan in which dependency information and partial order between actions are prescribed. This STRIPS plan is just a partial solution and it requires further processing to satisfy temporal requirements of the original TP problem.

To satisfy temporal constraints, we apply the Project Evaluation and Review Technique (PERT) [12], a critical path analysis chart for project management. At the heart of PERT is a network of actions needed to complete a project. Each path in the PERT chart starts with an initiation node from which the first action, or actions, originates. The path is complete when all final actions come together at the completion node in the chart. Given the actions and the dependency information, PERT aims at establishing a critical path in which the total time of actions is greater than any other path of actions.

After obtaining an optimal parallel plan by solving the STRIPS counterpart in max-SAT encoding, in order to minimize the total time of executing the actions, we apply PERT scheduling on this STRIPS plan as follow:

PERT scheduling on a STRIPS plan:

1. Record all partial orders between actions in the parallel plan, i.e. if two actions are not within the same time step in the parallel plan, we add a partial order that forces one action with an earlier time step to be ordered before the other action if some effect of the former one is a pre-condition of the latter one.
2. Generate a total order of actions, satisfying all the partial orders in step 1. (Although there are many possible total orders, the final result of total time from PERT scheduling will be the same.)
3. Process actions by the total order. For each action A, do:
   - Find all preceding actions before A in the partial order.
   - Set the beginning time of A to be the maximum of the ending time (or beginning time if the preceding actions can make immediate effects right after they start) of these preceding actions.
   - Find all preceding actions before A in the total order whose pre-conditions will be deleted by executing A.
   - Delay the beginning time of A until all such preceding actions have started (or have finished if the pre-conditions of such preceding actions should be held during the overall action period).
   - Set the ending time of A to be the beginning time plus the duration of A.

In essence, we treat the optimal parallel plan obtained from Phase I as a network specifying partial orders among actions at different time steps, and use the PERT technique to find the shortest temporal plan satisfying these partial orders given the parallel plan structure.

However, applying PERT to a STRIPS plan with the minimal parallel steps may not guarantee the optimality of the corresponding temporal plan. On the other hand, for most cases that we tested, the first STRIPS plan from solving max-SAT can often give rise to an optimal temporal plan. This observation makes PERT scheduling very attractive for finding an optimal or near-optimal temporal plan in practice. Furthermore, we have implemented an anytime search scheme that enumerates all the STRIPS plans within a time step bound and applies PERT to these plans to find the optimal temporal solution. We are currently investigating theoretical conditions under which an optimal parallel plan can be converted into an optimal temporal plan by PERT.

5   Experimental Evaluation and Analysis

Even though our main focus in this paper is TP, our new temporal planner embodies a strong STRIPS planner, an interesting contribution on its own. In this section, we experimentally evaluate the performance of these two new planners. We run all our experiments on a PC workstation with Intel Xeon (TM) 2.4 GHZ CPU and 2 GB memory.

5.1   Evaluation of the New STRIPS Planner

To fully appreciate our two-phase temporal planner, we first evaluated the performance of the STRIPS planner, SATPLAN04, with different combinations of our proposed methods. One of the objectives of this experimental study is to understand the strength of the two new methods, i.e., accumulative learning (AL) and goal-oriented variable selection (GOVS). In our current implementation, we used Zchaff as the SAT engine, because it is a top performer in many SAT competitions and its source code is available. Another competitive SAT solver, Siege, was not chosen since it is not an open-source software so that AL and GOVS cannot be incorporated.

To fully understand the strength of each component, we compared four versions of SATPLAN04, with Zchaff, Zchaff+GOVS, Zchaff+AL, and Zchaff+GOVS+AL, respectively, on some STRIPS domains from IPC3, IPC4 Competitions [1, 2], and the logistics-strips domains [7]. Since optimal plans are returned by these combinations of components, we only evaluated their running time. For all runs, we set a CPU time limit of 1,800 seconds. Table 1 summarizes the results on these domains.

Figures 2 and 3 show the running time of the four algorithmic combinations. Problems not solvable in 1,800 seconds were considered unsolvable and corresponded to missing points in the graphs. It is evident from Figures 2 and 3 that our proposed methods, GOVS heuristic and AL, can consistently improve the original SAT planner. We further compared the relative contributions of AL and GOVS. We observed that GOVS is particularly effective on Pipesworld-Notankage, Telegraph, Logistics, and Satellite domains. Zchaff+GOVS was able to solve several large problems in Pipesworld-Notankage, i.e., those numbered 21, 22, and 23, in 1711, 271, 1779 seconds, respectively, whereas Zchaff spent 521 seconds on number 22 and failed on number 21 and

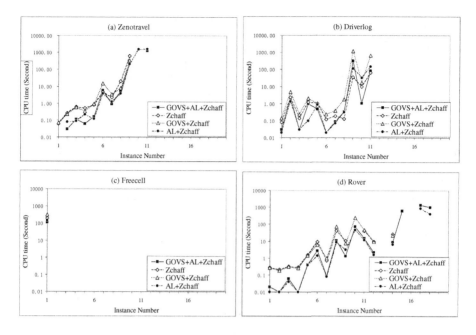

Fig. 2. A comparison of the performance of four algorithmic combinations on IPC3 STRIPS domains

Table 1. The number of instances each propositional planning method can solve for the IPC3, IPC4 STRIPS domains and logistics-strips domain. N is the total number of problem instances available in each domain. Highlighted in boxes are the methods that can solve the most number of problems.

Domains	N	Zchaff	GOVS	AL	GOVS+AL	Domains	N	Zchaff	GOVS	AL	GOVS+AL
Airport	50	27	26	26	27	PW-Notkg	50	9	11	7	7
PW-Tkg	50	3	4	4	4	Telegraph	14	7	13	13	13
Philosopher	29	29	29	29	29	PSR-small	50	46	46	47	47
Satellite	36	2	2	2	2	Logistics	30	20	20	20	24
Zenotravel	20	9	9	11	10	Driverlog	20	11	11	11	11
Freecell	20	1	1	1	1	Rover	20	12	13	15	16

23. For the Telegraph domain, Zchaff+GOVS was able to solve 13 problem instances, whereas Zchaff could only handle 6. Zchaff+GOVS was able to significantly reduce the running time of Zchaff on the largest problems in Logistics and Satellite.

The AL method was found very useful in the Telegraph, Philosopher, PSR, Rover, Driverlog, and Logistics domains. For these domains, Zchaff+AL can significantly and consistently improve upon Zchaff across all problem instances. The improvement made by AL can be more than one order of magnitude on the Telegraph, Philosopher, Rover, and Driverlog domains.

Table 2. The number of instances in the IPC4 temporal domains each temporal planner can solve. N is the total number of instances in each domain.

Domains	N	maxSATPLAN04 GOVS+AL	Siege	CPT	TP4	Domains	N	maxSATPLAN04 GOVS+AL	Siege	CPT	TP4
Airport	50	27	30	20	13	PW-Notankage	50	9	21	11	8
PW-Tankage	50	4	9	5	6	Satellite	36	2	15	8	4

Table 3. Comparison of solution quality and running time of CPT, TP4, and maxSATPLAN04 with GOVS+AL and Siege on Airport domain.

Instance	CPT		TP4		GOVS+AL		Siege		Instance	CPT		TP4		GOVS+AL		Siege	
1	64	0.1	64	0.1	64	0.1	64	0.3	2	185	0.1	185	0.1	185	0.1	185	0.3
3	200	0.1	200	0.4	200	0.1	200	0.4	4	127	0.1	127	0.2	127	0.1	127	0.6
5	227	0.1	227	0.4	227	0.1	227	0.8	6	232	0.4	232	6.7	232	0.1	232	1.2
7	232	0.4	232	5.5	232	0.1	232	1.1	8	394	237.0	—	—	394	0.8	394	3.7
9	402	935.5	—	—	402	1.7	402	6.2	10	126	0.1	126	0.2	126	0.1	126	0.6
11	228	0.2	228	0.7	228	0.1	228	0.9	12	228	0.6	228	1.4	*232	0.2	232	1.4
13	230	0.5	230	4.8	230	0.1	230	1.3	14	390	238.2	—	—	*394	1.0	394	4.8
15	262	41.6	262	50.5	262	0.5	262	2.9	16	393	1444.8	—	—	*397	3.5	397	9.1
17	399	190.4	—	—	*408	7.2	408	15.0	18	435	375.3	—	—	435	48.6	435	52.0
19	413	313.5	—	—	413	6.8	413	15.4	20	435	833.2	—	—	435	145.2	435	94.9
21	—	—	285	614.7	285	6.6	285	36.9	22	—	—	—	—	285	6.7	285	60.5
23	—	—	—	—	—	—	264	481.9	24	—	—	—	—	376	368.8	376	220.5
38	—	—	—	—	322	253.3	322	170.9									

Table 4. Comparison of solution quality and running time of CPT, TP4, and maxSATPLAN04 with GOVS+AL and Siege on PW-Notankage domain

Instance	CPT		TP4		GOVS+AL		Siege		Instance	CPT		TP4		GOVS+AL		Siege	
1	6	0.1	6	0.2	6	0.1	6	0.3	2	12	0.1	12	0.2	12	0.2	12	0.4
3	16	0.9	12	0.8	12	1.2	12	0.5	4	22	94.4	12	1.2	—	—	12	0.4
5	14	1.2	12	81.7	12	1.0	12	0.9	6	14	0.9	12	32.6	12	2.5	12	1.3
7	14	19.4	12	23.0	12	6.5	12	1.2	8	14	1.9	14	196.0	14	16.2	14	26.2
9	18	369.5	—	—	—	—	16	74.5	10	—	—	—	—	—	—	20	1009.2
11	—	—	—	—	—	—	6	292.6	12	—	—	—	—	—	—	7	820.9
13	6	403.8	—	—	—	—	6	278.0	14	8	543.1	—	—	—	—	8	1038.3
15	—	—	—	—	—	—	8	1170.1	17	—	—	—	—	—	—	6	625.2
21	—	—	—	—	—	—	13	649.5	23	—	—	—	—	—	—	16	499.6
24	—	—	—	—	—	—	26	901.6	31	—	—	—	—	—	—	24	1298.9

Table 5. Comparison of solution quality and running time of CPT, TP4, and maxSATPLAN04 with GOVS+AL and Siege on PW-tankage

Instance	CPT		TP4		GOVS+AL		Siege		Instance	CPT		TP4		GOVS+AL		Siege	
1	6	0.2	6	0.6	6	0.1	6	0.3	2	22	8.9	20	2.7	—	—	20	210.0
3	16	2087.9	12	447.3	12	176.6	12	4.3	4	—	—	12	721.1	—	—	12	17.9
5	14	200.0	12	65.5	12	25.4	12	8.9	6	14	214.3	12	282.8	12	34.6	12	18.8
7	—	—	—	—	—	—	12	444.5	8	—	—	—	—	—	—	14	1551.5
11	—	—	—	—	—	—	6	1187.6									

Table 6. Comparison of solution quality and running time of CPT, TP4, and maxSATPLAN04 with GOVS+AL and Siege on Satellite domain

Instance	CPT		TP4		GOVS+AL		Siege		Instance	CPT		TP4		GOVS+AL	Siege	
1	135.5	0.1	135.5	0.1	*174.9	33.3	174.9	19.4	2	156.3	1.0	156.2	7.1	—	—	259.5 341.8
3	65.2	0.2	65.2	7.3	*128.5	38.3	128.5	10.8	4	122.2	2.5	—	—	—	—	173.8 145.6
5	105.3	0.7	105.3	132.4	—	—	279.1	83.8	6	64.8	2.7	—	—	—	—	188.7 33.2
7	60.2	1.5	—	—	—	—	149.2	41.2	8	74.0	287.5	—	—	—	—	277.4 358.3
9	81.0	41.5	—	—	—	—	226.2	65.5	10	105.0	14.6	—	—	—	—	288.7 259.2
11	115.8	31.4	—	—	—	—	306.2	242.0	14	—	—	—	—	—	—	257.7 700.5
15	—	—	—	—	—	—	148.2	860.4	17	—	—	—	—	—	—	200.9 238.7

5.2 Evaluation of the New Temporal Planner

We integrated SATPLAN04 with PERT for temporal planning. We called the resulting algorithm *maxSATPLAN04*, which has a max-SAT solver as its core. Even though we do not have access to the source code of Siege, we can still apply it as a blackbox SAT solver for our max-SAT based STRIPS planner and combine it with PERT. We thus included this Siege-based temporal planner, along with the Zchaff-based temporal planner. In the Zchaff-based planner, we used GOVS heuristic and AL method. In our comparison, we also included CPT and TP4 [11], which are among the best optimal temporal planners in IPC4 Competition.

We considered the Airport, Pipesworld-Tankage, Pipesworld-Notankage, and Satellite temporal domains in IPC4 Competition, which are the only domains containing temporal features in the competition. Table 2 summarizes the results of different algorithms on four TP domains. When Siege was used as a blackbox SAT engine, our new temporal planner was able to solve more instances than CPT and TP4 on three out of the four domains. However, the new planner with Zchaff+GOVS+AL performed worse than CPT on these domains except Airport. This difference is evidently due to the SAT engines used. Siege is a SAT solver highly specialized for planning, while Zchaff is a generic SAT solver. Although our GOVS heuristic and AL method can significantly improve the performance of Zchaff, they were still unable to make up the gap between Zchaff and Siege. In our future work, we will try to re-implement Siege and use it to replace Zchaff.

Even though applying PERT to an optimal STRIPS plan cannot guarantee an optimal temporal plan, our experiments showed that it was able to produce optimal plans on most temporal benchmarks tested. Tables 3 to 6 list performance comparison between different temporal planners, CPT, TP4, maxSATPLAN04 with Zchaff+GOVS+AL as SAT engine, and maxSATPLAN04 with Siege as SAT engine. All the temporal plans from maxSATPLAN04 were produced by applying PERT to the optimal solutions from our new STRIPS planner. We highlight in the tables all the instances for which the *first* optimal STRIPS plans do not give rise to optimal temporal plans; for these instance, "*" means that further evaluating multiple STRIPS plans reached temporal optimality. In addition, "-" means no result for an algorithm after 1,800 seconds.

The results in Tables 3 to 6 show that our new temporal planner works very well on Airport, Notankage, and Tankage domains. For each algorithm, the left and right columns are the solution quality (makespan) and CPU time, respectively. As shown in the tables, for every instance in notankage and tankage domains, applying PERT to the

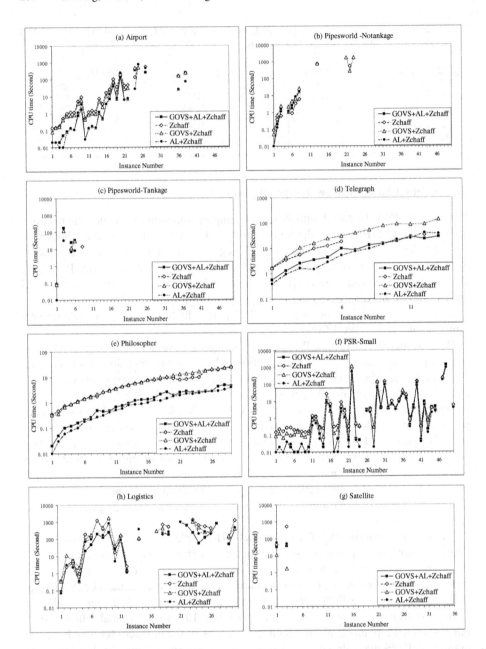

Fig. 3. A comparison of four algorithmic combinations on IPC4 STRIPS domains and logistics-strips domain

first optimal solution from our new STRIPS planner resulted in an optimal temporal plan. This is also true for most instances in the airport domain.

For instances 2, 3, 5 and 6 in Tankage domain and instances 3 to 9 in Notankage domain, TP4 and our maxSATPLAN04 generated solutions better than that from CPT. This result also revealed that CPT is not optimal, even though it was claimed to be so. Note that the results for our maxSATPLAN04 on these instances are consistent with that from TP4.

One interesting phenomenon is that our PERT-based strategy does not work very well on the Satellite domain. One possible reason is that the action durations of the problems in this domain are not integers. We plan to study this issue more carefully in our future work.

6 Additional Related Work and Discussions

Besides the related works - particularly planning as satisfibility and PERT - that we have already mentioned, we now briefly discuss some previous work about goal-oriented planning approaches and incremental learning.

Goal-oriented variable selection techniques have been exploited by heuristic planners. The backward chaining planning methods [22, 19] start from goal states and directly use the goal information to limit and focus search efforts. Although the forward chaining planning methods [5, 13, 21, 23] start from the initial state, the goal information is usually utilized in the heuristic function. For example, in the relaxed plan heuristic used by FF [13], the goal state information is used to extract a relaxed solution plan to estimate the solution length. However, unlike the SAT planners, heuristic planners generally have difficulty to find optimal plans or provide some degree of guarantee on the solution quality. In the optimization phase of the proposed Zchaff+GOVS algorithm, we essentially assign the completion time steps to different subgoals and try to minimize the solution length subject to the SAT constraints. Therefore, intelligent goal ordering information may be useful in this phase to prune infeasible or unreasonable orderings. A number of related approaches [18, 29] have been developed to provide some kind of goal ordering information by studying the intrinsic structure of the planning problems.

Several incremental learning approaches have been developed. For example, Katukam et al. proposed to learn explanation-based search control rules for a partial order planning scheme [15]. However, most of the learning schemes are proposed for heuristic search and partial order search methods. For incremental SAT learning methods, the most related research focuses on applying incremental SAT checker on bounded model checking [10, 3, 26].

7 Concluding Remarks

We developed a novel hybrid strategy for temporal planning, in which we synergistically combined many existing techniques from planning, constraint programming and Operations Research. We made four main contributions to temporal and STRIPS planning: 1) integration of planning as satisfiability (SATPlan) strategy and program evaluation and review technique (PERT); 2) a method of planning as max-SAT for STRIPS planning; 3) An accumulative learning scheme for incremental graph plan method; and 4) a goal-oriented heuristic for variable selection. In our research, we extensively exploited many characteristics of temporal planning and its underlying propositional planning problem. In addition to the new algorithms for temporal and STRIPS planning problems, our work also resulted in a novel maximum satisfiability (max-SAT) solver.

We implemented our new methods in SATPLAN04 and Zchaff, and performed extensive experimental analyses on benchmark problems from IPC4 and other domains. Experimental results show that our new approach significantly outperforms many state-of-the-art planners across different planning domains. We also performed an ablation study to analyze the relative contributions of the proposed methods. Our analysis shows that each of our new methods yields substantial performance improvements. In the future, we plan to study the relationship between optimal parallel plans and the temporal plans obtained from PERT and derive optimality conditions for the proposed strategy.

Acknowledgement

This research was supported in part by NSF grants ITR/EIA-0113618 and IIS-0535257.

References

1. http://planning.cis.strath.ac.uk/competition.
2. http://ls5-www.cs.uni-dortmund.de/ edelkamp/ipc-4.
3. M. Benedetti and S. Bernardini. Incremental compilation-to-SAT procedures. In *Proceedings of The Seventh International Conference on Theory and Applications of Satisfiability Testing (SAT04)*, 2004.
4. A. Blum and M.L. Furst. Fast planning through planning graph analysis. *Artificial Intelligence*, 90:281–300, 1997.
5. B. Bonet and H. Geffner. Planning as heuristic search. *Artificial Intelligence, Special issue on Heuristic Search*, 129(1), 2001.
6. http://www.cril.univ-artois.fr/ vidal/cpt.fr.html.
7. http://www.cs.washington.edu/homes/kautz/satplan/blackbox/blackbox-download.html.
8. S. Edelkamp. Mixed propositional and numerical planning in the model checking integrated planning system. In *Proceedings of AIPS 2002, Workshop on Planning for Temporal Domains*, pages 47–55, 2002.
9. R.E. Fikes and N.J. Nilsson. STRIPS: A new approach to the application of theorem proving to problem solving. *Artificial Intelligence*, 2:189–208, 1971.
10. M. Awedh H. Jin and F. Somenzi. CirCus: a satisfiability solver geared towards bounded model checking. In *CAV 2004*, pages 519–522, 2004.
11. P. Haslum. TP4'04 and HSP. In *Proceedings of IPC4, ICAPS*, pages 38–40, 2004.
12. F. Hillier and G. Lieberman. *Introduction to Operations Research*. McGraw-Hill, Boston, 7th edition, 2001.
13. J. Hoffmann and B. Nebel. The FF planning system: Fast plan generation through heuristic search. *Journal of Artificial Intelligence Research*, 14:253–302, 2001.
14. J.N. Hooker and V. Vinay. Branching rules for satisfiability. *Journal of Automated Reasoning*, 15:359–383, 1995.
15. S. Katukam and S. Kambhampati. Learning explanation-based search control rules for partial order planning. In *Proceedings of AAAI-94*, pages 582–587, 1994.

16. H. Kautz. SATPLAN04: Planning as satisfiability. In *Proceedings of IPC4, ICAPS*, 2004.
17. H. Kautz and B. Selman. Unifying SAT-based and graph-based planning. In *Proceedings of IJCAI-99*, pages 318–325, 1999.
18. J. Koehler and J. Hoffmann. On reasonable and forced goal orderings and their use in an agenda-driven planning algorithm. *Journal of Artificial Intelligence Research*, 12:338–386, 2000.
19. D. McDermott. Estimated-regression planning for interactions with web services. In *Proceedings of AIPS 2002*, pages 204–211, 2002.
20. M.W. Moskewicz, C.F. Madigan, Y. Zhao, L. Zhang, and S. Malik. Chaff: Engineering an efficient SAT solver. In *Proceedings of the 38th Design Automation Conference (DAC'01)*, 2001.
21. R. S. Nigenda, X. Nguyen, and S. Kambhampati. AltAlt: Combining the advantages of Graphplan and heuristic state search. Technical report, Arizona State University, 2000.
22. J. L. Pollock. The logical foundations of goal-regression planning in autonomous agents. *Artificial Intelligence*, 106(2):267–334, 1998.
23. I. Refanidis and I. Vlahavas. The GRT planner. *AI Magazine*, pages 63–66, 2001.
24. L. Ryan. Efficient algorithms for clause-learning SAT solvers. Master's thesis, Simon Fraser University, 2003.
25. B. Selman and H. Kautz. Planning as satisfiability. In *Proceedings ECAI-92*, pages 359–363, 1992.
26. O. Shtrichman. Tuning SAT checkers for bounded model checking. In *Computer Aided Verification*, pages 480–494, 2000.
27. V. Vidal and H. Geffner. CPT: An optimal temporal POCL planner based on constraint programming. In *Proceedings of IPC4, ICAPS*, pages 59–60, 2004.
28. S. Wolfman and D. Weld. Combining linear programming and satisfiability solving for resource planning. *The Knowledge Engineering Review*, 15(1), 2000.
29. C. Hsu Y. Chen and B. W. Wah. SGPlan: Subgoal partitioning and resolution in planning. In *Proceedings of IPC4, ICAPS*, pages 30–32, 2004.
30. L. Zhang, C. F. Madigan, M. W. Moskewicz, and S. Malik. Efficient conflict driven learning in boolean satisfiability solver. In *ICCAD*, pages 279–285, 2001.

Improved Algorithm for the Soft Global Cardinality Constraint

Alessandro Zanarini[1], Michela Milano[2], and Gilles Pesant[1]

[1] Département de génie informatique
École Polytechnique de Montréal
C.P. 6079, succ. Centre-ville
Montreal, Canada H3C 3A7
{azanarini, pesant}@crt.umontreal.ca
[2] D.E.I.S., Universitá di Bologna
Viale Risorgimento 2, 40136 Bologna, Italy
mmilano@deis.unibo.it

Abstract. We propose two algorithms achieving generalized arc consistency for the soft global cardinality constraint with variable-based violation and with value-based violation. They are based on graph theory and their complexity is $O(\sqrt{n}m)$ where n is the number of variables and m is the sum of the cardinalities of the domains. They improve previous algorithms that ran respectively in $O(n(m+n\log n))$ and $O((n+k)(m+n\log n))$ where k is the cardinality of the union of the domains.

1   Introduction

Many real-life problems are over-constrained. The tightness and the high number of constraints can make the problems become unfeasible. In these situations it is worth finding a solution that partially violates some constraints but that it is still interesting for the user. Constraints can be partitioned among *hard constraints* that cannot be violated, and *soft constraints* that can be (partially) violated. Hard constraints are used for modelling the inherent structure of the problem and soft constraints are more related to preferences that the user wishes to introduce to the model. Clearly, solutions satisfying a maximum of preferences are more interesting for the user. Different approaches deal with the concept of violation in different ways: some methods (MAX-CSP) try to minimize the number of violated constraints, others (Weighted-CSP [6] [7], Possibilistic-CSP [12], Fuzzy-CSP [2] [3]) propose more granular ways to measure the level of violation. Petit et al. in [9] proposed a new approach in which the over-constrained problem is translated to a constraint optimization problem. It is then worth trying to identify ad hoc filtering algorithms that can prune the variable domains on the basis of the cost (violation). Recent work started in that direction by exploring the area of soft global constraints. In particular van Hoeve et al. in [4] exploited Flow Theory and proposed filtering algorithms for the soft versions of the well known all-different, gcc, regular, and same constraints.

J.C. Beck and B.M. Smith (Eds.): CPAIOR 2006, LNCS 3990, pp. 288–299, 2006.
© Springer-Verlag Berlin Heidelberg 2006

In this paper we present an improved algorithm for achieving generalized arc consistency for the soft gcc (with variable based violation and value based violation) exploiting Matching Theory, with a better complexity. Intuitively the soft gcc constraint is violated when either

- too many variables are assigned to a value, exceeding its upper bound (producing an overflow) or
- too few variables are assigned to a value, violating its lower bound (producing an underflow) or
- both.

The idea of the paper is to compute separately the best possible overflow and underflow and, since we claim they are independent, find a class of solutions minimizing both overflow and underflow. On the basis of these best underflow and overflow we perform filtering.

The paper is organized as follows: in Section 2 we give a brief overview of basic notions about Constraint Satisfaction Problem and Matching Theory; in Section 3 we formally present the soft gcc constraint and the related violation measures; then we discuss the relationship between the violation measures and matching theory. In Section 4 we introduce the consistency theorems and the filtering algorithms for reaching generalized arc consistency. Finally in Section 5 conclusions are given.

2   Background

2.1   Constraint Satisfaction Problem

A *Constraint Satisfaction Problem* (CSP) consists of a finite set of variables $\mathcal{X} = \{x_1, x_2, \ldots, x_n\}$ with finite domains $\mathcal{D} = \{D_1, D_2, \ldots, D_n\}$ such that $x_i \in D_i$ for all i, together with a finite set of constraints $\mathcal{C}$, each on a subset of $\mathcal{X}$. A constraint $C \in \mathcal{C}$ is defined as a subset of the Cartesian product of the domains of the variables that are in C. A *solution* to a CSP is an assignment of a value to each variable that satisfies all the constraints.

A CSP is defined as inconsistent if no assignment that satisfies all the constraints exists. For these over-constrained problems it is natural to identify a subset of constraints, defined as soft constraints, that can be (partially) violated. The main objective is then to find a solution that minimizes the total violation according to some criteria. Petit et al. proposed in [9] to introduce a cost variable z representing the violation and an associated function that measures the violation of a constraint:

$$violation_C : D_1 \times D_2 \times \cdots \times D_n \to \mathbb{N}$$

Clearly, if the constraint is satisfied then $z = 0$, otherwise $z > 0$. A common accepted measure is the variable-based cost violation (see [9]) in which the violation is measured by the minimum number of variables that need to change their value in order to satisfy the constraint. There exists also violation measures that are specific to a particular constraint: this is the case of the value-based cost violation (introduced by van Hoeve et al. in [4]) that is applied in the soft global cardinality constraint. Our work covers both variable-based violation and value-based violation for the soft gcc constraint.

2.2   Matching Theory

In this section we recall the main results and definitions that will be used in the rest of the paper (see [1] for further explanations).

A graph is defined as $G = (V, E)$ where V is a set of vertices and E is a set of unordered pairs (edges) from V. A graph is called bipartite if V can be partitioned in two subset X and Y and all the edges are in the form $e = (v_i, v_k)$ where $v_i \in X$ and $v_j \in Y$ (i.e. there is no edge that joins two vertices of the same subset).

A path in a graph $G = (V, E)$ is a sequence of vertices $v_0, v_1, \ldots, v_k$ such that $(v_i, v_{i+1}) \in E$ with $i = 0, \ldots, k - 1$.

Definition 1 (Maximum Matching). *A subset of edges in a graph G is called matching if no two edges have a vertex in common. A matching of maximum cardinality is called maximum matching.*

Given a matching M in G, a vertex is called *free vertex* if it is not adjacent to any edge of the matching M.

An alternating path with respect to a matching M (*M-alternating path*) is defined as a path whose edges $e_i = (v_i, v_{i+1})$ belong alternatively to $E - M$ and to M.

An augmenting path with respect to a matching M (*M-augmenting path*) is defined as a path that starts from and ends to a free vertex, and its edges $e_i = (v_i, v_{i+1})$ belong alternatively to $E - M$ (odd edges) and to M (even edges); note that an augmenting path has an odd number of edges.

Intuitively, an augmenting path can be used to increase the number of edges that belong to a matching. Given a matching M and an M-augmenting path P, we can build M' as $M' = M \oplus P$ (the set operation $\oplus$ is defined as $A \oplus B = (A - B) \cup (B - A)$), that is the odd numbered edges are added to the matching and the even numbered edges are removed from the matching; the resulting matching increases its cardinality, $|M'| = |M| + 1$.

Theorem 1. *Let M be a matching, M is maximum if and only if there is no augmenting path relative to M.*

Theorem 2. *Let G be a graph and M a maximum matching in G. An edge belongs to a maximum matching in G if and only if it either belongs to M, or to an even M-alternating path starting from a free vertex, or to an even alternating circuit.*

Lemma 1. *Given a maximum matching M in G, for any edge $e = (v_i, v_j)$ in G, there exists a matching M_e such that $e \in M_e$ and $|M_e| \geq |M| - 1$.*

Proof. If e belongs to M then $M_e = M$; otherwise, starting from the matching M, we obtain M_e adding e and removing all the edges that belong to M and that are incident to v_i or v_j (at most one on each). The result is a matching of size $|M_e| \geq |M| + 1 - 2$.

We introduce the concept of degree $deg_M(v)$ of a vertex v as the number of edges adjacent to v that belongs to M (for the traditional definition of matching $deg_M(v) \in \{0, 1\}$).

Theorem 3. *Given a matching M in G, an M-augmenting path P and the matching $M' = M \oplus P$, each vertex v has $deg_{M'}(v) \geq deg_M(v)$.*

Proof. The degree of a vertex v decreases if and only if v is not free w.r.t M and the incident edge that belongs to M is removed from the matching. For every removed edge $e = (v, v_j)$, two new edges from P are added, incidents respectively to v and v_j, so $deg_M(v) = deg_{M'}(v)$.

Hopcroft and Karp (see [5]) described an algorithm based on Theorem 1 with a running time complexity of $O(\sqrt{n}m)$ where n is the number of vertices and m the sum of the cardinalities of the domains.

In [8], Quimper et al. generalized this algorithm maintaining the same complexity. In their generalization they associate to each vertex of the graph a capacity. Given a matching M, the capacity of a vertex v indicates the maximum number of edges in M adjacent to v.

Intuitively they build a duplicated graph G_d in which every vertex with a capacity greater than one is substituted by a number of vertices equal to the capacity, also the edges associated to these vertices are duplicated. In this way a traditional matching (in which all the capacities are equal to 1) in G_d corresponds to a matching on the original graph (in which the capacities can be greater than 1).

Quimper's approach is equivalent to the traditional one when all the capacities are set to 1.

3   Soft Global Cardinality Constraint

A *Global Cardinality Constraint* on a set of variables specifies the minimum and the maximum number of occurrences for each value in a solution.

Definition 2 (Global Cardinality Constraint)

$$gcc(X, l, u) = \{(d_1, d_2, \ldots, d_n) | d_i \in D_i, l_d \leq |\{d_i | d_i = d\}| \leq u_d \, \forall d \in D_X\}$$

A generic definition for a soft version of the gcc is:

Definition 3 (Soft Global Cardinality Constraint)

$$softgcc[*](X, l, u, z) =$$
$$\{(d_1, d_2, \ldots, d_n, d_z) | d_i \in D_i, d_z \in D_Z, violation_{softgcc[*]}(d_1, d_2, \ldots, d_n) \leq d_z\}$$

*where * defines a violation measure for the gcc.*

To calculate the violation measures van Hoeve et al. (see [4]) introduced the following definitions:

Definition 4. *Given a* $softgcc(X, l, u, z)$, *we define for all* $d \in D$

$$overflow(X, d) = max(|\{x_i \mid x_i = d\}| - u_d, 0)$$

$$underflow(X, d) = max(l_d - |\{x_i \mid x_i = d\}|, 0)$$

Definition 5 (Variable-based violation). *Given a constraint* C *and a solution* $\tilde{X}$, *the* variable-based violation *is defined as the number of variables that should change their value in order to satisfy* C.

Lemma 2 (SoftGCC Variable-based violation). *Given a softgcc, if* $\sum_{d \in D_X} l_d \leq |X| \leq \sum_{d \in D_X} u_d$ *then the* variable based violation *can be expressed as:*

$$violation_{[var]}(X) = max\left(\sum_{d \in D} overflow(X, d), \sum_{d \in D} underflow(X, d)\right)$$

Consider for example the variables x_1, x_2, x_3 and x_4 and the related domains $D_1 = \{1, 2\}$, $D_2 = \{1, 2\}$, $D_3 = \{1, 2\}$ and $D_4 = \{1, 2, 3\}$. Suppose we post the constraint $softgcc[var](\{x_1, x_2, x_3, x_4\}, \{l_1 = 0, l_2 = 1, l_3 = 2\}, \{u_1 = 1, u_2 = 1, u_3 = 2\}, z)$. A possible assignment is $(1, 1, 2, 3)$ which has an overflow equal to 1 and an underflow equal to 1; the variable-based violation is equal to 1.

Note that it is not always possible to calculate the variable based violation. To avoid this limitation van Hoeve et al. introduced the *value-based violation* (see [4]):

Definition 6 (Value-based violation). *Given a softgcc, the* value-based violation *is defined as:*

$$violation_{[val]}(X) = \sum_{d \in D} overflow(X, d) + \sum_{d \in D} underflow(X, d)$$

Consider again the example mentioned above. The assignment $(1, 1, 2, 3)$ which has unitary overflow and underflow, has a value-based violation equal to 2.

Van Hoeve et al. (see [4]) proposed two algorithms (one for variable-based violation and one for value-based violation) achieving generalized arc consistency both based on flow theory. In their solution they build a value graph (similarly to Régin in [10]) in which some arcs take into account the violations; a cost is associated to each of these arcs. A maximum flow with minimum cost in that graph is equivalent to a solution with minimum violation of the soft gcc.

Their algorithms have a complexity of $O(n(m + n \log n))$ for variable-based violation and of $O((n + k)(m + n \log n))$ for value-based violation (k is the cardinality of the union of the domains).

3.1 Soft gcc and Matching

The main idea of this paper is to exploit matching theory to calculate two assignments that minimize respectively the overflow and the underflow. We prove that it is possible to find a class of assignments that have overflow and underflow equal to the respective bounds. Then, we figure out how the violation cost of this class of assignments may change when we force an individual assignment $x_i = d$. Finally, we can perform filtering based on optimality reasoning.

Let $G(X \cup D, E)$ be an undirected bipartite graph (also called value graph) such that one partition represents the variable set and the other one the value set. There is an edge $(x_i, d) \in E$ if and only if $d \in D_i$.

Overflow. Let G_o be a value graph such that the capacities of value-vertices are set to $c(d) = u_d$ (variable-vertices have unitary capacity). Using the algorithm described in Section 2.2, we compute a maximum matching M_o in G_o. A maximum matching M_o corresponds to an assignment that should satisfy the upper bound constraint of the gcc. If $|M_o| = |X|$ then the matching corresponds to a consistent assignment (w.r.t. the upper bound constraint); if $|M_o| < |X|$ it means that some variables cannot be assigned to a value otherwise the upper bound constraint would be violated.

Exactly $|X| - |M_o|$ variables must be assigned to some values that have already reached the maximum number of occurrences so the overflow is exactly $|X| - |M_o|$.

Theorem 4. *Given a maximum matching M_o in the graph G_o, it is not possible to find an assignment with a total overflow less than $|X| - |M_o|$.*

Proof. Suppose that there exists an assignment X with an overflow equal to $OF < |X| - |M_o|$. We build the bipartite graph that represents X and we remove from this graph the OF edges that cause the overflow, therefore each value-vertex d has $deg(d) \le u_d$. The resulting graph can be seen as a feasible matching M' in G_o. Since $|M'| = |X| - OF$ then $|M'| > |M_o|$, i.e. M_o is not maximum.

Underflow. Analogously, we exploit matching theory to compute the underflow. In this case the graph G_u is built such that the capacities of value-vertices are set to $c(d) = l_d$ (variable-vertices have unitary capacity). Value-vertices with capacity equal to 0 are removed from the graph together with the related edges; in fact a value-vertex d with $c(d) = 0$ cannot cause underflow. A maximum matching M_u in G_u corresponds to a partial assignment that should satisfy the lower bound constraint of the gcc.

If $|M_u| = \sum_{d \in D} l_d$ then it means that for each value $deg_{M_u}(d) = l_d$, thus there exists at least one (partial) assignment that satisfies the lower bound constraint (i.e. there is no underflow and no violation w.r.t. the lower bound constraint). If $|M_u| < \sum_{d \in D} l_d$ then there are one or more values that do not reach the minimum number of requested occurrences (some value vartices are still free) and no variable can be assigned to these values.

Note that $l_d - deg_{M_u}(d) \geq 0$, hence by definition:

$$underflow(X, d) = l_d - deg_{M_u}(d)$$

and the total underflow is:

$$\sum_{d \in D} underflow(X, d) = \sum_{d \in D} l_d - deg_{M_u}(d) =$$

$$= \sum_{d \in D} l_d - \sum_{d \in D} deg_{M_u}(d) = \sum_{d \in D} l_d - |M_u|$$

Theorem 5. *Given a softgcc constraint and two maximum matchings M_o and M_u, respectively in G_o and G_u, it is possible to build a class of assignments with overflow equal to $BOF = |X| - |M_o|$ (best overflow) and underflow equal to $BUF = \sum_{d \in D} l_d - |M_u|$ (best underflow).*

Proof. We compute a maximum matching M_u in G_u whose underflow is equal to BUF. The matching M_u is clearly a feasible matching (probably not maximum) also in G_o because all the capacities of G_o are greater than those of G_u. Starting from M_u we compute the maximum matching M_o in G_o whose overflow is equal to BOF. As stated in Theorem 3, when we compute a matching, the degree of each vertex does not decrease, hence the underflow of M_o in G_o remains equal to BUF.

If $|M_o| < |X|$ then there exists a set X_{OF} of unassigned variables, that is, there is no edge in M_o adjacent to the variables in X_{OF}. These variables cause the overflow and, in the final solution, can be assigned to any value in their respective domain.

In order to develop a filtering algorithm, it is worth figuring out how overflow and underflow may change (w.r.t. the bounds of Theorem 5) when we try to force an individual assignment $x_i = d$. They change depending on whether the edge (x_i, d) belongs to a maximum matching in the graphs G_o and G_u or not; intuitively if it belongs to a maximum matching the overflow (or underflow) does not change otherwise it increases by 1 (see Lemma 1).

Theorem 6. *Given a softgcc constraint, an individual assignment $x_i = d$ and a solution $\tilde{X}$ with $x_i = d$ that minimizes the overflow (OF) and the underflow (UF) then $BOF \leq OF \leq BOF + 1$ and $BUF \leq UF \leq BUF + 1$ where BOF is the best overflow and BUF the best underflow.*

Proof. Let G_o and G_u be the overflow and underflow graphs and M_o and M_u the related maximum matchings. Suppose we remove from G_o (overflow graph) and G_u (underflow graph) the vertex x_i (and the related edges) and decrease u_d and l_d by 1; we call the resulting graph G'_o and G'_u. This is equivalent to forcing $x_i = d$ in the final assignment. Then we find the maximum matching M'_o in G'_o and M'_u in G'_u, clearly their cardinalities can be at most $|M'_o| = |M_o| - 1$ and $|M'_u| = |M_u| - 1$. Hence:

- if $|M'_o| = |M_o| - 1$ and $|M'_u| = |M_u| - 1$ then $x_i = d$ belongs to a maximum matching both in G_o and in G_u and the assignment has $OF = BOF$ and $UF = BUF$;
- if $|M'_o| = |M_o| - 1$ and $|M'_u| < |M_u| - 1$ then $x_i = d$ belongs to a maximum matching in G_o but not in G_u and the assignment has $OF = BOF$ and $UF = BUF + 1$ (equivalently if $x_i = d$ belongs to a maximum matching in G_u but not in G_o);
- if $|M'_o| < |M_o| - 1$ and $|M'_u| < |M_u| - 1$ then $x_i = d$ does not belong to a maximum matching in G_o nor in G_u and the assignment has $OF = BOF + 1$ and $UF = BUF + 1$.

In Figure 1 we give an example of the concepts explained above. Figure 1a shows the value graph of a global cardinality constraint; the variable domains are $D_1 = D_2 = D_3 = D_4 = D_5 = \{v_1, v_2\}$, $D_6 = \{v_3, v_4\}$, $D_7 = \{v_2, v_3\}$, $D_8 = D_9 = \{v_4, v_5\}$; for each value the minimum and the maximum number of occurences

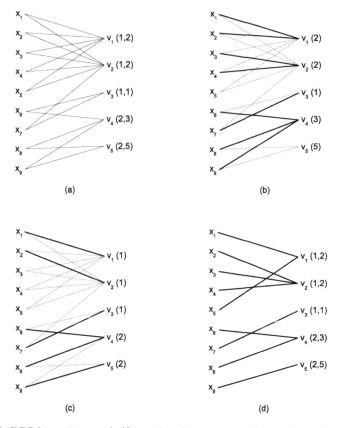

Fig. 1. (a) GCC bipartite graph (for each value, upper and lower bound are indicated between parenthesis). (b) Maximum Matching in G_o. (c) Maximum Matching in G_u. (d) Possible solution with minimum violation.

are indicated between parenthesis. Figure 1b and 1c show respectively G_o and G_u and the related maximum matchings. In details, the maximum matching M_o in G_o has an overflow equal to $|X| - |M_o| = 1$; as we can see x_5 causes the overflow since it is not assigned w.r.t. M_o. The maximum matching M_u in G_u has an underflow equal to $\sum_{d \in D} l_d - |M_u| = 1$ and the underflow is caused by the value v_5. Then it is possible to find a solution that minimizes overflow and underflow (figure (d)): this assignment has a variable-based violation equal to 1 and a value-based violation equal to 2.

4 Consistency and Filtering Algorithms

In this section we explain the basis to reach generalized arc consistency and we show the filtering algorithms for the variable-based and value-based violations. Our approach is similar to the one proposed by Petit et al. in [9] for the Soft All-Different constraint.

Briefly, we recall that the variable z represents the cost of the violation and D_z its domain; during the search $max\, D_z$ represents the maximum violation allowed; the objective is to minimize z in order to minimize the total violation.

Moreover, we recall that variable-based violation is equal to $max\left(\sum_{d \in D} overflow(X, d), \sum_{d \in D} underflow(X, d)\right)$ and that value-based violation is equal to $\sum_{d \in D} overflow(X, d) + \sum_{d \in D} underflow(X, d)$.

4.1 Variable Based Violation

Theorem 7. *Let G_o and G_u be the value graphs with respectively upper and lower bound capacities and let M_o and M_u be maximum matchings respectively in G_o and G_u; let BOF and BUF be respectively $BOF = |X| - |M_o|$ and $BUF = \sum_{d \in D} l_d - |M_u|$. The constraint $softgcc_{[var]}(X, l, u, z)$ is generalized arc consistent on X if and only if $min\, D_z \le max(BOF, BUF)$ and either:*

1. $max\left(\text{BOF}, \text{BUF}\right) \le (max\, D_z - 1)$ *or*

2. *if* $\left(\text{BOF} = max\, D_z\right)$ *and*

   $\left(\text{BUF} \le (max\, D_z - 1)\right)$ *and all edges in G_o belong to a maximum matching or*

3. *if* $\left(\text{BOF} \le (max\, D_z - 1)\right)$ *and*

   $\left(\text{BUF} = max\, D_z\right)$ *and all edges in G_u belong to a maximum matching or*

4. *if* $\left(\text{BOF} = \text{BUF} = max\, D_z\right)$ *and all edges in G_u and in G_o belong to a maximum matching in G_u*

Proof. In the first case we can build an assignment with $violation_{[var]}$ strictly less than $max\, D_z$; from Theorem 6 the change of a single variable can cause a unitary increase of the overflow and underflow hence the total violation is still less or equal to $max\, D_z$.

In the second case (resp. third case) if the overflow (resp. underflow) is equal to $max\,D_z$ then all the edges must belong to a maximum matching in G_o (resp. in G_u) such that there is no violation increase; from Theorem 6 we know that an edge that does not belong to a maximum matching would cause an overflow (resp. underflow) increase making it greater than $max\,D_z$.

In the last case the only way for not having a violation increase is that all edges belong to a maximum matching both in G_o and in G_u.

Filtering Algorithm. Firstly we compute the maximum matchings M_o in G_o and M_u in G_u. If the overflow or underflow is greater than $max\,D_z$ then we fail because the best possible solution is worse than the maximum allowed violation.

If $BOF = |X| - |M_o| < max\,D_z$ and $BUF = \sum_{d \in D} l_d - |M_u| < max\,D_z$ then all the values are consistent.

In the case of $|X| - |M_o| = max\,D_z$ we can remove all the edges that do not belong to a maximum matching in G_o; from matching theory (Theorem 2), we know that an edge can be part of a matching iff it belongs to a strongly connected component (alternating circuit) or it lies on an alternating path of even length starting from or leading to a free vertex. Analogously, if $\sum_{d \in D} l_d - |M_u| = max\,D_z$ we remove all the edges that do not belong to a maximum matching in G_u. Finally, we update the bound of the violation variable, if necessary $(min\,D_z = max(BOF\,,\,BUF))$.

The maximum matchings can be computed in $O(\sqrt{n}m)$ through Quimper's adaptation of Hopcroft-Karp's algorithm (where n is the number of variables and m the sum of the cardinalities of the domains); the running time for computing strongly connected components is $O(n + m)$ and for finding alternating paths is $O(m)$, hence the overall complexity can be bounded by $O(\sqrt{n}m)$.

Note that if all the values have u_d equal to 1 then the GCC is equivalent to the All-Different constraint; in that case the solution proposed is equivalent to Petit et al.'s solution for the Soft All-Different with variable based violation (see [9]).

Consider again the example given in Section 3 and suppose that $max\,D_z = 1$. We briefly recall that $D_1 = D_2 = D_3 = \{1,2\}$, $D_4 = \{1,2,3\}$ and that the values $(1,2,3)$ have lower bounds and upper bounds respectively of $(0,1,2)$ and $(1,1,2)$. Firstly we compute a maximum matching in G_o: $M_o = \{(x_1,1),(x_2,2),(x_4,3)\}$; thus the overflow is $OF = |X| - |M_o| = 1$. Then we compute a maximum matching in G_u: $M_u = \{(x_1,2),(x_4,3)\}$; the underflow is $UF = \sum_{d \in D} l_d - |M_u| = 1$. Since both the overflow and the underflow are equal to $max\,D_z$ then we prune all the edges that do not belong to a maximum matching in G_o and/or in G_u. In particular, all the edges belong to a maximum matching in G_o; the edges $(x_4,1)$ and $(x_4,2)$ do not belong to a maximum matching in G_u, so they can be pruned; in fact if x_4 would have been equal to 1 (or 2) then the underflow would have been equal to 2 (caused by the value 3).

4.2   Value Based Violation

Theorem 8. *Let G_o and G_u be the value graphs with respectively upper and lower bound capacities and let M_o and M_u be maximum matchings respectively in G_o and G_u; let BOF and BUF be respectively $BOF = |X| - |M_o|$ and*

$BUF = \sum_{d \in D} l_d - |M_u|$. The constraint $softgcc_{[val]}(X, l, u, z)$ is generalized arc consistent on X if and only if $min\,D_Z \leq BOF + BUF$ and either:

1. $BOF + BUF < (max\,D_Z - 1)$ or
2. if $BOF + BUF = (max\,D_Z - 1)$ and all edges belong to a maximum matching at least in one of G_o or G_u or
3. if $BOF + BUF = max\,D_Z$ and all edges belong to a maximum matching both in G_o and in sG_u

Proof. We start from the best solution found following Theorem 5. From this solution a single change of a variable can cause in the worst case a violation increase equal to 2 (Theorem 6). So in the worst case the total violation is less or equal to $max\,D_Z$ hence all the values are consistent.

If the overall violation is equal to $max\,D_Z - 1$ then we have to verify that all the edges belong to at least a maximum matching; for Theorem 6 the maximum violation increase would be at most equal to 1, hence the total violation remains less or equal to $max\,D_Z$.

If the overall violation is equal to $max\,D_Z$ and all the edges belong to a maximum matching in both G_o and G_u then there would be no increase in the total violation so the constraint remains feasible.

Filtering Algorithm. Firstly we compute the maximum matchings M_o in G_o and M_u in G_u. We denote with $S_{(OF,UF)}$ the sum of overflow and underflow.

If $S_{(OF,UF)}$ is greater than $max\,D_z$ then we fail because the best possible solution is worse than the best current solution found.

If $S_{(OF,UF)} < max\,D_z - 1$ then all the values are consistent. In the case of $S_{(OF,UF)} = max\,D_z - 1$ we can remove all the edges that belong neither to a maximum matching in G_o nor in G_u.

If $S_{(OF,UF)} = max\,D_z$ then we remove all the edges that do not belong to a maximum matching in G_o and/or in G_u.

Finally, we update $min\,D_z$, if necessary ($min\,D_z = S_{(OF,UF)}$).

The overall complexity is analogous to the variable-based algorithm, that is $O(\sqrt{n}m)$.

Following the example shown in Figure 1, suppose that $max\,D_z = 3$.

Instead, if we consider the value-based violation, we have to remove all the edges that belong neither to a maximum matching in G_o nor in G_u. In particular focusing on G_o, the edges $e_1 = (x_7, v_2)$ and $e_2 = (x_6, v_3)$ belong neither to an alternating circuit nor to an alternating path starting from or leading to a free vertex. This means that they do not belong to a maximum matching in G_o. Analyzing G_u, the situation is analogous. Hence, e_1 and e_2 cause an increase equal to 2 of the total violation (unitary increase of overflow and of underflow). Forcing e_1 (or e_2) to be in a solution, the resulting value-based violation is 4 then e_1 (resp. e_2) is inconsistent and can be pruned.

5   Conclusion

We have presented two algorithms for reaching generalized arc consistency in the Soft Global Cardinality Constraint with variable-based violation and value-based violation. They check the consistency of the constraint with a running time complexity of $O(\sqrt{n}m)$ and they prune inconsistent values in $O(m+n)$ where n is the cardinality of the set of variables and $m = \sum_i |D_i|$. We outperform previous algorithms that ran in $O(n(m + n \log n))$ (variable-based violation) and $O((n + k)(m + n \log n))$ (value-based violation) for constraint consistency check and in $O(\Delta(m + n \log n))$ for domain pruning where $\Delta = min(n, k)$ $(k = |\bigcup_i D_i|)$.

References

1. R.K. Ahuja, T.L. Magnanti, and J.B. Orlin (1993). Network Flows. Prentice Hall.
2. D. Dubois, H. Fargier and H. Prade. The calculus of fuzzy restrictions as a basis for flexible constraint satisfaction. *Proceedings of the Second IEEE International Conference on Fuzzy Systems*, volume 2, pp. 1131-1136, 1993.
3. H. Fargier, J. Lang and T. Schiex. Selecting preferred solutions in fuzzy constraint satisfaction problems. *Proceedings of the First European Congress on Fuzzy and Intelligent Technologies (EUFIT 93)*, Aachen, Vol. 3, pp. 1128-1134.
4. W. J. van Hoeve, G. Pesant and L.M. Rousseau. On Global Warming: Flow-Based Soft Global Constraints. *Journal of Heuristics*, to appear.
5. J.E. Hopcroft and R.M. Karp. An $n^{5/2}$ algorithm for maximum matchings in bipartite graphs. *SIAM Journal on Computing*, 2(4):225:231, 1973.
6. J. Larrosa. Node and Arc Consistency in Weighted CSP. *Proceedings of the Eighteenth National Conference on Artificial Intelligence*, pp. 48-53. AAAI Press, 2002.
7. J. Larrosa and T. Schiex. In the Quest of the best form of local consistency for Weighted CSP. *Proceedings of the Eighteenth International Joint Conference on Artificial Intelligence*, pp. 239-244. Morgan Kaufmann, 2003.
8. C-G. Quimper, Alejandro López-Ortiz, P. van Beek and Alexander Golynski. Improved Algorithms for the Global Cardinality Constraint. *Proceedings of the Tenth International Conference on Principles and Practice of Constraint Programming (CP 2004)*, Springer LNCS 3258: 542-556.
9. T. Petit, J-C. Régin and C. Bessière. Specific Filtering Algorithms for Over Constrained Problems. *Proceedings of the Seventh International Conference on Principles and Practice of Constraint Programming (CP 2001)*, Springer LNCS 2239: 451-463.
10. J-C. Régin. Arc Consistency for Global Cardinality Constraints with Costs. *Proceedings of the Fifth International Conference on Principles and Practice of Constraint Programming (CP 1999)*, Springer LNCS 1713: 390-404.
11. J-C. Régin. Generalized Arc Consistency for Global Cardinality Constraint. *Proceedings of the Thirteenth National Conference on Artificial Intelligence (AAAI-96)*, pp.209-215.
12. T. Schiex. Possibilistic Constraint Satisfaction Problems or "How to handle soft constraints?". *Proceedings of the 8th Annual Conference on Uncertainty in Artificial Intelligence*, pp.268-275. Morgan Kaufmann, 1992.

Author Index

Lecture Notes in Computer Science

For information about Vols. 1–3897

please contact your bookseller or Springer

Vol. 3947: Y.-C. Chung, J.E. Moreira (Eds.), Advances in Grid and Pervasive Computing. XXI, 667 pages. 2006.

Vol. 3946: T.R. Roth-Berghofer, S. Schulz, D.B. Leake (Eds.), Modeling and Retrieval of Context. XI, 149 pages. 2006. (Sublibrary LNAI).

Vol. 3945: M. Hagiya, P. Wadler (Eds.), Functional and Logic Programming. X, 295 pages. 2006.

Vol. 3944: J. Quiñonero-Candela, I. Dagan, B. Magnini, F. d'Alché-Buc (Eds.), Machine Learning Challenges. XIII, 462 pages. 2006. (Sublibrary LNAI).

Vol. 3943: N. Guelfi, A. Savidis (Eds.), Rapid Integration of Software Engineering Techniques. X, 289 pages. 2006.

Vol. 3942: Z. Pan, R. Aylett, H. Diener, X. Jin, S. Göbel, L. Li (Eds.), Technologies for E-Learning and Digital Entertainment. XXV, 1396 pages. 2006.

Vol. 3941: S.W. Gilroy, M.D. Harrison (Eds.), Interactive Systems. XI, 267 pages. 2006.

Vol. 3940: C. Saunders, M. Grobelnik, S. Gunn, J. Shawe-Taylor (Eds.), Subspace, Latent Structure and Feature Selection. X, 209 pages. 2006.

Vol. 3939: C. Priami, L. Cardelli, S. Emmott (Eds.), Transactions on Computational Systems Biology IV. VII, 141 pages. 2006. (Sublibrary LNBI).

Vol. 3936: M. Lalmas, A. MacFarlane, S. Rüger, A. Tombros, T. Tsikrika, A. Yavlinsky (Eds.), Advances in Information Retrieval. XIX, 584 pages. 2006.

Vol. 3935: D. Won, S. Kim (Eds.), Information Security and Cryptology - ICISC 2005. XIV, 458 pages. 2006.

Vol. 3934: J.A. Clark, R.F. Paige, F.A. C. Polack, P.J. Brooke (Eds.), Security in Pervasive Computing. X, 243 pages. 2006.

Vol. 3933: F. Bonchi, J.-F. Boulicaut (Eds.), Knowledge Discovery in Inductive Databases. VIII, 251 pages. 2006.

Vol. 3931: B. Apolloni, M. Marinaro, G. Nicosia, R. Tagliaferri (Eds.), Neural Nets. XIII, 370 pages. 2006.

Vol. 3930: D.S. Yeung, Z.-Q. Liu, X.-Z. Wang, H. Yan (Eds.), Advances in Machine Learning and Cybernetics. XXI, 1110 pages. 2006. (Sublibrary LNAI).

Vol. 3929: W. MacCaull, M. Winter, I. Düntsch (Eds.), Relational Methods in Computer Science. VIII, 263 pages. 2006.

Vol. 3928: J. Domingo-Ferrer, J. Posegga, D. Schreckling (Eds.), Smart Card Research and Advanced Applications. XI, 359 pages. 2006.

Vol. 3927: J. Hespanha, A. Tiwari (Eds.), Hybrid Systems: Computation and Control. XII, 584 pages. 2006.

Vol. 3925: A. Valmari (Ed.), Model Checking Software. X, 307 pages. 2006.

Vol. 3924: P. Sestoft (Ed.), Programming Languages and Systems. XII, 343 pages. 2006.

Vol. 3923: A. Mycroft, A. Zeller (Eds.), Compiler Construction. XIII, 277 pages. 2006.

Vol. 3922: L. Baresi, R. Heckel (Eds.), Fundamental Approaches to Software Engineering. XIII, 427 pages. 2006.

Vol. 3921: L. Aceto, A. Ingólfsdóttir (Eds.), Foundations of Software Science and Computation Structures. XV, 447 pages. 2006.

Vol. 3920: H. Hermanns, J. Palsberg (Eds.), Tools and Algorithms for the Construction and Analysis of Systems. XIV, 506 pages. 2006.

Vol. 3918: W.K. Ng, M. Kitsuregawa, J. Li, K. Chang (Eds.), Advances in Knowledge Discovery and Data Mining. XXIV, 879 pages. 2006. (Sublibrary LNAI).

Vol. 3917: H. Chen, F.Y. Wang, C.C. Yang, D. Zeng, M. Chau, K. Chang (Eds.), Intelligence and Security Informatics. XII, 186 pages. 2006.

Vol. 3916: J. Li, Q. Yang, A.-H. Tan (Eds.), Data Mining for Biomedical Applications. VIII, 155 pages. 2006. (Sublibrary LNBI).

Vol. 3915: R. Nayak, M.J. Zaki (Eds.), Knowledge Discovery from XML Documents. VIII, 105 pages. 2006.

Vol. 3914: A. Garcia, R. Choren, C. Lucena, P. Giorgini, T. Holvoet, A. Romanovsky (Eds.), Software Engineering for Multi-Agent Systems IV. XIV, 255 pages. 2006.

Vol. 3911: R. Wyrzykowski, J. Dongarra, N. Meyer, J. Waśniewski (Eds.), Parallel Processing and Applied Mathematics. XXIII, 1126 pages. 2006.

Vol. 3910: S.A. Brueckner, G.D.M. Serugendo, D. Hales, F. Zambonelli (Eds.), Engineering Self-Organising Systems. XII, 245 pages. 2006. (Sublibrary LNAI).

Vol. 3909: A. Apostolico, C. Guerra, S. Istrail, P. Pevzner, M. Waterman (Eds.), Research in Computational Molecular Biology. XVII, 612 pages. 2006. (Sublibrary LNBI).

Vol. 3908: A. Bui, M. Bui, T. Böhme, H. Unger (Eds.), Innovative Internet Community Systems. VIII, 207 pages. 2006.

Vol. 3907: F. Rothlauf, J. Branke, S. Cagnoni, E. Costa, C. Cotta, R. Drechsler, E. Lutton, P. Machado, J.H. Moore, J. Romero, G.D. Smith, G. Squillero, H. Takagi (Eds.), Applications of Evolutionary Computing. XXIV, 813 pages. 2006.

Vol. 3906: J. Gottlieb, G.R. Raidl (Eds.), Evolutionary Computation in Combinatorial Optimization. XI, 293 pages. 2006.

Vol. 3905: P. Collet, M. Tomassini, M. Ebner, S. Gustafson, A. Ekárt (Eds.), Genetic Programming. XI, 361 pages. 2006.

Vol. 3904: M. Baldoni, U. Endriss, A. Omicini, P. Torroni (Eds.), Declarative Agent Languages and Technologies III. XII, 245 pages. 2006. (Sublibrary LNAI).

Vol. 3903: K. Chen, R. Deng, X. Lai, J. Zhou (Eds.), Information Security Practice and Experience. XIV, 392 pages. 2006.

Vol. 3902: R. Kronland-Martinet, T. Voinier, S. Ystad (Eds.), Computer Music Modeling and Retrieval. XI, 275 pages. 2006.

Vol. 3901: P.M. Hill (Ed.), Logic Based Program Synthesis and Transformation. X, 179 pages. 2006.

Vol. 3900: F. Toni, P. Torroni (Eds.), Computational Logic in Multi-Agent Systems. XVII, 427 pages. 2006. (Sublibrary LNAI).

Vol. 3899: S. Frintrop, VOCUS: A Visual Attention System for Object Detection and Goal-Directed Search. XIV, 216 pages. 2006. (Sublibrary LNAI).

Vol. 3898: K. Tuyls, P.J. 't Hoen, K. Verbeeck, S. Sen (Eds.), Learning and Adaption in Multi-Agent Systems. X, 217 pages. 2006. (Sublibrary LNAI).